MACHINE LEARNING WITH NEURAL NETWORKS

This modern and self-contained book offers a clear and accessible introduction to the important topic of machine learning with neural networks. In addition to describing the mathematical principles of the topic, and its historical evolution, strong connections are drawn with underlying methods from statistical physics and current applications within science and engineering. Closely based around a well-established undergraduate course, this pedagogical text provides a solid understanding of the key aspects of modern machine learning with artificial neural networks, for students in physics, mathematics, and engineering. Numerous exercises expand and reinforce key concepts within the book and allow students to hone their programming skills. Frequent references to current research develop a detailed perspective on the state-of-the-art in machine learning research.

BERNHARD MEHLIG is Professor in Complex Systems at the University of Gothenburg, Sweden. His research is focused on statistical physics of complex systems and fluid mechanics, and he has published extensively in these areas. In 2010, he was awarded the prestigious Göran Gustafsson Prize in physics for his research in statistical physics. He has taught a course on machine learning for more than 15 years at Chalmers University of Technology and at the University of Gothenburg.

MACHINE LEARNING WITH NEURAL NETWORKS

An Introduction for Scientists and Engineers

BERNHARD MEHLIG
University of Gothenburg, Sweden

CAMBRIDGE
UNIVERSITY PRESS

CAMBRIDGE
UNIVERSITY PRESS

University Printing House, Cambridge CB2 8BS, United Kingdom

One Liberty Plaza, 20th Floor, New York, NY 10006, USA

477 Williamstown Road, Port Melbourne, VIC 3207, Australia

314–321, 3rd Floor, Plot 3, Splendor Forum, Jasola District Centre, New Delhi – 110025, India

103 Penang Road, #05–06/07, Visioncrest Commercial, Singapore 238467

Cambridge University Press is part of the University of Cambridge.

It furthers the University's mission by disseminating knowledge in the pursuit of education, learning, and research at the highest international levels of excellence.

www.cambridge.org
Information on this title: www.cambridge.org/9781108494939
DOI: 10.1017/9781108860604

First published 2022

Printed in the United Kingdom by TJ Books Limited, Padstow Cornwall

A catalogue record for this publication is available from the British Library.

Library of Congress Cataloging-in-Publication Data
Names: Mehlig, Bernhard, 1964- author.
Title: Machine learning with neural networks : an introduction for scientists and engineers / Bernhard Mehlig, University of Gothenburg, Sweden.
Description: Cambridge, United Kingdom ; New York, NY : Cambridge University Press, 2021. | Includes bibliographical references and index.
Identifiers: LCCN 2021027339 | ISBN 9781108494939 (hardback) | ISBN 9781108860604 (ebook)
Subjects: LCSH: Neural networks (Computer science) | Machine learning. | BISAC: SCIENCE / Physics / Mathematical & Computational | SCIENCE / Physics / Mathematical & Computational
Classification: LCC QA76.87 .M447 2021 | DDC 006.3/2–dc23
LC record available at https://lccn.loc.gov/2021027339

ISBN 978-1-108-49493-9 Hardback

The cover image shows an input pattern designed to maximise the output of neurons corresponding to one feature map in a given convolution layer of a deep convolutional neural network. (See Section 8.7 and references cited there.) Image by Hampus Linander. Reproduced with permission.

Contents

Acknowledgements

This textbook is based on lecture notes for the course Artificial Neural Networks that I have given at Gothenburg University and at Chalmers Technical University in Gothenburg, Sweden. When I prepared my lectures, my main source was *Introduction to the Theory of Neural Computation* by Hertz, Krogh, and Palmer [1]. Other sources were *Neural Networks: A Comprehensive Foundation* by Haykin [2]; Horner's lecture notes from Heidelberg [3]; *Deep Learning* by Goodfellow, Bengio, and Courville [4]; and the online book *Neural Networks and Deep Learning* by Nielsen [5].

I thank Martin Čejka for typesetting the first version of my handwritten lecture notes, Erik Werner and Hampus Linander for their help in preparing Chapter 8, Kristian Gustafsson for his detailed feedback on Chapter 11, Nihat Ay for his comments on Section 4.5, and Mats Granath for discussions about autoencoders. I would also like to thank Juan Diego Arango, Oleksandr Balabanov, Anshuman Dubey, Johan Fries, Phillip Gräfensteiner, Navid Mousavi, Marina Rafajlovic, Jan Schiffeler, Ludvig Storm, and Arvid Wenzel Wartenberg for implementing algorithms described in this book. Many figures are based on their results. Oleksandr Balabanov, Anshuman Dubey, Jan Meibohm, and in particular Johan Fries and Marina Rafajlovic contributed exam questions that became exercises in this book. Finally, I would like to express my gratitude to Stellan Östlund for his encouragement and criticism. Last but not least, a large number of colleagues and students – past and present – pointed out misprints and errors and suggested improvements. I thank them all.

1

Introduction

The term *neural networks* historically refers to networks of neurons in the mammalian brain. Neurons are its fundamental units of computation, and they are connected together in networks to process data. This can be a very complex task. The dynamics of such neural networks in response to external stimuli is therefore often quite intricate. Inputs and outputs of each neuron vary as functions of time in the form of spike trains, but the network itself also changes over time: we learn and improve our data-processing capabilities by establishing new connections between neurons.

Neural-network algorithms for machine learning are inspired by the architecture and the dynamics of networks of neurons in the brain. The algorithms use highly idealised neuron models. Nevertheless, the fundamental principle is the same: artificial neural networks learn by changing the connections between their neurons. Such networks can perform a multitude of information-processing tasks.

Neural networks can for instance learn to recognise structures in a data set and, to some extent, generalise what they have learnt. A *training set* contains a list of input patterns together with a list of corresponding labels, or target values, that encode the properties of the input patterns the network is supposed to learn. Artificial neural networks can be trained to classify such data very accurately by adjusting the connection strengths between their neurons, and can learn to generalise the result to other data sets – provided that the new data is not too different from the training data. A prime example for a problem of this type is object recognition in images, for instance in the sequence of camera images taken by a self-driving car. Recent interest in machine learning with neural networks is driven in part by the success of neural networks in *visual object recognition*.

Another task at which neural networks excel is *machine translation* with dynamical or recurrent networks. Such networks take sentences as inputs. As one feeds word after word, the network outputs the words in the translated sentence. Recurrent networks can be efficiently trained on large training sets of input sentences and their translations. Google translate works in this way. Recurrent networks have also been used with considerable success to predict chaotic dynamics. These are all examples of *supervised* learning, where the networks are trained to associate a certain target, or label, with each input.

Artificial neural networks are also good at analysing large sets of unlabelled, often high-dimensional data – where it may be difficult to determine a priori which questions are most relevant and rewarding to ask. *Unsupervised*-learning algorithms organise the unlabelled input data in different ways. They can, for instance, detect familiarity and similarity (clusters) of input patterns, or other structures in the input data. Unsupervised-learning algorithms work well when there is redundancy in the input data, and they are particularly useful for large, high-dimensional data sets, where it may be a challenge to detect clusters or other data structures by inspection.

Many problems lie between these two extremes of supervised and unsupervised learning. Consider how an agent may learn to navigate a complex environment, in order to get from one location to another as quickly as possible, or expending as little energy as possible. The method of *reinforcement learning* allows the agent to do just that, by optimising its behaviour in response to environmental cues in the shape of penalties and rewards. In short, the agent learns to act in such a way that it receives positive feedback (reward) more often than a penalty.

The tools for machine learning with neural networks were developed long ago, most of them during the second half of the last century. In 1943, McCulloch and Pitts [6] analysed how networks of neurons can process information. Using an abstract model for a neuron, they demonstrated how such units can be coupled together to represent logical functions (Figure 1.1). Their analysis and conclusions are formulated using Carnap's logical syntax [7], not in terms of algebraic equations as we are used to today. Nevertheless, their neuron model is essentially the binary threshold unit, closely related to the fundamental building block of most neural-network algorithms for machine learning to this date. In this book, we therefore refer to this model as the *McCulloch-Pitts neuron*. The purpose of this early

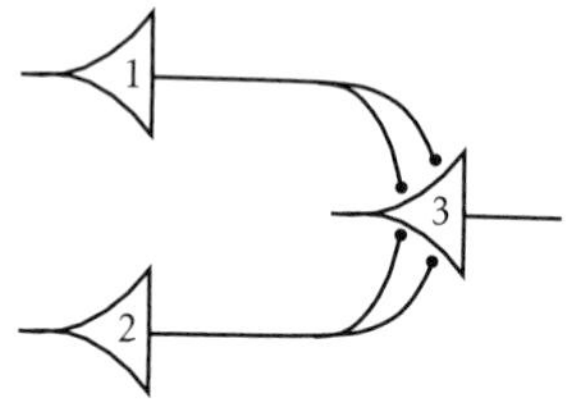

Figure 1.1 Logical OR function represented by three neurons. Neuron 3 fires actively if at least one of the neurons 1 and 2 is active. After Figure 1b by McCulloch and Pitts [6]

research on neural networks was to explain neuro-physiological mechanisms [8]. Perhaps the most significant advance was Hebb's learning principle, describing how neural networks learn by strengthening connections between neurons that are active simultaneously. The principle is described in Hebb's book *The Organization of Behavior: A Neuropsychological Theory* [9], published in 1949.

About 10 years later, research in artificial neural networks had intensified, sparked by the influential work of Rosenblatt. In 1958, he formulated a learning rule for the McCulloch-Pitts neuron, related to Hebb's rule, and demonstrated that the rule converges to the correct solution for all problems this model can solve [10]. He coined the term *perceptron* for layered networks of McCulloch-Pitts neurons and showed that such neural networks could in principle solve tasks that a single McCulloch-Pitts neuron could not. However, there was no general learning rule for perceptrons at the time. The work of Minsky and Papert [11] emphasised the geometric aspects of learning. This allowed them to prove which kinds of problems perceptrons could solve, and which not. In 1969, they summarised these results in their elegant book *Perceptrons: An Introduction to Computational Geometry*.

Perceptrons are classifiers that output a label for each input pattern. A perceptron represents a mathematical function, an input-output mapping. A breakthrough in perceptron learning was the paper by Rumelhart et al. [12]. The authors demonstrated in 1986 that perceptrons can be trained by gradient descent. This means that the connection strengths between the neurons are iteratively changed in small steps, to eventually minimise the output error. For a single McCulloch-Pitts neuron, this gives essentially Hebb's rule. The important point is that gradient descent allows one to efficiently train perceptrons with many layers (*backpropagation* for multilayer perceptrons). A second contribution of Rumelhart et al. is the idea to use local feature maps for object recognition with neural networks. The corresponding mathematical operation is a convolution. Therefore such architectures are now known as *convolutional networks*.

The work of Hopfield popularised an entirely different approach, also based on Hebb's rule. In 1982, Hopfield analysed the properties of a dynamical, or recurrent, network that functions as a memory [13]: the dynamics is designed to find stored patterns by converging to a corresponding steady state. Such *Hopfield networks* were especially popular amongst physicists because there are close connections to the statistical physics of spin glasses that made it possible to derive a precise mathematical understanding of such artificial neural networks. Hopfield networks are the basis for important developments in computer science. More general recurrent networks, for example, are trained like perceptrons for language processing. Hopfield networks with hidden neurons, so-called Boltzmann machines [14], are generative models that allow one to sample from a distribution the neural network learned. The training algorithm for Boltzmann machines with many hidden layers [15], published in 1986, is one of the first algorithms for training networks with many layers, so-called *deep* networks.

An important problem in behavioural psychology is to understand how we learn from experience. One hypothesis is that desirable behaviours are reinforced by positive feedback. Around the same time as researchers began to analyse perceptrons, a different line of neural-network research emerged: to find learning rules that allow neural networks to learn by reinforcement. In many problems of this kind, the positive feedback or reward is not immediate but is received at some time in the future, as in a board game, for example. Therefore it is necessary to understand how to estimate the future reward for a certain behaviour, and how to find strategies that optimise the future reward. *Reinforcement* learning [16] is designed for this purpose. In 1995, an early application of this method demonstrated how a neural network could learn to play the board game backgammon [17].

A related research field originated from the neuro-physiological question: how do we learn to map visual or sensory stimuli to spatio-temporal patterns of neural activity in the brain? In 1992, Kohonen described a *self-organising map* [18] that suggests how neurons might create meaningful geometric representations of inputs. At the same time, Kohonen's algorithm is one of the first methods for non-linear dimensionality reduction for large data sets.

There are many connections between neural-network algorithms for machine learning and methods used in mathematical statistics, such as for instance Markov-chain Monte-Carlo algorithms and simulated-annealing methods. Certain unsupervised learning algorithms are related to principal-component analysis, others to clustering algorithms such as K-means clustering. Supervised learning with deep networks is essentially regression analysis, trying to fit an input-output function to the training data. In other words, this is just function fitting – and usually with a very large number of fitting parameters. Recent convolutional neural networks have millions of parameters. To determine so many parameters requires very large and accurate data sets. This makes it clear that neural networks are not a *solution for everything*. One of the difficult problems is to understand when machine learning with neural networks is called for and when it is not. Therefore we need a detailed understanding of how the algorithms work, and in particular when and how they fail.

There were some early successes of machine learning with neural networks, but these methods were not widely used in the last century. During the past decade, by contrast, machine learning with neural networks has become increasingly successful and popular. For many applications, neural-network-based algorithms are now regarded as the method of choice, for example for predicting how proteins fold [19]. What caused this paradigm shift? After all, the methods are essentially those developed forty or more years ago. A reason for the new success is perhaps that industry, in acute need of machine-learning algorithms for large data sets, has invested time and effort into generating larger and more accurate training sets than previously available. Computer hardware has improved too, so that networks with many layers containing many neurons can now be efficiently trained, making the recent progress possible.

This book is based on notes for lectures on artificial neural networks I have given for more than 15 years at Gothenburg University and Chalmers University of Technology in Gothenburg, Sweden. When I prepared these lectures, my primary source was *Introduction to the Theory of Neural Computation* by Hertz, Krogh, and Palmer [1]. The material is organised into three parts: Hopfield networks, supervised learning of labelled data, and learning for unlabelled data sets (unsupervised and reinforcement learning). One reason for starting with Hopfield networks is that there is an elegant mathematical theory that describes how these neural networks learn, making it possible to understand their strengths and weaknesses from first principles. This is not the case for most of the other algorithms discussed in this book. The analysis of Hopfield networks sets the scene for the later parts of the book. Part II describes supervised learning with multilayer perceptrons and convolutional neural networks, starting from the simple geometrical picture emphasised by Minsky and Papert, and leading to the recent successes of convolutional networks in object recognition and recurrent networks in language processing. Part III explains what neural networks can learn about data that is not labelled, with particular emphasis on reinforcement learning. The overall goal is to explain the fundamental principles that allow neural networks to learn, emphasising ideas and concepts that are common to all three parts.

1.1 Neural Networks

Different regions in the mammalian brain perform different tasks. The *cerebral cortex* is the outer part of the mammalian brain, one of its largest and best developed segments. We can think of the cerebral cortex as a thin sheet (about 2 to 5 mm thick) that folds upon itself to form a layered structure with a large surface area that can accommodate large numbers of nerve cells, *neurons*. The human cerebral cortex contains about 10^{10} neurons. They are linked together by nerve strands (*axons*) that branch and end in *synapses*. These synapses are the connections to other neurons. The synapses connect to *dendrites*, branches extending from the neural cell body that are designed to collect input from other neurons in the form of electrical signals. A neuron in the human brain may have thousands of synaptic connections with other neurons. The resulting network of connected neurons in the cerebral cortex is responsible for processing visual, audio, and sensory data.

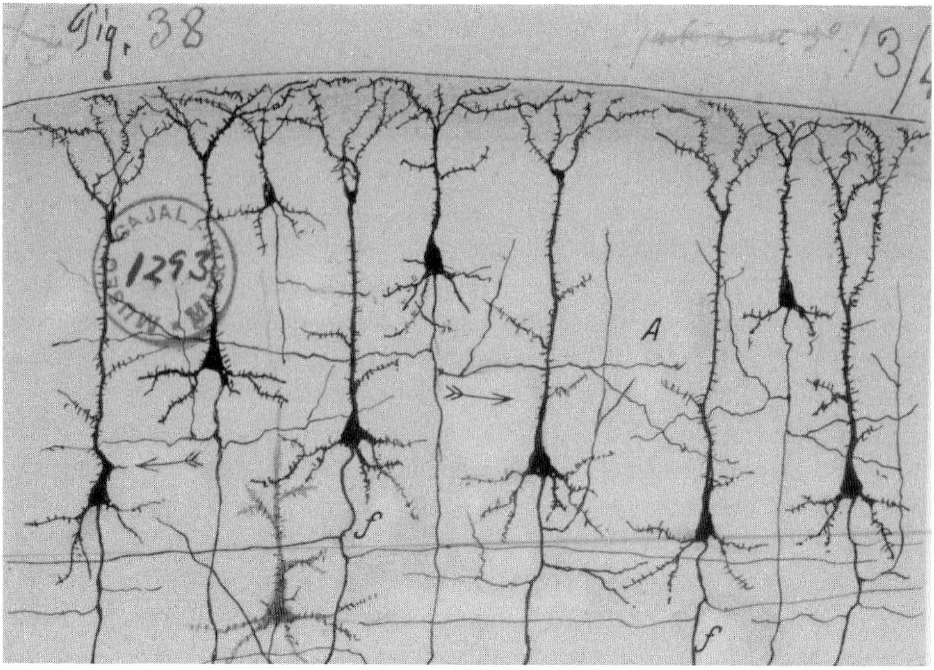

Figure 1.2 Neurons in the *cerebral cortex*, a part of the mammalian brain. Drawing by Santiago Ramón y Cajal, the Spanish neuroscientist who received the Nobel Prize in Physiology and Medicine in 1906 together with Camillo Golgi 'in recognition of their work on the structure of the nervous system' [20]. Courtesy of the Cajal Institute, 'Cajal Legacy', Spanish National Research Council (CSIC), Madrid, Spain. Original in colour

Figure 1.2 shows neurons in the cerebral cortex. This drawing was made by Santiago Ramón y Cajal more than 100 years ago. By microscope he studied the structure of neural networks in the brain and documented his observations by ink-on-paper drawings like the one reproduced in Figure 1.2. One can distinguish the cell bodies of the neural cells, their axons (f), and their dendrites. The axons of some neurons connect to the dendrites of other neurons, forming a neural network (see Ref. [21] for a slightly more detailed description of this drawing).

A schematic image of a neuron is drawn in Figure 1.3. Information is processed from left to right. On the left are the dendrites that receive signals and connect to the cell body of the neuron where the signal is processed. The right part of the figure shows the axon, through which the output is sent to other neurons. The axon connects to their dendrites via synapses.

Information is transmitted as an electrical signal. Figure 1.4 shows an example of the time series of the electric potential for a *pyramidal* neuron in fish [22]. The time

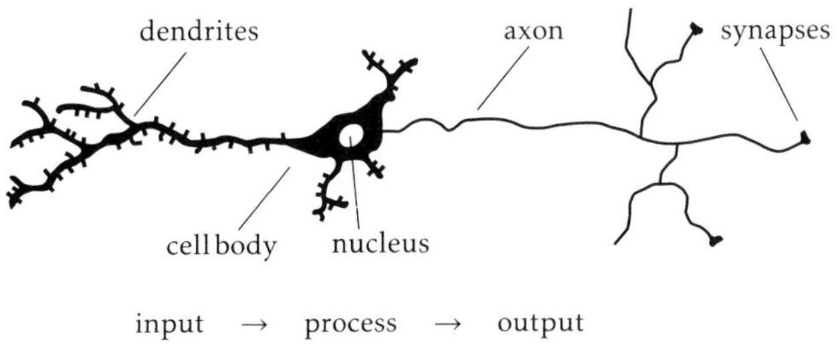

Figure 1.3 Schematic image of a neuron. Dendrites receive input in the form of electrical signals, via synapses. The signals are processed in the cell body of the neuron. The cell nucleus is shown as a white blob. The output travels from the neural cell body along the axon which connects via synapses to other neurons

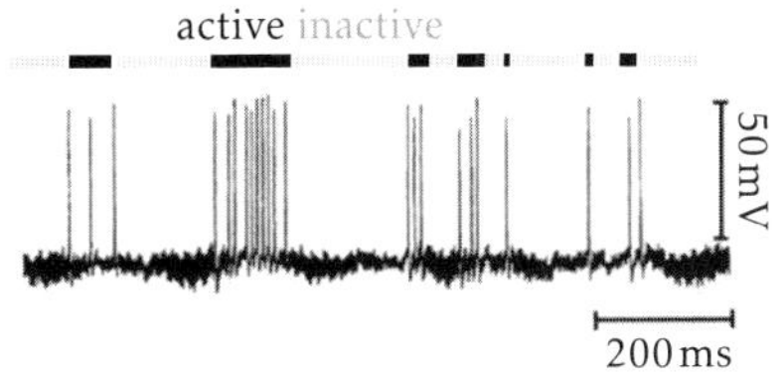

Figure 1.4 Spike train in electro-sensory pyramidal neuron of a fish. The time series is from Ref. [22]. It is reproduced by permission of the publisher. The labels were added

series consists of an intermittent series of electrical-potential spikes. Quiescent periods without spikes occur when the neuron is *inactive*, during spike-rich periods we say that the neuron is *active*.

1.2 McCulloch-Pitts Neurons

In artificial neural networks, the ways in which information is processed and signals are transferred are highly idealised. McCulloch and Pitts [6] modelled the neuron, the computational unit of the neural network, as a *binary threshold unit*. It has only two possible outputs, or *states*: active or inactive. To compute the output, the unit sums the weighted inputs. If the sum exceeds a given threshold, the state of the neuron is said to be active, otherwise inactive. A slightly more general model than the original one is illustrated in Figure 1.5. The model performs repeated computations in discrete time steps $t = 0, 1, 2, 3, \ldots$. The state of neuron number j at time step t is denoted by

$$s_j(t) = \begin{cases} -1 & \text{inactive,} \\ 1 & \text{active.} \end{cases} \tag{1.1}$$

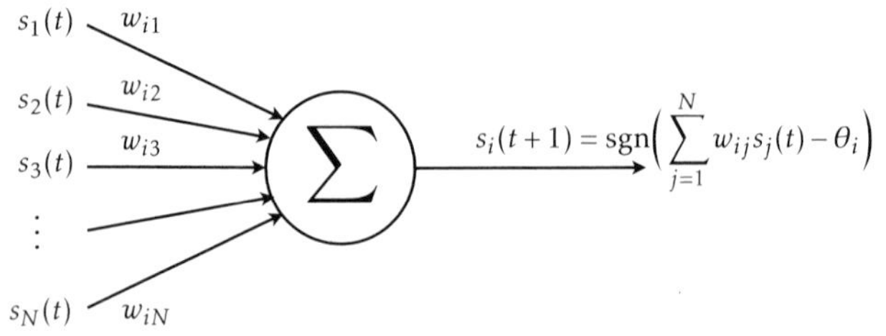

Figure 1.5 Schematic diagram of a McCulloch-Pitts neuron. The index of the neuron is i; it receives inputs from N neurons. The strength of the connection from neuron j to neuron i is denoted by w_{ij}. The *threshold* value for the neuron is denoted by θ_i. The index $t = 0, 1, 2, 3, \ldots$ labels the discrete time sequence of computation steps, and $\text{sgn}(b)$ stands for the signum function [Figure 1.6 and Equation (1.3)]

Given the states $s_j(t)$, neuron number i computes

$$s_i(t+1) = \text{sgn}\Big(\sum_{j=1}^{N} w_{ij}s_j(t) - \theta_i\Big) \equiv \text{sgn}[b_i(t)]. \tag{1.2}$$

Here $\text{sgn}(b)$ is the signum function (Figure 1.5):

$$\text{sgn}(b) = \begin{cases} -1, & b < 0, \\ +1, & b \geq 0. \end{cases} \tag{1.3}$$

The argument of the signum function,

$$b_i(t) = \sum_{j=1}^{N} w_{ij}s_j(t) - \theta_i, \tag{1.4}$$

is called the *local field*. We see that the neuron performs a weighted average of the inputs $s_j(t)$. The parameters w_{ij} are called *weights*. Here the first index, i, refers to the neuron that does the computation, and j labels the neurons that connect to neuron i. In general weights between different pairs of neurons assume different numerical values, reflecting different strengths of the synaptic couplings. Weights can be positive or negative, and we say that there is no connection when $w_{ij} = 0$.

In this book, we refer to the model described in Figure 1.5 as the *McCulloch-Pitts neuron*, although their original model had additional constraints on the weights. The threshold[1] for neuron i is denoted by θ_i.

Finally, note that the computation (1.2) is performed for all neurons i in parallel, given the states $s_j(t)$ at time step t. The outputs $s_i(t+1)$ are the inputs to all neurons

[1] In the *deep-learning* literature [4], the thresholds are called *biases*, defined as the negative of θ_i, with a plus sign in Equation (1.4). In this book, we use the convention (1.4), with the minus sign.

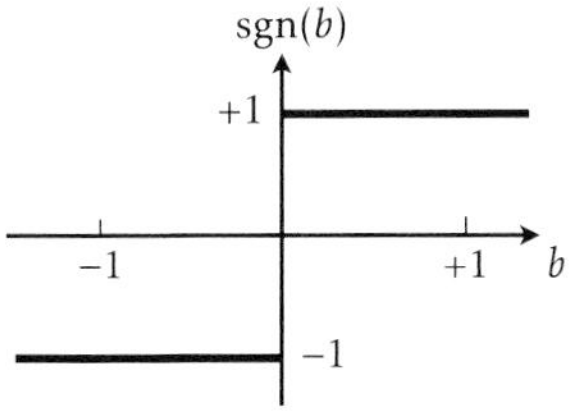

Figure 1.6 Signum function [Equation (1.3)]

at the next time step. Therefore the outputs have the time argument $t + 1$. These steps are repeated many times, resulting in time series of the activity levels of all neurons in the network.

1.3 Activation Functions

The McCulloch-Pitts model approximates the patterns of spiking activity in Figure 1.4 in terms of two states, -1 and $+1$, representing the inactive and active periods shown in the figure. For many computation tasks this is sufficient, and for our purposes it does not matter that the dynamics of electrical signals in the cortex is quite different in detail. The aim after all is not to model the neural dynamics in the brain but to construct computation models inspired by these dynamics.

It will become apparent later that the simplest model described above must be generalised somewhat for certain tasks and questions. For example, the jump in the signum function at $b = 0$ may cause large fluctuations in the activity levels of a network of neurons, caused by infinitesimal changes of the local fields across $b = 0$. To dampen this effect, one allows the neuron to respond continuously to its inputs, replacing Equation (1.2) by

$$s_i(t+1) = g\Big(\sum_j w_{ij} s_j(t) - \theta_i\Big). \tag{1.5}$$

Here $g(b)$ is a continuous *activation function*. It could just be a linear function, $g(b) \propto b$. But we shall see that many tasks require non-linear activation functions, such as $\tanh(b)$ (Figure 1.7). When the activation function is continuous, the neuron states assume continuous values too, not just the discrete values -1 and $+1$ given in Equation (1.1).

Alternatively, one may use a piecewise linear activation function (Figure 1.8). This is motivated in part by the response curve of the *leaky integrate-and-fire* neuron, a model for the relation between the electrical current I through the cell membrane into the neuron cell, and the membrane potential U. The simplest model for the dynamics of the membrane potential represents the neuron as a capacitor. In the

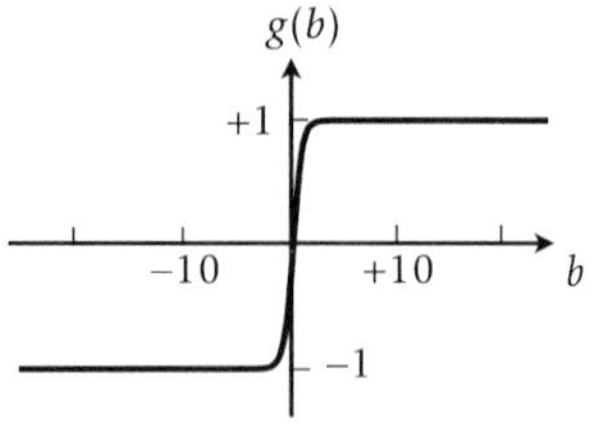

Figure 1.7 Continuous activation function $g(b) = \tanh(b)$

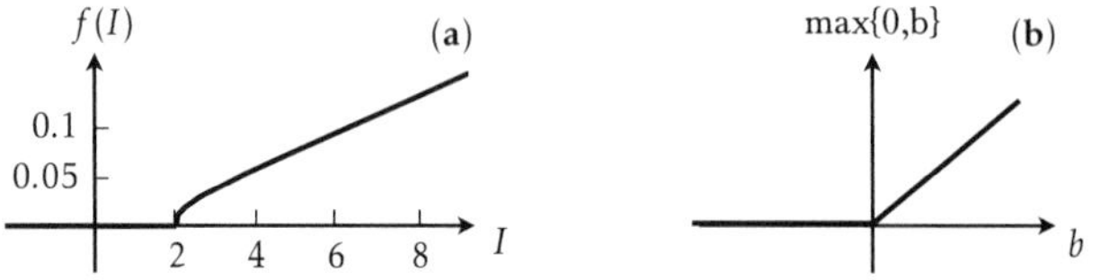

Figure 1.8 (**a**) Firing rate $f(I)$ of a *leaky integrate-and-fire* neuron as a function of the electrical current I through the cell membrane, Equation (1.7) for $\tau = 25$ and $U_c/R = 2$. (**b**) Piecewise linear activation function, $g(b) = \max\{0, b\}$

leaky integrate-and-fire neuron, leakage is modelled by adding a resistor R in parallel with the capacitor C, so that

$$I = \frac{U}{R} + C\frac{\mathrm{d}U}{\mathrm{d}t}\,. \tag{1.6}$$

For a constant current, the membrane potential grows from zero as a function of time, $U(t) = RI[1 - \exp(-t/\tau)]$, where $\tau = RC$ is the time constant of the model. One says that the neuron produces a *spike* when the membrane potential exceeds a critical value, U_c. Immediately after, the membrane potential is set to zero (and begins to grow again). In this model, the *firing rate* $f(I)$ is thus given by t_c^{-1}, where t_c is the solution of $U(t) = U_c$. It follows that the firing rate exhibits a threshold behaviour. In other words, the system works like a rectifier:

$$f(I) = \begin{cases} 0 & \text{for} \quad I \le U_c/R\,, \\ \left[\tau \log\left(\frac{RI}{RI - U_c}\right)\right]^{-1} & \text{for} \quad I > U_c/R\,. \end{cases} \tag{1.7}$$

This response curve is illustrated in Figure 1.8 (**a**). The main point is that there is a threshold below which the response is strictly zero. The response function looks qualitatively like the piecewise linear function

$$g(b) = \max\{0, b\}\,, \tag{1.8}$$

shown in panel (**b**). Neurons with this activation function are called *rectified linear units*, and the activation function (1.8) is called the *ReLU* function.

1.4 Asynchronous Updates

Equations (1.2) and (1.5) are called *synchronous* update rules, because all neurons are updated in parallel: at time step t all inputs $s_j(t)$ are stored. Then all neurons i are simultaneously updated using the stored inputs. An alternative is to update only a single neuron per iteration, for example the one with index m:

$$s_i(t+1) = \begin{cases} g\big(\sum_j w_{mj} s_j(t) - \theta_m\big) & \text{for } i = m, \\ s_i(t) & \text{otherwise.} \end{cases} \tag{1.9}$$

This is called an *asynchronous* update rule. Different schemes for choosing neurons are used in asynchronous updating. One possibility is to arrange the neurons into a two-dimensional array and to update them one by one, in a certain order. In the *typewriter scheme*, for example, one updates the neurons in the top row of the array first, from left to right, then the second row from left to right, and so forth. A second possibility is to choose randomly which neuron to update.

If there are N neurons, then one synchronous step corresponds to N asynchronous steps, on average. This difference in time scales is not the only difference between synchronous and asynchronous updating. The asynchronous dynamics can be shown to converge to a definite state in certain cases, while the synchronous dynamics may fail to do so, resulting in periodic cycles that persist forever.

1.5 Summary

Artificial neural networks use a highly idealised model for the fundamental computation unit: the McCulloch-Pitts neuron (Figure 1.5) is a binary threshold unit, very similar to the model introduced originally by McCulloch and Pitts [6]. The units are linked together by weights w_{ij}, and each unit computes a weighted average of its inputs. The network performs these computations in sequence. Most neural-network algorithms are built using the model described in this chapter.

1.6 Further Reading

Two accounts of the history of artificial neural networks are especially recommended. First, the early history of the field is summarised in the introduction to the second edition of *Perceptrons: An Introduction to Computational Geometry* by Minsky and Papert [11], which came out in 1988. This book also contains a concise bibliography, with comments by Minsky and Papert. Second, in a short note, Kanal [23] reviews the work of Rosenblatt and puts it into context.

Part I
Hopfield Networks

Hopfield networks [13, 24] are artificial neural networks that can recognise or reconstruct images, for instance the binary images of digits shown in Figure 2.1. The images are *stored* in the network by choosing its weights w_{ij} according to *Hebb's rule* [9]. One feeds a distorted digit (Figure 2.2) to the network by assigning the initial states of its neurons to the bits of the distorted digit. The idea is that the neural-network dynamics converges to the closest stored digit. In this way, the network can recognise the input as a distorted version of the correct pattern, it can *discern* the correct digit. Hopfield networks recognise patterns with many bits quite reliably, and in the past, such networks were used to perform pattern-recognition tasks. Today there are more efficient algorithms for this purpose (Chapter 8).

Nevertheless, Hopfield networks exemplify fundamental principles of machine learning with neural networks. For a start, most neural-network algorithms discussed in this book are built from similar building blocks and use learning rules related to Hebb's rule. Moreover, Hopfield networks are examples of *recurrent networks*, their neurons are updated following a dynamical rule. Widely used algorithms for machine translation and time-series prediction are based on this principle.

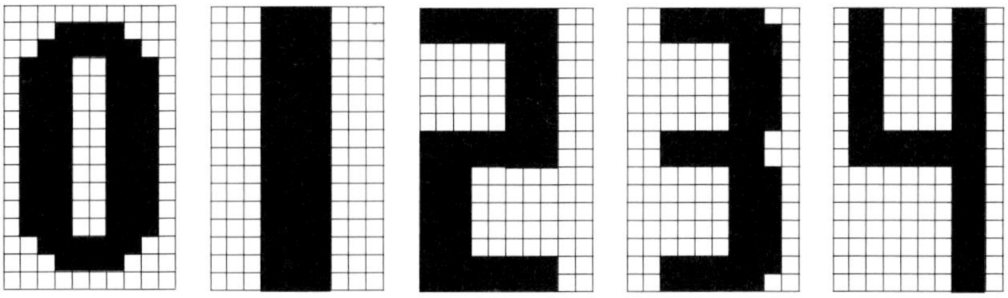

Figure 2.1 Binary representation of the digits 0 to 4. Each pattern has 10×16 pixels. Adapted from Figure 14.17 in Ref. [2]. The slightly peculiar shapes help the Hopfield network to distinguish the patterns [28]

Furthermore, *restricted Boltzmann machines* are closely related to Hopfield networks. These machines use hidden neurons to learn distributions of input patterns. This makes it possible to generate image textures and to complete partially obscured images [25]. Generalisations of these machines, *deep-belief networks*, are examples of the first deep network architectures for machine learning. Restricted Boltzmann machines have been developed into more efficient generative models, *Helmholtz machines*, to sample new patterns similar to those in a given data distribution. The training algorithm for recent generative models, *variational autoencoders*, is based on the same principles as the learning rule for Helmholtz machines.

The dynamics of Hopfield networks is closely related to stochastic *Markov-chain Monte-Carlo algorithms* used in a wide range of problems in the natural sciences. Hopfield networks highlight the role of stochasticity in neural-network dynamics. A certain degree of noise, not too much, can substantially improve the performance of the Hopfield network. In engineering problems, it is usually better to avoid stochasticity, when it is due to multiplicative or additive noise that diminishes the performance of the system. In neural-network dynamics, by contrast, stochasticity is often helpful, because it allows one to explore a wider range of configurations or actions and thus helps the dynamics to converge to a better solution. In general, it is challenging to analyse the stochastic dynamics of neural networks. But for the Hopfield network, much is known. The reason is that Hopfield networks are closely related to stochastic systems studied in statistical physics, so-called *spin glasses* [26, 27].

2

Deterministic Hopfield Networks

2.1 Pattern Recognition

As an example of a pattern-recognition task, consider p images (*patterns*), for instance the digits shown in Figure 2.1. The different patterns are indexed by $\mu = 1, \ldots, p$. Here p is the number of patterns ($p = 5$ in Figure 2.1). The bits of pattern μ are denoted by $x_i^{(\mu)}$. The index $i = 1, \ldots, N$ labels the different bits of a given pattern, and N is the number of bits per pattern ($N = 160$ in Figure 2.1). The bits are *binary*: they can take only the values -1 and $+1$. To determine the generic properties of the algorithm, one often turns to *random patterns* where each bit $x_i^{(\mu)}$ is chosen randomly, taking either value with probability equal to $\frac{1}{2}$. It is convenient to gather the bits of a pattern in a column vector like this:

$$\boldsymbol{x}^{(\mu)} = \begin{bmatrix} x_1^{(\mu)} \\ x_2^{(\mu)} \\ \vdots \\ x_N^{(\mu)} \end{bmatrix} . \tag{2.1}$$

In this book, vectors are written in a bold math font, as in Equation (2.1).

The task of the neural network is to recognise distorted patterns, to determine for instance that the pattern on the right in Figure 2.2 is a perturbed version of the digit on the left of this figure. To this end, one *stores* p patterns in the network, presents it with a distorted version of one of these patterns, and asks the network to find the stored pattern that is most similar to the distorted one.

The formulation of the problem makes it necessary to define how similar a distorted pattern $\boldsymbol{x}$ is to any of the stored patterns, say $\boldsymbol{x}^{(\nu)}$. One possibility is to quantify the distance d_ν between the patterns $\boldsymbol{x}$ and $\boldsymbol{x}^{(\nu)}$ by the mean squared error:

$$d_\nu = \frac{1}{4N} \sum_{i=1}^{N} \left(x_i - x_i^{(\nu)}\right)^2 . \tag{2.2}$$

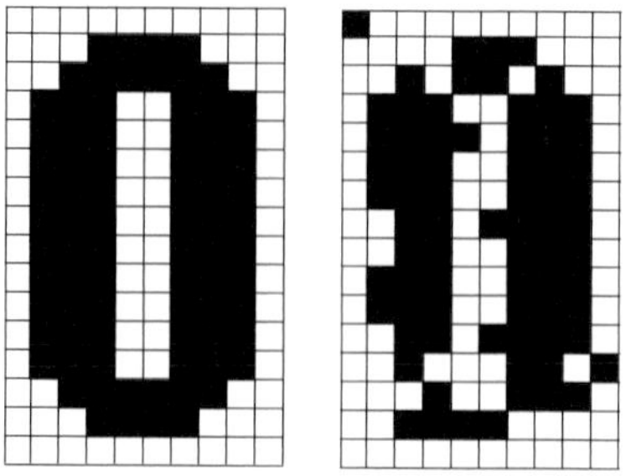

Figure 2.2 Binary image ($N = 160$) of the digit 0, and a distorted version of the same image. There are $N = 160$ bits x_i, $i = 1, \ldots, N$, and ■ stands for $x_i = +1$ while □ denotes $x_i = -1$

The prefactor is chosen so that the distance equals the fraction of bits by which two ± 1-patterns differ. Note that the distance (2.2) does not refer to distortions by translations, rotations, or shearing. An improved measure of distance might take the minimum distance between the patterns subject to all possible translations, rotations, and so forth.

2.2 Hopfield Networks and Hebb's Rule

Hopfield networks [13, 24] are networks of McCulloch-Pitts neurons designed to solve the pattern-recognition task described in the previous section. The bits of the patterns to be recognised are encoded in a particular choice of weights called Hebb's rule, as explained in the following.

All possible states of the McCulloch-Pitts neurons in the network,

$$\boldsymbol{s} = \begin{bmatrix} s_1 \\ s_2 \\ \vdots \\ s_N \end{bmatrix}, \tag{2.3}$$

form the *configuration space* of the network. The components of the states $\boldsymbol{s}$ are updated either with the synchronous rule (1.2):

$$s_i(t+1) = \operatorname{sgn}\big[b_i(t)\big] \quad \text{with local field} \quad b_i(t) = \sum_{j=1}^{N} w_{ij} s_j(t) - \theta_i , \tag{2.4}$$

or with the asynchronous rule

$$s_i(t+1) = \begin{cases} \operatorname{sgn}\big[b_m(t)\big] & \text{for } i = m, \\ s_i(t) & \text{otherwise.} \end{cases} \tag{2.5}$$

To recognise a distorted pattern, one feeds its bits x_i into the network by assigning the initial states of the neurons to the pattern bits,

$$s_i(t=0) = x_i\,. \tag{2.6}$$

The idea is to choose a set of weights w_{ij} so that the network dynamics (2.4) or (2.5) converges to the correct stored pattern. The choice of weights must depend on all p patterns, $\boldsymbol{x}^{(1)}, \ldots, \boldsymbol{x}^{(p)}$. We say that these patterns are *stored* in the network by assigning appropriate weights. For example, if $\boldsymbol{x}$ is a distorted version of $\boldsymbol{x}^{(\nu)}$ (Figure 2.2), then we want the network to converge to this pattern:

$$\text{if}\quad \boldsymbol{s}(t=0) = \boldsymbol{x} \approx \boldsymbol{x}^{(\nu)} \quad\text{then}\quad \boldsymbol{s}(t) \to \boldsymbol{x}^{(\nu)} \quad\text{as}\quad t \to \infty\,. \tag{2.7}$$

Equation (2.7) means that the network corrects the errors in the distorted pattern $\boldsymbol{x}$. If this works, the stored pattern $\boldsymbol{x}^{(\nu)}$ is said to be an *attractor* of the network dynamics.

But convergence is not guaranteed. If the initial distortion is too large, the network may converge to another pattern, or to some other state that bears no or little relation to the stored patterns, or it may not converge at all. The region around pattern $\boldsymbol{x}^{(\nu)}$ in which all patterns $\boldsymbol{x}$ converge to $\boldsymbol{x}^{(\nu)}$ is called the *region of attraction* of $\boldsymbol{x}^{(\nu)}$. The size of this region depends in an intricate way upon the ensemble of stored patterns, and there is no general convergence proof.

Therefore we ask a simpler question first: if one feeds one of the undistorted patterns, for instance $\boldsymbol{x}^{(\nu)}$, does the network recognise it as one of the stored, undistorted ones? The network should not make any changes to $\boldsymbol{x}^{(\nu)}$ because all bits are correct:

$$\text{if}\quad \boldsymbol{s}(t=0) = \boldsymbol{x}^{(\nu)} \quad\text{then}\quad \boldsymbol{s}(t) = \boldsymbol{x}^{(\nu)} \quad\text{for all}\quad t = 0, 1, 2, \ldots\,. \tag{2.8}$$

How can we choose weights and thresholds to ensure that Equation (2.8) holds? Let us consider a simple case first, where there is only one pattern, $\boldsymbol{x}^{(1)}$, to recognise. In this case, a suitable choice of weights and thresholds is

$$w_{ij} = \frac{1}{N} x_i^{(1)} x_j^{(1)} \quad\text{and}\quad \theta_i = 0\,. \tag{2.9}$$

This *learning rule* reminds of a relation between activation patterns of neurons and their coupling, postulated by Hebb [9] more than 70 years ago:

When an axon of cell A is near enough to excite a cell B and repeatedly or persistently takes part in firing it, some growth process or metabolic change takes place in one or both cells such that A's efficiency, as one of the cells firing B, is increased.

This is a mechanism for learning through establishing connections: the coupling between neurons tends to increase if they are active at the same time. Equation (2.25), expresses an analogous principle. Together with Equation (2.7), it says that

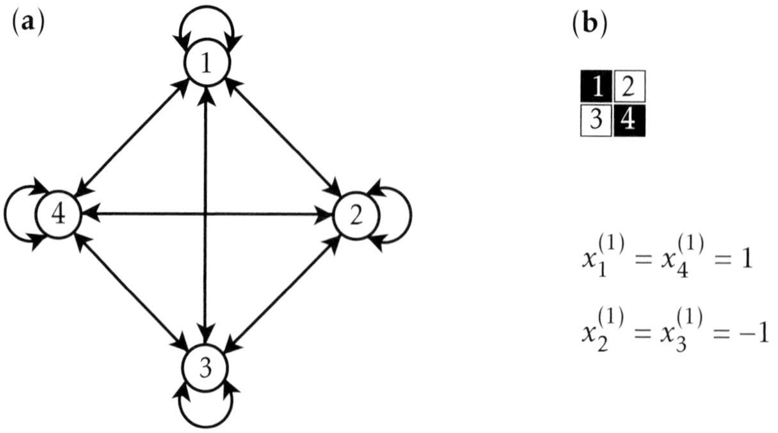

Figure 2.3 Hopfield network with $N = 4$ neurons. (**a**) Network layout. Neuron i is represented as $\textcircled{i}$. Arrows indicate symmetric connections. (**b**) Pattern $\boldsymbol{x}^{(1)}$

the coupling w_{ij} between two neurons is positive if they are both active ($s_i = s_j = 1$); if their states differ, the coupling is negative. Therefore the rule (2.25) is called Hebb's rule. Hopfield networks are networks of McCulloch-Pitts neurons with weights determined by Hebb's rule.

Does a Hopfield network recognise the pattern $\boldsymbol{x}^{(1)}$ stored in this way? To check that the rule (2.9) does the trick, we feed the pattern to the network by assigning $s_j(t = 0) = x_j^{(1)}$, and evaluate Equation (2.4):

$$\sum_{j=1}^{N} w_{ij} x_j^{(1)} = \frac{1}{N} \sum_{j=1}^{N} x_i^{(1)} x_j^{(1)} x_j^{(1)} = \frac{1}{N} \sum_{j=1}^{N} x_i^{(1)} = x_i^{(1)} . \tag{2.10}$$

The second equality follows because $x_j^{(1)}$ can only take the values ± 1, so that $[x_j^{(1)}]^2 = 1$. Using $\text{sgn}(x_i^{(1)}) = x_i^{(1)}$, we obtain

$$\text{sgn}\Big(\sum_{j=1}^{N} w_{ij} x_j^{(1)} \Big) = x_i^{(1)} . \tag{2.11}$$

Comparing Equation (2.11) with the update rule (2.4) shows that the bits $x_j^{(1)}$ of the pattern $\boldsymbol{x}^{(1)}$ remain unchanged under the update, as required by Equation (2.8). The network recognises the pattern as a stored one, so Equation (2.9) does what we asked for. Note that we obtain the same result if we leave out the factor of N^{-1} in Equation (2.9).

But does the network correct errors? In other words, is the pattern $\boldsymbol{x}^{(1)}$ an attractor [Equation (2.7)]? This question cannot be answered in general. Yet, in

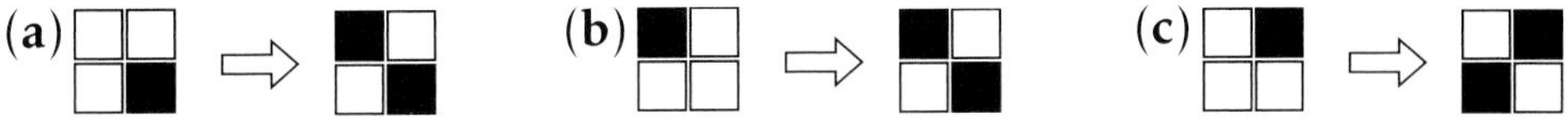

Figure 2.4 Reconstruction of a distorted pattern. Under synchronous updating (2.4), the first two distorted patterns (**a**) and (**b**) converge to the stored pattern $\boldsymbol{x}^{(1)}$, but pattern (**c**) does not

practice, Hopfield networks work often quite well. It is a fundamental insight that neural networks may perform well although no proof exists that their dynamics converges on the correct solution.

To illustrate the difficulties, consider an example: a Hopfield network with $N = 4$ neurons (Figure 2.3). Store the pattern $\boldsymbol{x}^{(1)}$ shown in Figure 2.3 by assigning the weights w_{ij} using Equation (2.9). Now consider a distorted pattern $\boldsymbol{x}$ that has a non-zero distance to $\boldsymbol{x}^{(1)}$ [Figure 2.4 (**a**)]:

$$d_1 = \frac{1}{16}\sum_{i=1}^{4}\left(x_i - x_i^{(1)}\right)^2 = \frac{1}{4}\,. \tag{2.12}$$

To feed the pattern to the network, we set $s_j(t = 0) = x_j$. Then we iterate the dynamics using the synchronous rule (2.4). Results for different distorted patterns are shown in Figure 2.4. We see that the first two distorted patterns converge to the stored pattern, cases (**a**) and (**b**). But the third distorted pattern does not [case (**c**)].

To understand this behaviour, we analyse the synchronous dynamics (2.4) using the *weight matrix*

$$\mathbb{W} = \frac{1}{N}\boldsymbol{x}^{(1)}\boldsymbol{x}^{(1)\mathsf{T}}\,. \tag{2.13}$$

Here $\boldsymbol{x}^{(1)\mathsf{T}}$ denotes the *transpose* of the column vector $\boldsymbol{x}^{(1)}$, so that $\boldsymbol{x}^{(1)\mathsf{T}}$ is a row vector. The standard rules for matrix multiplication apply also to column and row vectors, they are just $N \times 1$ and $1 \times N$ matrices. So the product on the r.h.s. of Equation (2.13) is an $N \times N$ matrix. In the following, matrices with elements A_{ij} or B_{ij} are written as $\mathbb{A}$, $\mathbb{B}$, and so forth. The product in Equation (2.13) is also referred to as an *outer product*. If we change the order of $\boldsymbol{x}^{(1)}$ and $\boldsymbol{x}^{(1)\mathsf{T}}$ in the product, we obtain a number instead:

$$\boldsymbol{x}^{(1)\mathsf{T}}\boldsymbol{x}^{(1)} = \sum_{j=1}^{N}\left[x_j^{(1)}\right]^2 = N\,. \tag{2.14}$$

The product (2.14) is called the *scalar product*. It is also denoted by $\boldsymbol{a}\cdot\boldsymbol{b} = \boldsymbol{a}^{\mathsf{T}}\boldsymbol{b}$ and equals $|\boldsymbol{a}||\boldsymbol{b}|\cos\varphi$, where φ is the angle between the vectors $\boldsymbol{a}$ and $\boldsymbol{b}$, and $|\boldsymbol{a}|$ is the magnitude of $\boldsymbol{a}$. We use the same notation for multiplying a transposed vector with a matrix from the left: $\boldsymbol{x}\cdot\mathbb{A} = \boldsymbol{x}^{\mathsf{T}}\mathbb{A}$. An excellent source for those not familiar with

Figure 2.5 Reproduced from xkcd.com/1838 under the creative commons attribution-noncommercial 2.5 license

these terms from linear algebra (Figure 2.5) is Chapter 6 of *Mathematical Methods of Physics* by Mathews and Walker [29].

Using Equation (2.14), we see that $\mathbb{W}$ *projects* onto the vector $\boldsymbol{x}^{(1)}$:

$$\mathbb{W}\boldsymbol{x}^{(1)} = \boldsymbol{x}^{(1)} . \tag{2.15}$$

In addition, it follows from Equation (2.14) that the matrix (2.13) is *idempotent*:

$$\mathbb{W}^t = \mathbb{W} \quad \text{for} \quad t = 1, 2, 3, \ldots . \tag{2.16}$$

Equations (2.15) and (2.16) together with $\operatorname{sgn}\left(x_i^{(1)}\right) = x_i^{(1)}$ imply that the network recognises the pattern $\boldsymbol{x}^{(1)}$ as the stored one. This example illustrates the general proof, Equations (2.10) and (2.11).

Now consider the distorted pattern (**a**) in Figure 2.4. We feed this pattern to the network by assigning

$$\boldsymbol{s}(t=0) = \begin{bmatrix} -1 \\ -1 \\ -1 \\ 1 \end{bmatrix} . \tag{2.17}$$

To compute one step in the synchronous dynamics (2.4), we apply $\mathbb{W}$ to $\boldsymbol{s}(t=0)$. This is done in two steps, using the outer-product form (2.13) of the weight matrix. We first multiply $\boldsymbol{s}(t=0)$ with $\boldsymbol{x}^{(1)\mathsf{T}}$ from the left

$$\boldsymbol{x}^{(1)\mathsf{T}}\boldsymbol{s}(t=0) = \begin{bmatrix}1, & -1, & -1, & 1\end{bmatrix}\begin{bmatrix}-1\\-1\\-1\\1\end{bmatrix} = 2\,, \tag{2.18}$$

and then we multiply this result with $\boldsymbol{x}^{(1)}$. This results in

$$\mathbb{W}\boldsymbol{s}(t=0) = \tfrac{1}{2}\boldsymbol{x}^{(1)}\,. \tag{2.19}$$

The signum of the i-th component of the vector $\mathbb{W}\boldsymbol{s}(t=0)$ yields $s_i(t=1)$:

$$s_i(t=1) = \operatorname{sgn}\Big(\sum_{j=1}^{N} w_{ij}s_j(t=0)\Big) = x_i^{(1)}\,. \tag{2.20}$$

We conclude that the state of the network converges to the stored pattern, in one synchronous update. Since $\mathbb{W}$ is idempotent, the network stays there: the pattern $\boldsymbol{x}^{(1)}$ is an attractor. Case (**b**) in Figure 2.4 works in a similar way.

Now look at case (**c**), where the network fails to converge to the stored pattern. We feed this pattern to the network by assigning $\boldsymbol{s}(t=0) = [-1,1,-1,-1]^{\mathsf{T}}$. For one iteration of the synchronous dynamics, we evaluate

$$\boldsymbol{x}^{(1)\mathsf{T}}\boldsymbol{s}(0) = \begin{bmatrix}1, & -1, & -1, & 1\end{bmatrix}\begin{bmatrix}-1\\1\\-1\\-1\end{bmatrix} = -2\,. \tag{2.21}$$

It follows that

$$\mathbb{W}\boldsymbol{s}(t=0) = -\tfrac{1}{2}\boldsymbol{x}^{(1)}\,. \tag{2.22}$$

Using the update rule (2.4), we find

$$\boldsymbol{s}(t=1) = -\boldsymbol{x}^{(1)}\,. \tag{2.23}$$

Equation (2.16) implies that

$$\boldsymbol{s}(t) = -\boldsymbol{x}^{(1)} \quad \text{for} \quad t \geq 1\,. \tag{2.24}$$

Thus the network shown in Figure 2.3 has two attractors, the pattern $\boldsymbol{x}^{(1)}$ as well as the *inverted* pattern $-\boldsymbol{x}^{(1)}$. As we shall see in Section 2.5, this is a general property of Hopfield networks: if $\boldsymbol{x}^{(1)}$ is an attractor, then the pattern $-\boldsymbol{x}^{(1)}$ is an attractor too. In the next section, we discuss what happens when more than one pattern is stored in the Hopfield network.

2.3 The Cross-Talk Term

When there are more patterns than just one, we need to generalise Equation (2.9). One possibility is to simply sum Equation (2.9) over the stored patterns [13]:

$$w_{ij} = \frac{1}{N}\sum_{\mu=1}^{p} x_i^{(\mu)} x_j^{(\mu)} \quad \text{and} \quad \theta_i = 0\,. \tag{2.25}$$

Equation (2.25) generalises Hebb's rule to p patterns. Because of the sum over μ, the relation to Hebb's learning hypothesis is less clear, but we nevertheless refer to Equation (2.25) as Hebb's rule. At any rate, we see that the weights are proportional to the second moments of the pattern bits. It is plausible that a neural network based upon the rule (2.25) can recognise properties of the patterns $\boldsymbol{x}^{(\mu)}$ that are encoded in two-point correlations of their bits.

Note that the weights are symmetric, $w_{ij} = w_{ji}$. Also, note that the prefactor N^{-1} in Equation (2.25) is not important. It is chosen to simplify the large-N analysis of the model (Chapter 3). An alternative version of Hebb's rule [2, 13] sets the diagonal weights to zero:

$$w_{ij} = \frac{1}{N}\sum_{\mu=1}^{p} x_i^{(\mu)} x_j^{(\mu)} \quad \text{for} \quad i \neq j, \quad w_{ii} = 0, \quad \text{and} \quad \theta_i = 0. \tag{2.26}$$

In this section, we use the form (2.26) of Hebb's rule. If we assign the weights in this way, does the network recognise the stored patterns? The question is whether

$$\operatorname{sgn}\underbrace{\Big(\frac{1}{N}\sum_{j\neq i}\sum_{\mu} x_i^{(\mu)} x_j^{(\mu)} x_j^{(\nu)}\Big)}_{=b_i^{(\nu)}} = x_i^{(\nu)} \tag{2.27}$$

holds or not. To check this, we repeat the calculation described in the previous section. As a first step, we evaluate the local field

$$b_i^{(\nu)} = \big(1 - \tfrac{1}{N}\big)x_i^{(\nu)} + \frac{1}{N}\sum_{j\neq i}\sum_{\mu\neq\nu} x_i^{(\mu)} x_j^{(\mu)} x_j^{(\nu)}\,. \tag{2.28}$$

Here we split the sum over the patterns into two contributions. The first term corresponds to $\mu = \nu$, where ν refers to the pattern that was fed to the network, the one we want the network to recognise. Condition (2.27) is satisfied if the second term in (2.28) does not affect the sign of the r.h.s. of this equation. This second term is called the *cross-talk* term.

Whether adding the cross-talk term to $\boldsymbol{x}^{(\nu)}$ affects $\operatorname{sgn}(b_i^{(\nu)})$ or not depends on the stored patterns. Since the cross-talk term contains a sum over $\mu = 1, \ldots, p$, we expect that this term does not matter if p is small enough. The fewer patterns we store, the more likely it is that all of them are recognised. Furthermore, by analogy with the example described in the previous section, it is plausible that the stored patterns are then also attractors, so that slightly distorted patterns converge to the correct stored pattern.

For a more quantitative analysis of the effect of the cross-talk term, we store patterns with random bits (*random patterns*). Different bits are assigned ± 1 independently with equal probability:

$$\text{Prob}(x_i^{(\nu)} = \pm 1) = \tfrac{1}{2}\,. \tag{2.29}$$

This means in particular that different patterns are *uncorrelated*, because their *covariance* vanishes:

$$\langle x_i^{(\mu)} x_j^{(\nu)} \rangle = \delta_{ij}\delta_{\mu\nu}\,. \tag{2.30}$$

Here $\langle \cdots \rangle$ denotes an average over many realisations of random patterns, and δ_{ij} is the *Kronecker delta*, equal to unity if $i = j$ but zero otherwise. Note that $\langle x_j^{(\mu)} \rangle = 0$. This follows from Equation (2.29).

Given an ensemble of random patterns, what is the probability that the cross-talk term changes $\text{sgn}(b_i^{(\nu)})$? In other words, what is the probability that the network produces a wrong bit in one asynchronous update if all bits were initially correct? The magnitude of the cross-talk term does not matter when it has the same sign as $x_i^{(\nu)}$. If it has a different sign, then the cross-talk term may matter. To determine when this is the case, one defines

$$C_i^{(\nu)} \equiv -x_i^{(\nu)} \underbrace{\frac{1}{N}\sum_{j\neq i}\sum_{\mu\neq\nu} x_i^{(\mu)} x_j^{(\mu)} x_j^{(\nu)}}_{\text{cross-talk term}}\,. \tag{2.31}$$

If $C_i^{(\nu)} < 0$, then the cross-talk term has same sign as $x_i^{(\nu)}$, so that the cross-talk term does not make a difference: adding this term does not change the sign of $x_i^{(\nu)}$. If $0 < C_i^{(\nu)} < 1$, it does not matter either, only when $C_i^{(\nu)} > 1$. The network produces an error in bit i of pattern ν if $C_i^{(\nu)} > 1$ (we approximated $1 - 1/N \approx 1$ in Equation (2.28), assuming that N is large).

2.4 One-Step Error Probability

The one-step *error probability* $P_{\text{error}}^{t=1}$ is defined as the probability that an error occurs in one attempt to update a bit, given that initially all bits were correct:

$$P_{\text{error}}^{t=1} = \text{Prob}(C_i^{(\nu)} > 1)\,. \tag{2.32}$$

Since patterns and bits are identically distributed, $\text{Prob}(C_i^{(\nu)} > 1)$ does not depend on i or ν. Therefore $P_{\text{error}}^{t=1}$ does not carry any indices.

How does $P_{\text{error}}^{t=1}$ depend on the parameters of the problem, p and N? When both p and N are large, we can use the *central-limit theorem* [29, 30] to answer this question. Since different bits/patterns are independent, we can think of $C_i^{(\nu)}$ as a sum of independent random numbers c_m that take the values -1 and $+1$ with equal probabilities,

$$C_i^{(\nu)} = -\frac{1}{N}\sum_{j\neq i}\sum_{\mu\neq\nu} x_i^{(\mu)} x_j^{(\mu)} x_j^{(\nu)} x_i^{(\nu)} = -\frac{1}{N}\sum_{m=1}^{(N-1)(p-1)} c_m\,. \tag{2.33}$$

There are $M = (N-1)(p-1)$ terms in the sum on the r.h.s. because terms with $\mu = \nu$ are excluded, and also those with $j = i$ [Equation (2.26)]. If we use the rule (2.25) instead, then there is a correction to Equation (2.33) from the diagonal weights. For $p \ll N$, this correction is small.

When p and N are large, the sum $\sum_{m=1}^{M} c_m$ contains a large number of independently identically distributed random numbers with mean zero and variance unity. It follows from the central-limit theorem [29, 30] that $\sum_{m=1}^{M} c_m$ is Gaussian distributed with mean zero and variance M.

Since the central-limit theorem plays an important role in the analysis of neural-network algorithms, it is worth discussing this theorem in a little more detail. To begin with, note that the sum $\sum_{m=1}^{M} c_m$ equals $2k - M$, where k is the number of occurrences of $c_m = +1$ in the sum. Choosing c_m randomly to equal either -1 or $+1$ is called a *Bernoulli trial* [30], and the probability $P_{k,M}$ of drawing k times $+1$ and $M-k$ times -1 is given by the *binomial* distribution [30]. In our case, the probability of $c_m = \pm 1$ equals $\frac{1}{2}$, so that

$$P_{k,M} = \binom{M}{k}\left(\tfrac{1}{2}\right)^k\left(\tfrac{1}{2}\right)^{M-k}. \tag{2.34}$$

Here $\binom{M}{k} = M!/[k!\,(M-k)!]$ denotes the number of ways in which k occurrences of $+1$ can be distributed over M places.

We want to show that $P_{k,M}$ approaches a Gaussian distribution for large M, with mean zero and with variance M. Since the variance diverges as $M \to \infty$, it is convenient to use the variable $z = (2k - M)/\sqrt{M}$. The central-limit theorem implies that z is Gaussian with mean zero and unit variance in the limit of large M. To prove that this is the case, we substitute $k = \frac{M}{2} + \frac{\sqrt{M}}{2} z$ into Equation (2.34) and take the limit of large M using Stirling's approximation

$$n! \approx \mathrm{e}^{n\log n - n + \frac{1}{2}\log 2\pi n}\,. \tag{2.35}$$

Expanding $P_{k,M}$ to leading order in M^{-1} assuming that z remains of order unity gives $P_{k,M} = \sqrt{2/(\pi M)}\exp(-z^2/2)$. Now one changes variables from k to z. This stretches local neighbourhoods $\mathrm{d}k$ to $\mathrm{d}z$. Conservation of probability implies that $P(z)\mathrm{d}z = P(k)\mathrm{d}k$. It follows that $P(z) = (\sqrt{M}/2)P(k)$, so that $P(z) = (2\pi)^{-1/2}$

$\exp(-z^2/2)$. In other words, the distribution of z is Gaussian with zero mean and unit variance, as we intended to show.

Returning to Equation (2.33), we conclude that $C_i^{(\nu)}$ is Gaussian distributed

$$P(C) = (2\pi\sigma_C^2)^{-1/2} \exp[-C^2/(2\sigma_C^2)]\,, \tag{2.36}$$

with zero mean, as illustrated in Figure 2.6, and with variance

$$\sigma_C^2 = \frac{M}{N^2} \approx \frac{p}{N}\,. \tag{2.37}$$

Here we used $M \approx Np$ for large N and p.

Another way to compute this variance is to square Equation (2.33) and to average over random patterns:

$$\sigma_C^2 = \frac{1}{N^2}\Big\langle\Big(\sum_{m=1}^{M} c_m\Big)^2\Big\rangle = \frac{1}{N^2}\sum_{n=1}^{M}\sum_{m=1}^{M}\langle c_n c_m\rangle. \tag{2.38}$$

Here $\langle\cdots\rangle$ denotes the average over random realisations of c_m. Since the random numbers c_m are independent for different indices and because $\langle c_m^2\rangle = 1$, we have that $\langle c_n c_m\rangle = \delta_{nm}$. So only the diagonal terms in the double sum contribute, summing to $M \approx Np$. This yields Equation (2.37).

To determine $P_{\rm error}^{t=1}$ [Equation (2.32)], we must integrate the distribution of C from 1 to ∞:

$$P_{\rm error}^{t=1} = \frac{1}{\sqrt{2\pi}\sigma_C}\int_1^\infty {\rm d}C\ {\rm e}^{-\frac{C^2}{2\sigma_C^2}} = \frac{1}{2}\Big[1 - {\rm erf}\Big(\sqrt{\frac{N}{2p}}\Big)\Big]\,. \tag{2.39}$$

Here erf is the *error function* defined as [31]

$$\operatorname{erf}(z) = \frac{2}{\sqrt{\pi}}\int_0^z {\rm d}x\ {\rm e}^{-x^2}\,. \tag{2.40}$$

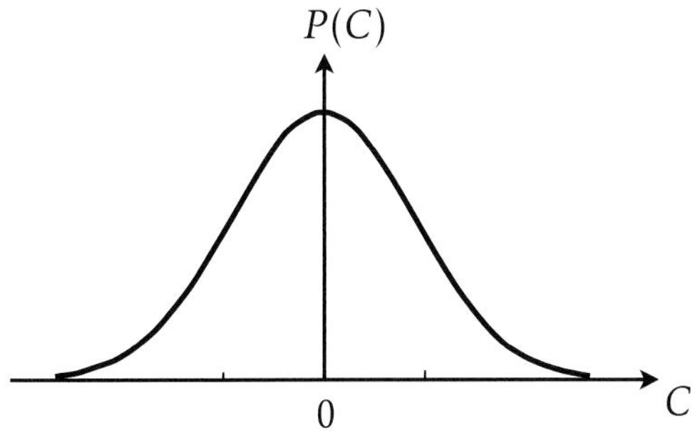

Figure 2.6 Gaussian distribution of the quantity C defined in Equation (2.31)

Since erf(z) increases monotonically as z increases, we conclude that $P_{\text{error}}^{t=1}$ increases as p increases, or as N decreases. This is expected: it is more difficult for the network to distinguish stored patterns when there are more of them. On the other hand, it is easier to differentiate stored patterns if they have more bits. We also see that the one-step error probability depends on p and N only through the combination

$$\alpha \equiv \frac{p}{N}. \tag{2.41}$$

The parameter α is called the *storage capacity* of the network. Figure 2.7 shows how $P_{\text{error}}^{t=1}$ depends on the storage capacity. For $\alpha = 0.2$, for example, the one-step error probability is slightly larger than 1%.

In the derivation of Equation (2.39), we assumed that the stored patterns are random with independent bits. Realistic patterns are not random. We nevertheless expect that $P_{\text{error}}^{t=1}$ describes the typical one-step error probability of the Hopfield network when p and N are large. However, it is straightforward to construct counterexamples. Consider for example *orthogonal patterns*:

$$\boldsymbol{x}^{(\mu)} \cdot \boldsymbol{x}^{(\nu)} = 0 \quad \text{for} \quad \mu \neq \nu\,. \tag{2.42}$$

For such patterns, the crosstalk term vanishes in the limit of large N (Exercise 2.2), so that $P_{\text{error}}^{t=1} = 0$.

More importantly, the error probability defined in this section refers only to the initial update, the first iteration. What happens in the next iteration, and after many iterations? Numerical experiments show that the error probability can be much higher in later iterations, because an error tends to increase the probability of making another error later on. So the estimate $P_{\text{error}}^{t=1}$ is only a lower bound for the probability of observing errors in the long run.

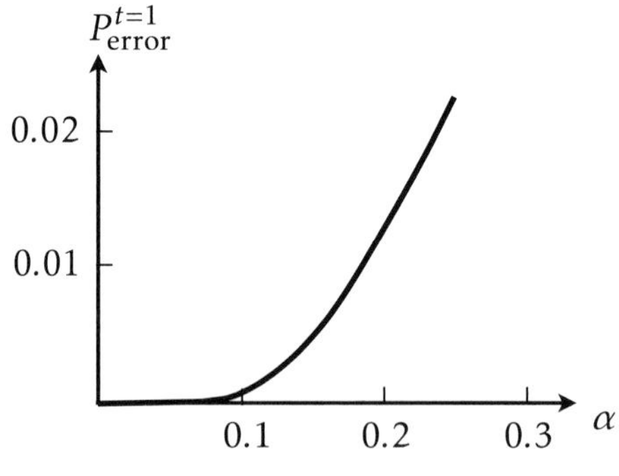

Figure 2.7 Dependence of the one-step error probability on the storage capacity α according to Equation (2.39)

2.5 Energy Function

Consider the long-time limit $t \to \infty$. Does the Hopfield dynamics converge, as required by Equation (2.7)? This is an important question in the analysis of neural-network algorithms, because an algorithm that does not converge to a meaningful solution is useless.

The standard way of analysing the convergence of neural-network algorithms is to define an *energy function* $H(\boldsymbol{s})$ that has a minimum at the desired solution, say $\boldsymbol{s} = \boldsymbol{x}^{(\nu)}$. We monitor how the energy function changes as we iterate, and keep track of the smallest values of H encountered, to find the minimum. If we store only one pattern, $p = 1$, then a suitable energy function is

$$H = -\frac{1}{2N}\Big(\sum_{i=1}^{N} s_i x_i^{(1)}\Big)^2 . \tag{2.43}$$

This function is minimal when $\boldsymbol{s} = \boldsymbol{x}^{(1)}$, *i.e.*, when $s_i = x_i^{(1)}$ for all i. It is customary to insert the factor $1/(2N)$; this does not change the fact that H is minimal at $\boldsymbol{s} = \boldsymbol{x}^{(1)}$.

A crucial point is that the asynchronous McCulloch-Pitts dynamics (2.5) *converges* to the minimum [13]. This follows from the fact that H cannot increase under the update rule (2.5). To prove this important property, we begin by evaluating the expression on the r.h.s. of Equation (2.43):

$$H = -\frac{1}{2}\sum_{ij}^{N}\Big(\frac{1}{N}x_i^{(1)}x_j^{(1)}\Big)s_i s_j . \tag{2.44}$$

Using Hebb's rule (2.9), we find that the energy function (2.43) becomes

$$H = -\frac{1}{2}\sum_{ij} w_{ij} s_i s_j . \tag{2.45}$$

This function has the same form as the energy function (or *Hamiltonian*) for certain physical models of magnetic systems consisting of interacting spins [32], where the interaction energy between spins s_i and s_j is $\frac{1}{2}(w_{ij} + w_{ji})s_i s_j$. Note that Hebb's rule (2.9) yields symmetric weights: $w_{ij} = w_{ji}$, and $w_{ii} > 0$. Note also that setting the diagonal weights to zero does not change the fact that H is minimal at $\boldsymbol{s} = \boldsymbol{x}^{(1)}$, because $s_i^2 = 1$. The diagonal weights just give a constant contribution to H, independent of $\boldsymbol{s}$.

The second step is to show that H cannot increase under the asynchronous McCulloch-Pitts dynamics (2.5). In this case, we say that the energy function is a *Lyapunov function*, or *loss function*. To demonstrate that the energy function is a Lyapunov function, choose a neuron m and update it according to Equation (2.5).

We denote the updated state of neuron m by s'_m:

$$s'_m = \operatorname{sgn}\Big(\sum_j w_{mj} s_j\Big). \tag{2.46}$$

All other neurons remain unchanged. There are two possibilities, either $s'_m = s_m$ or $s'_m = -s_m$. In the first case, H remains the same, $H' = H$. Here H' refers to the value of the energy function after the update (2.46). When $s'_m = -s_m$, by contrast, the energy function changes by the amount

$$\begin{aligned} H' - H &= -\frac{1}{2}\sum_{j\neq m}(w_{mj} + w_{jm})(s'_m s_j - s_m s_j) - \frac{1}{2} w_{mm}(s'_m s'_m - s_m s_m) \\ &= \sum_{j\neq m}(w_{mj} + w_{jm}) s_m s_j\,. \end{aligned} \tag{2.47}$$

The sum goes over all neurons j that are connected to the neuron m, the one to be updated in Equation (2.46). Now if the weights are symmetric, $H' - H$ equals

$$H' - H = 2\sum_{j\neq m} w_{mj} s_m s_j = 2\sum_j w_{mj} s_m s_j - 2w_{mm}\,. \tag{2.48}$$

Since the sign of $\sum_j w_{mj} s_j$ is that of $s'_m = -s_m$ and if $w_{mm} \geq 0$, it follows that

$$H' - H < 0\,. \tag{2.49}$$

In other words, the value of H must decrease when the state of neuron m changes, $s'_m \neq s_m$. In summary,[1] H either remains constant under the asynchronous McCulloch-Pitts dynamics ($s'_m = s_m$) or its value decreases ($s'_m \neq s_m$). Note that this does not hold for the synchronous dynamics (2.4); see Exercise 2.9. Since the energy function cannot increase under the asynchronous McCulloch-Pitts dynamics, it must converge to a minimum of the energy function. For the energy function (2.43) this implies that the dynamics must either converge to the stored pattern or to its inverse. Both are attractors.

We assumed the thresholds to vanish, but the proof also works when the thresholds are not zero, in this case for the energy function

$$H = -\frac{1}{2}\sum_{ij} w_{ij} s_i s_j + \sum_i \theta_i s_i \tag{2.50}$$

in conjunction with the update rule $s'_m = \operatorname{sgn}\big(\sum_j w_{mj} s_j - \theta_m\big)$.

[1] The derivation outlined here did not use the specific form of Hebb's rule (2.9), only that the weights are symmetric, and that $w_{mm} \geq 0$. However, the derivation fails when $w_{mm} < 0$. In this case, it is still true that H assumes a minimum at $\boldsymbol{s} = \boldsymbol{x}^{(1)}$, but H can increase under the update rule, so that convergence is not guaranteed. We therefore require that the diagonal weights are not negative.

Up until now, we considered only one stored pattern, $p = 1$. If we store more than one pattern [Hebb's rule (2.25)], the proof that (2.45) cannot increase under the McCulloch-Pitts dynamics works in the same way because no particular form of the weights w_{ij} was assumed, only that they must be symmetric, and that the diagonal weights must not be negative. Therefore it follows in this case too that the minima of the energy function must correspond to attractors, as illustrated schematically in Figure 2.8. The configuration space of the network, corresponding to all possible choices of $\boldsymbol{s} = [s_1, \dots s_N]^{\mathsf{T}}$, is drawn as a single axis, the x-axis. But when N is large, the configuration space is really very high dimensional.

However, some stored patterns may not be attractors when $p > 1$. This follows from our analysis of the cross-talk term in Section 2.2. If the cross-talk term causes errors for a certain stored pattern, then this pattern is not located at a minimum of the energy function. Another way to see this is to combine Equations (2.25) and (2.45) to give

$$H = -\frac{1}{2N} \sum_{\mu=1}^{p} \Big(\sum_{i=1}^{N} s_i x_i^{(\mu)} \Big)^2 . \tag{2.51}$$

While the energy function defined in Equation (2.43) has a minimum at $\boldsymbol{x}^{(1)}$, Equation (2.51) need not have a minimum at $\boldsymbol{x}^{(1)}$ (or at any other stored pattern), because a maximal value of $\big(\sum_{i=1}^{N} s_i x_i^{(1)} \big)^2$ may be compensated by terms stemming from other patterns. This happens rarely when p is small (Section 2.2).

Conversely, there may be minima that do not correspond to stored patterns. Such states are referred to as *spurious states*. The network may converge to spurious states. This is undesirable, but it occurs even when there is only one stored pattern, as we saw in Section 2.2: the McCulloch-Pitts dynamics may converge to the inverted pattern. This follows also from Equation (2.51): if $\boldsymbol{s} = \boldsymbol{x}^{(1)}$ is a local

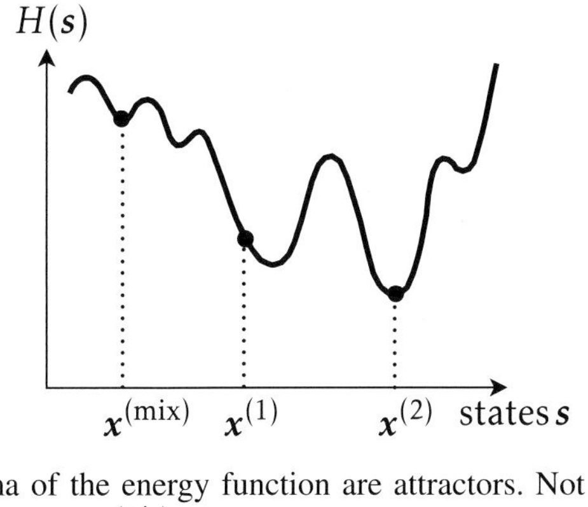

Figure 2.8 Minima of the energy function are attractors. Not all minima correspond to stored patterns ($\boldsymbol{x}^{(\text{mix})}$ is a mixed state; see the text), and stored patterns need not correspond to minima

minimum of H, then so is $\boldsymbol{s} = -\boldsymbol{x}^{(1)}$. This is a consequence of the invariance of H under $\boldsymbol{s} \to -\boldsymbol{s}$. There are other types of spurious states besides inverted patterns. An example are *mixed states*, *superpositions* of an odd number $2n+1$ of patterns [1]. For $n = 1$, for example, the bits of a mixed state read:

$$x_i^{(\text{mix})} = \text{sgn}\left(\pm x_i^{(1)} \pm x_i^{(2)} \pm x_i^{(3)}\right). \tag{2.52}$$

The number of mixed states increases as n increases. There are $2^{2n+1}\binom{p}{2n+1}$ mixed states that are superpositions of $2n+1$ out of p patterns, for $n = 1, 2, \ldots$ (Exercise 2.4). Mixed states such as (2.52) are sometimes recognised by the network (Exercise 2.5); therefore it may happen that the network converges to these states. Finally, there are spurious states that are not related in any way to the stored patterns $x_j^{(\mu)}$. Such *spin-glass* states are discussed in detail in Refs. [27, 33, 34], and also by Hertz, Krogh, and Palmer [1].

2.6 Summary

Hopfield networks are networks of McCulloch-Pitts neurons that recognise patterns (Algorithm 1). Their layout is defined by connection strengths, or weights, chosen according to Hebb's rule. The weights w_{ij} are symmetric, and the network is in general fully connected. Hebb's rule ensures that stored patterns are recognised, at least most of the time if the number of patterns is not too large. Convergence of the McCulloch-Pitts dynamics is analysed in terms of an energy function, which cannot increase under the asynchronous McCulloch-Pitts dynamics.

A single-step estimate for the error probability of the network dynamics was derived in Section 2.2. If one iterates several steps, the error probability is usually much larger, but it is difficult to evaluate in general. For stochastic Hopfield networks, the steady-state error probability can be estimated more easily, because the dynamics converges to a steady state (Chapter 3).

Algorithm 1 Pattern recognition with a deterministic Hopfield network

store patterns $\boldsymbol{x}^{(\mu)}$ using Hebb's rule;
feed distorted pattern $\boldsymbol{x}$ into network by assigning $\boldsymbol{s}(t=0) \leftarrow \boldsymbol{x}$;
for $t = 1, \ldots, T$ **do**
  choose a value of m and update $s_m(t) \leftarrow \text{sgn}\big(\sum_{j=1}^N w_{mj} s_j(t-1)\big)$;
end for
read out pattern $\boldsymbol{s}(T)$;

2.7 Exercises

2.1 Modified Hebb's rule. Show that the modified Hebb's rule (2.26) satisfies Equation (2.8) if we store only one pattern, $p = 1$.

2.2 Orthogonal patterns. For Hebb's rule (2.25), show that the cross-talk term vanishes for orthogonal patterns, so that $P_{\text{error}}^{t=1} = 0$. For the modified Hebb's rule (2.26), the cross-talk term is non-zero for orthogonal patterns. Show that it becomes negligible in the limit of large N.

2.3 Cross-talk term. Expression (2.33) for the cross-talk term was derived using modified Hebb's rule, Equation (2.26). How does Equation (2.33) change if you use the rule (2.25) instead? Show that the distribution of $C_i^{(\nu)}$ then acquires a non-zero mean, obtain an estimate for this mean value, and compute the one-step error probability. Show that your result approaches (2.39) for small values of α. Explain why your result is different from (2.39) for large α.

2.4 Mixed states. Explain why there are no mixed states that are superpositions of an even number of stored patterns. Show that there are $2^{2n+1}\binom{p}{2n+1}$ mixed states that are superpositions of $2n + 1$ out of p patterns, for $n = 1, 2, \ldots$.

2.5 Recognising mixed states. Store p random patterns in a Hopfield network with $N = 50$ and 100 neurons using Hebb's rule (2.25). Using computer simulations, determine the probability that the network recognises bit $x_i^{(\text{mix})}$ of the mixed state $\boldsymbol{x}^{(\text{mix})}$ with bits

$$x_i^{(\text{mix})} = \text{sgn}\big(x_i^{(1)} + x_i^{(2)} + x_i^{(3)}\big). \tag{2.53}$$

Show that the one-step error probability tends to zero as $\alpha \to 0$ for large N, by analysing how often $\text{sgn}\big(\frac{1}{N}\sum_{\mu=1}^{p}\sum_{j=1}^{N} x_i^{(\mu)} x_j^{(\mu)} x_j^{(\text{mix})}\big) = x_i^{(\text{mix})}$ holds. *Hint:* Think of $\frac{1}{N}\sum_j$ as an average of $x_j^{(\mu)} x_j^{(\text{mix})}$ over random bits and evaluate this average. Then apply the signum function.

2.6 Energy function. Figure 2.9 shows a network with two neurons with asymmetric weights, $w_{12} = 2$ and $w_{21} = -1$. Show that the energy function $H = -\frac{w_{12}+w_{21}}{2} s_1 s_2$ can increase under the asynchronous McCulloch-Pitts rule.

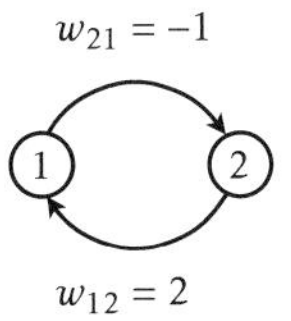

Figure 2.9 Two neurons with asymmetric connections (Exercise 2.6)

2.7 Higher-order Hopfield networks. Determine under which conditions the energy function $H = -\frac{1}{2}\sum_{ij} w_{ij}^{(2)} s_i s_j - \frac{1}{6}\sum_{ijk} w_{ijk}^{(3)} s_i s_j s_k$ is a Lyapunov function for the asynchronous dynamics $s'_m = \text{sgn}(b_m)$ with $b_m = \partial H/\partial s_m$.

2.8 Hebb's rule and energy function for 0/1 units. Suppose that the state of a neuron takes the values 0 (inactive) and 1 (active). The corresponding asynchronous update rule is $n'_m = \theta_{\rm H}\left(\sum_j w_{mj} n_j - \mu_m\right)$ with threshold μ_m. The activation function $\theta_{\rm H}(b)$ is the Heaviside function, equal to 0 if $b < 0$ and equal to 1 if $b \geq 0$ (Figure 2.10). Write down Hebb's rule for such 0/1 units and show that if one stores only one pattern, then this pattern is recognised. Show that $H = -\frac{1}{2}\sum_{ij} w_{ij} n_i n_j + \sum_i \mu_i n_i$ cannot increase under the asynchronous update rule (it is assumed that the weights are symmetric, and that $w_{ii} \geq 0$). See Ref. [13].

2.9 Energy function and synchronous dynamics. Analyse how the energy function (2.45) changes under the synchronous dynamics (2.4). Show that the energy function can increase, even though the weights are symmetric and the diagonal weights are zero.

2.10 Continuous Hopfield network. Hopfield [35] also analysed a version of his model with continuous-time dynamics. Consider $\tau \frac{\mathrm{d}}{\mathrm{d}t} n_i = -n_i + g\left(\sum_j w_{ij} n_j - \theta_i\right)$ with $g(b) = (1 + \mathrm{e}^{-b})^{-1}$ (this dynamical equation is slightly different from the one used by Hopfield [35]). Show that the energy function $E = -\frac{1}{2}\sum_{ij} w_{ij} n_i n_j + \sum_i \theta_i n_i + \sum_i \int_0^{n_i} \mathrm{d}n g^{-1}(n)$ cannot increase under the network dynamics if the weights are symmetric. It is not necessary to assume that $w_{ii} \geq 0$.

2.11 Hopfield network with four neurons. The pattern shown in Figure. 2.11 is stored in a Hopfield network using Hebb's rule $w_{ij} = \frac{1}{N} x_i^{(1)} x_j^{(1)}$. There are 2^4 four-bit patterns. Apply each of these to the Hopfield network, and perform one synchronous update. List the patterns you obtain and discuss your results.

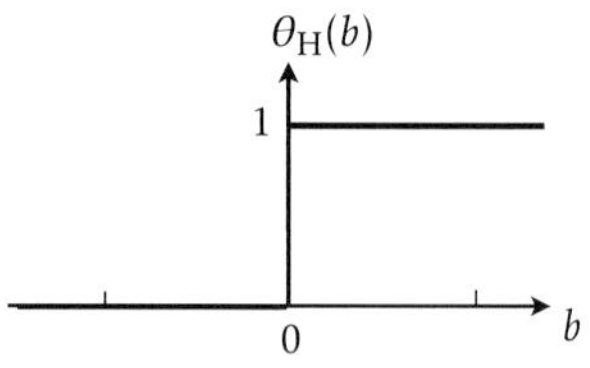

Figure 2.10 Heaviside function (Exercise 2.8)

Figure 2.11 The pattern $\boldsymbol{x}^{(1)}$ has $N = 4$ bits, $x_1^{(1)} = 1$, and $x_i^{(1)} = -1$ for $i = 2, 3, 4$. See Exercise 2.11

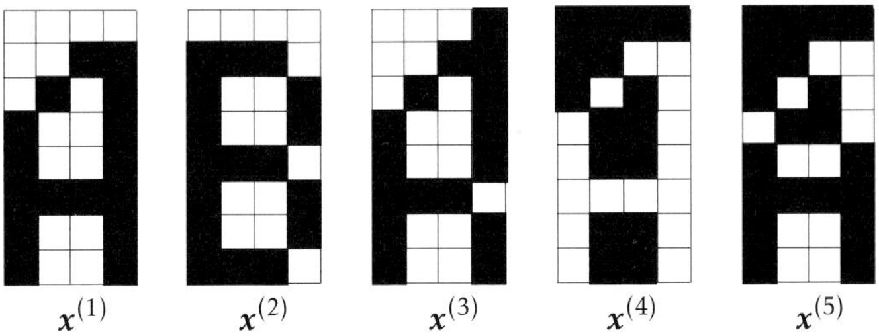

Figure 2.12 Each of the five patterns consists of 32 bits $x_i^{(\mu)}$. A black pixel i in pattern μ corresponds to $x_i^{(\mu)} = 1$, a white one to $x_i^{(\mu)} = -1$. See Exercise 2.12

2.12 Recognising letters with a Hopfield network. The five patterns in Figure 2.12 each have $N = 32$ bits. Store the patterns $\boldsymbol{x}^{(1)}$ and $\boldsymbol{x}^{(2)}$ in a Hopfield network using Hebb's rule $w_{ij} = \frac{1}{N}\sum_{\mu=1}^{2} x_i^{(\mu)} x_j^{(\mu)}$. Which of the patterns in Figure 2.12 remain unchanged after one synchronous update with $s_i' = \operatorname{sgn}\left(\sum_{j=1}^{N} w_{ij} s_j\right)$? *Hint*: read off $\sum_{j=1}^{N} x_j^{(\mu)} x_j^{(\nu)}$ from the *Hamming* distance between the two patterns, equal to the number of bits by which the patterns differ. Use this quantity to express the local fields $b_i^{(\mu)}$ as linear combinations of $x_i^{(1)}$ and $x_i^{(2)}$.

2.13 XOR function. The Boolean XOR function takes two binary inputs. For the inputs $[-1, -1]$ and $[1, 1]$ the function evaluates to -1, for the other two inputs to $+1$. Try to encode the XOR function in a Hopfield network with three neurons by storing the patterns $[-1, -1, -1]$, $[1, 1, -1]$, $[-1, 1, 1]$, and $[1, -1, 1]$ using Hebb's rule. Test whether the patterns are recognised or not. Discuss your findings.

2.14 Distance as a measure of convergence. The distance $d = \frac{1}{4N}\sum_i (s_i - x_i^{(1)})^2$ [Equation (2.2)] has a minimum at $\boldsymbol{s} = \boldsymbol{x}^{(1)}$. How are d and H [Equation (2.43)] related? What is the advantage of using H instead of d as a measure of convergence?

3
Stochastic Hopfield Networks

Two related problems became apparent in the previous chapter. First, the Hopfield dynamics may get stuck in local minima. In fact, if there is a local minimum down-hill from a given initial state, between this state and the correct attractor, then the dynamics arrests in the local minimum, so that the algorithm fails to converge to the correct attractor. Second, the energy function usually is a strongly varying function over a high-dimensional configuration space. Therefore it is difficult to predict the first local minimum encountered by the down-hill dynamics of the network.

Both problems are solved by introducing an element of stochasticity into the dynamics. This is a trick that works for many neural-network algorithms. In general, it is quite challenging to analyse the stochastic dynamics. For the Hopfield network, by contrast, much is known. The reason is that the stochastic Hopfield network is closely related to systems studied in statistical mechanics, so-called spin glasses. Like these systems – and many other physical systems – the stochastic Hopfield network exhibits an *order-disorder transition.* This transition becomes sharp in the limit of a large number of neurons. This has important consequences. Suppose that the network produces satisfactory results for a given number of patterns with a certain number of bits. If one tries to store just one more pattern, the network may fail to recognise anything. The goal of this chapter is to explain why this occurs and how it can be avoided.

3.1 Stochastic Dynamics

The asynchronous update rule (2.5) is called *deterministic* because a given set of states s_j determines the outcome of the update of neuron m. To introduce noise, one replaces the rule (2.5) by an asynchronous *stochastic* rule [36]:

$$s'_m = \begin{cases} +1 & \text{with probability} \quad p(b_m)\,, \\ -1 & \text{with probability} \quad 1 - p(b_m)\,. \end{cases} \tag{3.1a}$$

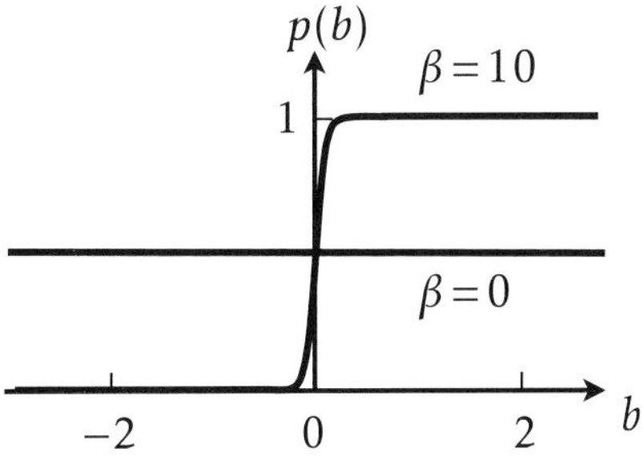

Figure 3.1 Probability function (3.1b) used in the definition of the stochastic rule (3.1). Here it is plotted for $\beta = 10$ and $\beta = 0$

A neuron with update rule (3.1a) is called *binary stochastic neuron*. As before, $b_m = \sum_j w_{mj} s_j - \theta_m$ is the local field, and the probability $p\,(b)$ is given by

$$p\,(b) = \frac{1}{1 + \mathrm{e}^{-2\beta b}}\,. \tag{3.1b}$$

The function $p\,(b)$ is plotted in Figure 3.1. The parameter β is the noise parameter. When β is large, the *noise level* is small. As β tends to infinity, the function $p\,(b)$ approaches zero if b is negative, and it tends to unity if b is positive. So for $\beta \to \infty$, the stochastic update rule (3.1) converges to the deterministic rule (2.5). In the opposite limit, when $\beta = 0$, the update probability $p\,(b)$ simply equals $\frac{1}{2}$. In this case, s_i is updated to -1 or $+1$ randomly, with equal probability. The dynamics does not depend upon the stored patterns contained in the local field $\boldsymbol{b}$.

The idea is to keep a small but finite noise level. Then the network dynamics is very similar to the deterministic Hopfield dynamics analysed in the previous chapter. But the noise allows the system to escape local minima. However, since the dynamics is stochastic, we must rephrase the convergence criterion (2.7). This is discussed next.

3.2 Order Parameters

If we feed one of the stored patterns, $\boldsymbol{x}^{(1)}$ for example, then we want the stochastic dynamics to stay in the vicinity of $\boldsymbol{x}^{(1)}$. This can only work if the noise is weak enough, and even then it is not guaranteed. At time step t, bit i is correct if $s_i(t)x_i^{(1)} = 1$. All bits are correct when $\sum_{i=1}^N s_i(t)x_i^{(1)} = N$; otherwise, the sum takes a value smaller than N. One measures success by averaging $\frac{1}{N}\sum_{i=1}^N s_i(t)x_i^{(1)}$ over the asynchronous stochastic dynamics of the network from $t = 0$ to $t = T$, for given bits $x_i^{(\mu)}$:

$$m_\mu(T) = \frac{1}{T}\sum_{t=1}^{T}\left(\frac{1}{N}\sum_{i=1}^{N} s_i(t)x_i^{(\mu)}\right). \tag{3.2a}$$

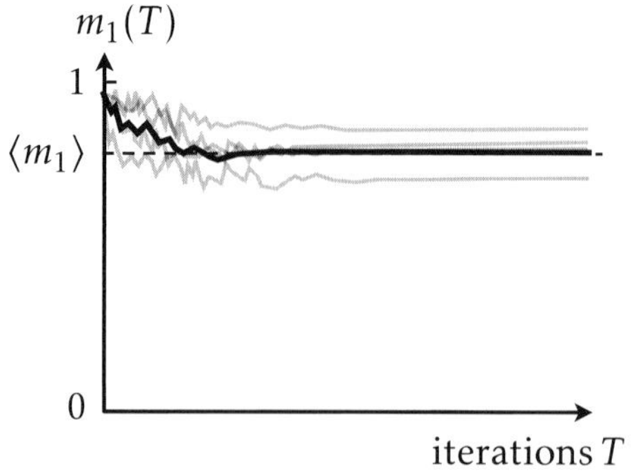

Figure 3.2 The finite-time average $m_1(T)$ [Equation (3.2a)] depends upon the total iteration time T. The light-gray lines show results for $m_1(T)$ for different realisations of random patterns stored in the network, at a large but finite N. The black line is the average of $m_1(T)$ over the different realisations of random patterns

If we feed pattern $\boldsymbol{x}^{(1)}$ to the network, we have $m_1(t=0) = 1$. We want $m_1(t)$ to remain close to unity so that the network recognises the pattern $\boldsymbol{x}^{(1)}$. In practice, the quantity $\frac{1}{N}\sum_{i=1}^{N} s_i(t)x_i^{(1)}$ settles into a *steady state*, where it fluctuates around a mean value with a definite distribution that becomes independent of the iteration number t. If the network works well, the finite-time average $m_1(T)$ converges to a value of order unity after an initial *transient* (Figure 3.2). The limiting value

$$m_1 \equiv \lim_{T\to\infty} m_1(T) \equiv \frac{1}{N}\sum_{i=1}^{N} \langle s_i \rangle x_i^{(1)} \tag{3.2b}$$

is called the *order parameter*. Since there is noise, the order parameter m_1 is usually smaller than unity. The last equality in Equation (3.2b) defines the time average $\langle s_i \rangle$ over the stochastic network dynamics.

Figure 3.2 also illustrates a subtlety. For finite values of N, the order parameter m_1 depends upon the stored patterns. Different realisations $\boldsymbol{x}^{(1)}, \ldots, \boldsymbol{x}^{(p)}$ of random patterns yield different values of m_1. In the limit of $N \to \infty$, this problem does not occur and the order parameter m_1 becomes independent of the stored patterns. We say that the system is *self-averaging* in this limit. To obtain definite value for the order parameter for finite values of N, one usually averages m_1 over different realisations of random patterns stored in the network (thick solid line in Figure 3.2). The dashed line in Figure 3.2 shows this average, $\langle m_1 \rangle$.

The other components, $m_\mu = \lim_{T\to\infty} m_\mu(T)$ for $\mu > 1$, are expected to be small. This is certainly true for random patterns with many independent bits. If $s_i(t) \approx x_i^{(1)}$, the individual terms in the sum over i in Equation (3.2) cancel approximately upon summation, because the bits of the patterns $\boldsymbol{x}^{(2)}$ to $\boldsymbol{x}^{(p)}$ are independent from those of $\boldsymbol{x}^{(1)}$. In summary, if we feed pattern $\boldsymbol{x}^{(1)}$, and if the network works well, we expect in the limit of large N:

$$m_\mu \approx \begin{cases} 1 & \text{if} \quad \mu = 1, \\ 0 & \text{otherwise.} \end{cases} \tag{3.3}$$

Whether this is the case or not depends on the values of p, N, and β. In the next sections, we determine how m_1 depends on these parameters.

3.3 Mean-Field Theory

The order parameter is defined as an average over the stochastic dynamics of the network in its steady state (Figure 3.2). It is a challenging task to compute this average because all neurons interact with each other in a non-linear fashion. Consider neuron number i. The fate of s_i is determined by its local field b_i, through Equation (3.1). The difficulty is that the local field in turn depends on the states s_j of all other neurons in the network:[1]

$$b_i(t) = \sum_{j=1}^{N} w_{ij} s_j(t)\,. \tag{3.4}$$

When N is large, we may assume that $b_i(t)$ remains essentially constant in the steady state, independent of t, because fluctuations of $s_j(t)$ average out when summing over j:

$$b_i(t) = \langle b_i \rangle + \text{fluctuations}\,. \tag{3.5}$$

Since $b_i(t)$ is given by a sum over many random numbers, we appeal to the central-limit theorem and argue that the fluctuations of $b_i(t)$ are of order $\sqrt{N}$. Since $\langle b_i(t)\rangle \sim N$, we ignore the fluctuations in the limit of large N and write

$$b_i(t) \approx \langle b_i \rangle = \sum_{j=1}^{N} w_{ij} \langle s_j \rangle = \frac{1}{N} \sum_{\mu} \sum_{j \neq i} x_i^{(\mu)} x_j^{(\mu)} \langle s_j \rangle\,, \tag{3.6}$$

using Hebb's rule (2.26) for given patterns $\boldsymbol{x}^{(\mu)}$. The time-averaged local field $\langle b_i \rangle$ is called the *mean field*. Theories that neglect the fluctuations in Equation (3.5) are called *mean-field theories*. They require a self-consistent solution, because the average $\langle s_j \rangle$ on the r.h.s. of Equation (3.6) depends on the mean field. Using the stochastic update rule (3.1), we find

$$\langle s_i \rangle = \text{Prob}(s_i = +1) - \text{Prob}(s_i = -1) = p(\langle b_i \rangle) - [1 - p(\langle b_i \rangle)] = \tanh(\beta \langle b_i \rangle)\,. \tag{3.7}$$

[1] We set the thresholds to zero, as assumed in Hebb's rule (2.26).

Equations (3.6) and (3.7) yield a set of N non-linear self-consistent equations for $\langle s_i \rangle$,

$$\langle s_i \rangle = \tanh(\beta \langle b_i \rangle) \quad \text{with} \quad \langle b_i \rangle = \frac{1}{N} \sum_{\mu} \sum_{j \neq i} x_i^{(\mu)} x_j^{(\mu)} \langle s_j \rangle \,. \tag{3.8}$$

Recall that the averages $\langle \cdots \rangle$ are time averages, evaluated for given patterns $\boldsymbol{x}^{(\mu)}$.

An equivalent yet slightly different derivation of the mean-field equations (3.8) is this: suppose we average s_i over the dynamics (3.1) at fixed $s_j \neq s_i$, and then we average all s_j over the dynamics. This gives $\langle s_i \rangle = \langle \tanh(\beta b_i) \rangle$. Comparing with Equation (3.8), we see that the mean-field approximation corresponds to approximating $\langle \tanh(\beta b_i) \rangle \approx \tanh(\beta \langle b_i \rangle)$.

Now, in order to calculate the order parameters,

$$m_\mu = \frac{1}{N} \sum_{j=1}^{N} \langle s_j \rangle x_j^{(\mu)} \,, \tag{3.9}$$

we must solve the mean-field equations (3.8) to obtain the time averages $\langle s_i \rangle$ in Equation (3.9). To this end, we express the mean field $\langle b_i \rangle$ in terms of the order parameters m_μ:

$$\langle b_i \rangle = \frac{1}{N} \sum_{\mu=1}^{p} \sum_{j \neq i} x_i^{(\mu)} x_j^{(\mu)} \langle s_j \rangle \approx \sum_{\mu=1}^{p} x_i^{(\mu)} m_\mu \,. \tag{3.10}$$

The last equality is only approximate because the j-sum in the definition of m_μ contains the term $j = i$. Whether or not to include this term makes only a small difference in the limit of large N.

Let us first calculate m_1 assuming Equation (3.3), neglecting terms with $\mu \neq 1$ in Equation (3.10). To make sure that these small $\mu \neq 1$-terms do not add up to a substantial correction to the first term, the storage capacity must be small enough. For large values of N, we assume

$$\alpha \ll 1 \,. \tag{3.11}$$

In this case, it is sufficient to keep only the first term on the r.h.s. of Equation (3.10) [37]. This approximation yields together with Equation (3.8):

$$\langle s_i \rangle = \tanh(\beta \langle b_i \rangle) \approx \tanh\left(\beta m_1 x_i^{(1)}\right) . \tag{3.12}$$

Applying the definition (3.9) of the order parameter, one finds

$$m_1 = \frac{1}{N} \sum_{i=1}^{N} \tanh\left(\beta m_1 x_i^{(1)}\right) x_i^{(1)} \,. \tag{3.13}$$

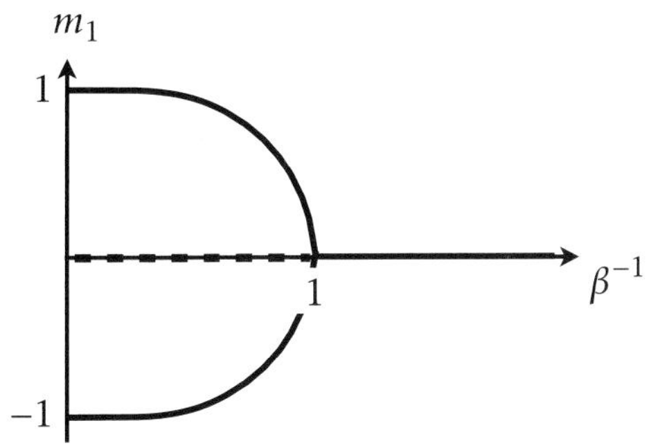

Figure 3.3 Solutions of the mean-field equation (3.14) are represented by solid lines. The dashed line corresponds to an unstable solution. Its stability changes at the critical noise level, $\beta_c = 1$.

Using that $\tanh(z) = -\tanh(-z)$ as well as the fact that the bits $x_i^{(\mu)}$ can only assume the values ± 1, one obtains:

$$m_1 = \tanh(\beta m_1)\,. \tag{3.14}$$

This is a self-consistent equation for m_1. For $\beta \to 0$, it has the solution $m_1 = 0$. This is not the desired solution because $m_1 = 0$ means that $\boldsymbol{x}^{(1)}$ is not recognised. For $\beta \to \infty$, by contrast, there are three solutions, $m_1 = 0, \pm 1$. Figure 3.3 shows results of the numerical evaluation of Equation (3.14) for intermediate values of β. For β larger the critical value

$$\beta_c = 1\,, \tag{3.15}$$

the three solutions persist. The solution $m_1 = 0$ is *unstable*. This can be shown by computing the derivatives of the *free energy* of the Hopfield network [1]. In other words, if we start with an initial condition that corresponds to $m_1 = 0$, the network dynamics does not stay there. The other two solutions are *stable*: when the network is initialised close to $\boldsymbol{x}^{(1)}$, then it converges to $m_1 > 0$, as long as $\beta > \beta_c$.

The symmetry of the problem dictates that there must also be a solution with $-m_1$. This solution corresponds to the inverted pattern $-\boldsymbol{x}^{(1)}$ (Section 2.5). If we start in the vicinity of $\boldsymbol{x}^{(1)}$, then the network is unlikely to converge to $-\boldsymbol{x}^{(1)}$, provided that N is large enough. The probability of the dynamical transition $\boldsymbol{x}^{(1)} \to -\boldsymbol{x}^{(1)}$ vanishes very rapidly as N increases and as the noise level decreases. If this transition occurs in a simulation, in this limit, the network stays near $-\boldsymbol{x}^{(1)}$ for a very long time. Consider the limit where T tends to ∞ at a finite but large value of N. Then the network jumps back and forth between $\boldsymbol{x}^{(1)}$ and $-\boldsymbol{x}^{(1)}$ at a very small rate. As a result, the order parameter averages to zero. This shows that the limits of large N and large T do not commute:

$$\lim_{T\to\infty}\lim_{N\to\infty} m_1(T) \neq \lim_{N\to\infty}\lim_{T\to\infty} m_1(T)\,. \tag{3.16}$$

In practice, the interesting limit is the left one – that of a large network run for a time T much longer than the initial transient, but not infinite. This is precisely where the mean-field theory applies. It corresponds to taking the limit $N \to \infty$ first, at finite but large T. This describes simulations where the transition $\boldsymbol{x}^{(1)} \to -\boldsymbol{x}^{(1)}$ does not occur.

In summary, Equation (3.14) predicts that the order parameter converges to a definite value, m_1, independent of the stored patterns in the limit $N \to \infty$. When N is finite, the limiting value of the order parameter does depend on the stored patterns (Figure 3.2). In this case, one also averages over different realisations of the stored patterns, as mentioned previously. The value of this average, $\langle m_1\rangle$, determines the average error probability $P^{t=\infty}_{\text{error}}$ in the steady state, the average fraction of wrong bits. The steady-state average number of correct bits is given by

$$\Big\langle \frac{1}{2}\sum_{i=1}^{N}\Big(1+\langle s_i\rangle x_i^{(1)}\Big)\Big\rangle = \frac{N}{2}(1+\langle m_1\rangle)\,, \tag{3.17}$$

because $\frac{1}{2}\big(1+s_i x_i^{(1)}\big) = 1$ if $x_i^{(1)}$ is correct, and equal to zero otherwise. The outer average is over different realisations of random patterns (the inner average is over the network dynamics). The second equality follows from Equation (3.2b). Since the l.h.s of Equation (3.17) equals N times $1 - P^{t=\infty}_{\text{error}}$, we deduce

$$P^{t=\infty}_{\text{error}} = \tfrac{1}{2}(1-\langle m_1\rangle)\,. \tag{3.18}$$

Since $m_1 \to 1$ as $\beta\to\infty$, the steady-state error probability tends to zero in this limit. This is expected since the stored patterns $\boldsymbol{x}^{(\mu)}$ are recognised for small enough values of α in the deterministic limit, when the cross-talk term is negligible. But note that the stochastic dynamics *slows down* as the noise level tends to zero. The lower the noise level, the longer the network remains stuck in local minima, so that it takes longer time to reach the steady state, and to sample the steady-state statistics of H. In the opposite limit, $\beta \to 0$, the steady-state error probability tends to $\frac{1}{2}$, because $m_1 \to 0$. In this noise-dominated limit the stochastic network ceases to function. If one were to assign N bits entirely randomly, then half of them would be correct, on average, $P^{t=\infty}_{\text{error}} = \frac{1}{2}$.

It is important to note that noise can help in another way: it may prevent the network dynamics from converging to mixed states (Section 2.5). This can be seen as follows [1, 33]. To derive the above mean-field result, we assumed $m_1 \approx 1$ and $m_\mu \approx 0$ for $\mu \neq 1$. Mixed states correspond to solutions where an odd number of components of $\boldsymbol{m}$ is non-zero – for example,

$$\boldsymbol{m}^{(\text{mix})} = \begin{bmatrix} m \\ m \\ m \\ 0 \\ \vdots \end{bmatrix} . \tag{3.19}$$

Neglecting the cross-talk term, the mean-field equation reads

$$\langle s_i \rangle = \tanh\Big(\beta \sum_{\mu=1}^{p} m_\mu^{(\text{mix})} x_i^{(\mu)}\Big) . \tag{3.20}$$

In the limit of $\beta \to \infty$, the averages $\langle s_i \rangle$ converge to the mixed states (2.52) when $\boldsymbol{m}^{(\text{mix})}$ is given by Equation (3.19). Averaging over the bits of the random patterns, one finds

$$m_\mu^{(\text{mix})} = \Big\langle x_i^{(\mu)} \tanh\Big(\beta \sum_{\nu=1}^{p} m_\nu^{(\text{mix})} x_i^{(\nu)}\Big)\Big\rangle . \tag{3.21}$$

The numerical solution of Equation (3.21) shows that there is a non-zero solution for $\beta^{-1} < \beta_{\rm c}^{-1} = 1$. Yet this solution is unstable for $0.46 < \beta^{-1} < 1$ [33]. In other words, the mixed states have a lower critical noise level than the stored patterns, equal to 0.46. For noise levels larger than that, but still smaller than unity, the network can recognise the stored patterns, and it does not converge to mixed states.

However, these results were obtained assuming that only one (or a few) order parameters are not zero. This corresponds to the limit of $\alpha = p/N \to 0$, where the cross-talk term (Section 2.3) is negligible. The next section describes a mean-field theory that remains accurate for larger values of α.

3.4 Critical Storage Capacity

The analysis in the previous section kept only the first term in the sum on the r.h.s. of Equation (3.10), $x_i^{(1)} m_1$. This can only work when p/N is small enough. Now we discuss how to proceed when p/N is not small.

Note that the analysis in Section 2.2 did not assume that p/N is small, but it yielded only the one-step error probability $P_{\text{error}}^{t=1}$, and we discussed the storage capacity $\alpha = p/N$ in relation to the one-step error probability. As the network dynamics is iterated, however, the number of errors tends to increase, at least when α is large enough so that the cross-talk term matters. Now we describe how to compute $P_{\text{error}}^{t=\infty}$ for general values of the storage capacity α, in order to demonstrate how the errors multiply as one iterates, causing the network to fail.

As before, we store p patterns in the network using Hebb's rule (2.26) and feed pattern $\boldsymbol{x}^{(1)}$ to the network. The aim is to determine the order parameter m_1 and the corresponding error probability in the steady state for $p \sim N$, so that α remains finite as $N \to \infty$. In this case, we can no longer approximate the sum in Equation (3.10) just by its first term, because the other terms for $\mu > 1$ may sum up to a contribution that is of the same order as m_1. Instead, we must evaluate all m_μ to compute the mean field $\langle b_i \rangle$.

The relevant calculation is summarised in Chapter 4 of Ref. [38]. It is also outlined in Section 2.5 of Hertz, Krogh, and Palmer [1]. The remainder of this section follows this outline quite closely. One starts by rewriting the mean-field equations (3.8) in terms of the order parameters m_μ. Using

$$\langle s_i \rangle = \tanh\Big(\beta \sum_\mu x_i^{(\mu)} m_\mu\Big), \tag{3.22}$$

we find

$$m_\nu = \frac{1}{N}\sum_i x_i^{(\nu)} \langle s_i \rangle = \frac{1}{N}\sum_i x_i^{(\nu)} \tanh\Big(\beta \sum_\mu x_i^{(\mu)} m_\mu\Big). \tag{3.23}$$

This coupled set of p non-linear equations is equivalent to the mean-field equations (3.8) in the limit of large N.

Now feed pattern $\boldsymbol{x}^{(1)}$ to the network. We assume that the network stays close to the pattern $\boldsymbol{x}^{(1)}$ in the steady state, so that m_1 remains of order unity. The other m_μ remain small. When p is large, however, we cannot simply approximate the sum over μ on the r.h.s. of Equation (3.23) by its first term only, because the sum of the remaining (small) terms might not be negligible. Therefore we need to estimate these terms, the other order parameters m_μ for $\mu \neq 1$.

The trick is to assume that the pattern bits are random, uncorrelated with mean zero [Equations (2.29) and (2.30)]. In this case, the order parameters m_μ, $\mu = 2, \ldots, p$, become random numbers that fluctuate around zero with variance $\langle m_\mu^2 \rangle$. (This average is over random patterns.) We use Equation (3.23) to compute the variance approximately.

In the μ-sum on the r.h.s of Equation (3.23), we must treat the term $\mu = \nu$ separately, because the index ν appears also on the l.h.s. of this equation. Also, the term $\mu = 1$ must be treated separately, as before, because $\mu = 1$ is the index of the pattern that is fed to the network. As a consequence, the calculations of m_1 and m_ν for $\nu \neq 1$ proceed slightly differently. We begin with the first case. Given that $x_i^{(\mu)} = \pm 1$, and that $\tanh(z)$ is an odd function, Equation (3.23) simplifies to

$$m_1 = \frac{1}{N}\sum_i \tanh\Big(\beta m_1 + \beta \sum_{\mu \neq 1} x_i^{(\mu)} x_i^{(1)} m_\mu\Big). \tag{3.24}$$

The next steps are similar to the analysis of the cross-talk term in Section 2.2. One assumes that the patterns are random, that their bits $x_i^{(\mu)} = \pm 1$ are independently and identically distributed. In the limit of large N and p, the sums in Equation (3.24) can then be estimated using the central-limit theorem. For random patterns, the variable

$$z \equiv \sum_{\mu \neq 1} x_i^{(\mu)} x_i^{(1)} m_\mu \tag{3.25}$$

is a sum of many independent, identically distributed random numbers with mean zero and finite variance. The variable z is therefore approximately Gaussian distributed, with mean zero. As a consequence, the distribution of z is entirely determined by its variance σ_z^2, and it is independent of i.

Returning to Equation (3.24), one approximates the sum $\frac{1}{N}\sum_i$ as an average over the Gaussian distributed variable z. This yields

$$m_1 = \int \frac{\mathrm{d}z}{\sqrt{2\pi\sigma_z^2}} e^{-\frac{z^2}{2\sigma_z^2}} \tanh(\beta m_1 + \beta z)\,. \tag{3.26}$$

Equation (3.26) is the desired result, a self-consistent equation for m_1 replacing the mean-field equation (3.14).

In order to determine m_1, we need to estimate the variance σ_z^2 featuring in Equation (3.26). To this end, one squares Equation (3.25), and averages the resulting double sum over pattern indices. Since the bits $x_i^{(\mu)}$ and $x_i^{(\mu')}$ are independent when $\mu \neq \mu'$, only the diagonal terms in this double sum contribute to the average:

$$\sigma_z^2 = \sum_{\mu \neq 1} \langle m_\mu^2 \rangle \approx p \langle m_\mu^2 \rangle \quad \text{for any} \quad \mu \neq 1\,. \tag{3.27}$$

Here we assumed that p is large, and approximated $p - 1 \approx p$. To evaluate the variance further, it is necessary to estimate the remaining order parameters. One starts again from Equation (3.23) and writes for $\nu \neq 1$

$$\begin{aligned} m_\nu &= \frac{1}{N}\sum_i x_i^{(\nu)} \tanh\Big(\beta x_i^{(1)} m_1 + \beta x_i^{(\nu)} m_\nu + \beta \sum_{\substack{\mu \neq 1 \\ \mu \neq \nu}} x_i^{(\mu)} m_\mu\Big) \\ &= \frac{1}{N}\sum_i x_i^{(\nu)} x_i^{(1)} \tanh\Big(\underbrace{\beta m_1}_{①} + \underbrace{\beta x_i^{(1)} x_i^{(\nu)} m_\nu}_{②} + \underbrace{\beta \sum_{\substack{\mu \neq 1 \\ \mu \neq \nu}} x_i^{(\mu)} x_i^{(1)} m_\mu}_{③}\Big)\,. \end{aligned} \tag{3.28}$$

Consider the three terms in the argument of $\tanh(\ldots)$. The term ① is of order unity, it is independent of N. The term ③ may be of the same order, because the sum over μ contains $\sim pN$ terms. The term ②, by contrast, is small for large values of N. Therefore it is a good approximation to Taylor-expand tanh(...) as follows:

$$\tanh\left(①+②+③\right) \approx \tanh\left(①+③\right) + ②\frac{\mathrm{d}}{\mathrm{d}x}\tanh\Big|_{①+③} + \ldots . \quad (3.29)$$

Using $\frac{\mathrm{d}}{\mathrm{d}x}\tanh(x) = 1 - \tanh^2(x)$, one obtains

$$m_\nu = \frac{1}{N}\sum_i x_i^{(\nu)} x_i^{(1)} \tanh\Big(\underbrace{\beta m_1}_{①} + \underbrace{\beta \sum_{\substack{\mu\neq 1\\ \mu\neq\nu}} x_i^{(\mu)} x_i^{(1)} m_\mu}_{③}\Big)$$
$$+ \frac{1}{N}\sum_i x_i^{(\nu)} x_i^{(1)} \underbrace{\beta x_i^{(1)} x_i^{(\nu)} m_\nu}_{②}\Big[1 - \tanh^2\Big(\beta m_1 + \beta \sum_{\substack{\mu\neq 1\\ \mu\neq\nu}} x_i^{(\mu)} x_i^{(1)} m_\mu\Big)\Big]. \quad (3.30)$$

Using the fact that $x^{(\mu)} = \pm 1$ and thus $[x_i^{(\mu)}]^2 = 1$, this expression simplifies as follows:

$$m_\nu = \frac{1}{N}\sum_i x_i^{(\nu)} x_i^{(1)} \tanh\Big(\beta m_1 + \beta \sum_{\substack{\mu\neq 1\\ \mu\neq\nu}} x_i^{(\mu)} x_i^{(1)} m_\mu\Big)$$
$$+ \beta m_\nu \frac{1}{N}\sum_i \Big[1 - \tanh^2\Big(\beta m_1 + \beta \sum_{\substack{\mu\neq 1\\ \mu\neq\nu}} x_i^{(\mu)} x_i^{(1)} m_\mu\Big)\Big]. \quad (3.31)$$

The goal is now to solve for m_ν. Approximating the sum $\frac{1}{N}\sum_i$ in the second line as an average over the Gaussian distributed variable z [Equation (3.25)] gives

$$\beta m_\nu \int_{-\infty}^{\infty} \mathrm{d}z \frac{1}{\sqrt{2\pi}\sigma_z} \mathrm{e}^{-\frac{z^2}{2\sigma_z^2}} \left[1 - \tanh^2(\beta m_1 + \beta z)\right]. \quad (3.32)$$

Defining the parameter q

$$q \equiv \int_{-\infty}^{\infty} \mathrm{d}z \frac{1}{\sqrt{2\pi}\sigma_z} \mathrm{e}^{-\frac{z^2}{2\sigma_z^2}} \tanh^2(\beta m_1 + \beta z), \quad (3.33)$$

one can write Equation (3.32) as

$$\beta m_\nu \left[1 - \int_{-\infty}^{\infty} \mathrm{d}z \frac{1}{\sqrt{2\pi}\sigma_z} \mathrm{e}^{-\frac{z^2}{2\sigma_z^2}} \tanh^2(\beta m_1 + \beta z)\right] = \beta m_\nu (1-q). \quad (3.34)$$

Returning to Equation (3.31), we see that it takes the form

$$m_\nu = \frac{1}{N}\sum_i x_i^{(\nu)} x_i^{(1)} \tanh\Big(\beta m_1 + \beta \sum_{\substack{\mu\neq 1\\ \mu\neq\nu}} x_i^{(\mu)} x_i^{(1)} m_\mu\Big) + (1-q)\beta m_\nu . \quad (3.35)$$

Solving for m_ν one finds for $\nu \neq 1$:

$$m_\nu = \frac{\frac{1}{N}\sum_i x_i^{(\nu)} x_i^{(1)} \tanh\Big(\beta m_1 + \beta \sum_{\substack{\mu\neq 1\\ \mu\neq\nu}} x_i^{(\mu)} x_i^{(1)} m_\mu\Big)}{1-\beta(1-q)}. \tag{3.36}$$

This expression allows us to compute the variance σ_z, defined by Equation (3.27). Equation (3.36) shows that the average $\langle m_\nu^2\rangle$ contains a double sum over the bit index, i. Since the bits are independent, only the diagonal terms contribute, so that

$$\langle m_\nu^2\rangle \approx \frac{\frac{1}{N^2}\sum_i \tanh^2\Big(\beta m_1 + \beta \sum_{\substack{\mu\neq 1\\ \mu\neq\nu}} x_i^{(\mu)} x_i^{(1)} m_\mu\Big)}{[1-\beta(1-q)]^2} \tag{3.37}$$

for $\nu \neq 1$, but otherwise independent of ν. The numerator is just q/N, from Equation (3.33). So the variance evaluates to

$$\sigma_z^2 = \frac{\alpha q}{[1-\beta(1-q)]^2}. \tag{3.38}$$

In summary, there are three coupled equations, for m_1, q, and σ_z, Equations (3.26), (3.34), and (3.38). They must be solved together to determine how m_1 depends on β and α.

In order to compare with the results described in Section 2.2, we must take the deterministic limit, $\beta \to \infty$. In this limit, q approaches unity, yet $\beta(1-q)$ remains finite [1]. Setting $q = 1$ in Equation (3.38) but retaining $\beta(1-q)$, one finds

$$\sigma_z^2 = \frac{\alpha}{[1-\beta(1-q)]^2}. \tag{3.39a}$$

The deterministic limits of Equations (3.34) and (3.26) become [1]:

$$\beta(1-q) = \sqrt{\frac{2}{\pi\sigma_z^2}}\,e^{-\frac{m_1^2}{2\sigma_z^2}}, \tag{3.39b}$$

$$m_1 = \operatorname{erf}\Big(\frac{m_1}{\sqrt{2\sigma_z^2}}\Big). \tag{3.39c}$$

Recall expression (3.18) for the steady-state error probability. Inserting Equation (3.39c) for m_1 into this expression, we find in the deterministic limit

$$P_{\text{error}}^{t=\infty} = \frac{1}{2}\Big[1-\operatorname{erf}\Big(\frac{m_1}{\sqrt{2\sigma_z^2}}\Big)\Big]. \tag{3.40}$$

Compare this with Equation (2.39) for the one-step error probability in the same limit. That equation was derived for only one step of the network dynamics, while Equation (3.40) describes the limit of many steps, the long-time or steady-state limit.

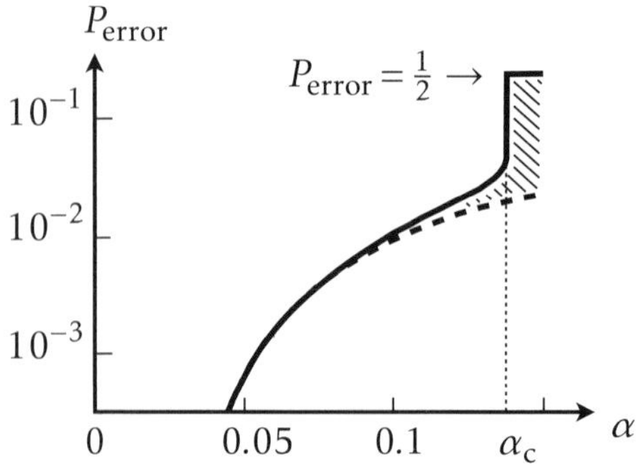

Figure 3.4 Error probability as a function of the storage capacity α in the deterministic limit. The one-step error probability $P_{\rm error}^{t=1}$ [Equation (2.39)] is shown as a dashed line; the steady-state error probability $P_{\rm error}^{t=\infty}$ [Equation (3.40)] is shown as a solid line. In the hashed region, error avalanches increase the error probability. After Figure 1 in Ref. [34]

Yet it turns out that Equation (3.40) reduces to (2.39) as $\alpha \to 0$. To see this, one solves the set of Equations (3.39) by introducing the variable $y = m_1/\sqrt{2\sigma_z^2}$. One obtains the following one-dimensional equation for y [1, 34]:

$$y\big(\sqrt{2\alpha} + (2/\sqrt{\pi})\, {\rm e}^{-y^2}\big) = {\rm erf}(y)\,. \tag{3.41}$$

The relevant solutions are those satisfying $0 \leq {\rm erf}(y) \leq 1$, because the order parameter is restricted to this range (transitions to $-m_1$ do not occur in the limit $N \to \infty$). Figure 3.4 shows the steady-state error probability obtained from Equations (3.40) and (3.41). Also shown is the one-step error probability

$$P_{\rm error}^{t=1} = \frac{1}{2}\Big[1 - {\rm erf}\Big(\frac{1}{\sqrt{2\alpha}}\Big)\Big]$$

derived in Section 2.2. As stated above, $P_{\rm error}^{t=\infty}$ approaches $P_{\rm error}^{t=1}$ for small α. We conclude that in this limit, for small α, the error probability does not increase significantly as one iterates the network dynamics. Errors in earlier iterations have little effect on the probability that later errors occur.

The situation is different at larger values of α. In that case, $P_{\rm error}^{t=1}$ significantly underestimates the steady-state error probability. In the hashed region, errors in the dynamics increase the probability of errors in subsequent steps, giving rise to *error avalanches*. Figure 3.4 illustrates that the steady-state error probability tends to $\frac{1}{2}$ as the parameter α increases beyond a critical value, α_c. Equation (3.41) yields

$$\alpha_c \approx 0.1379 \tag{3.42}$$

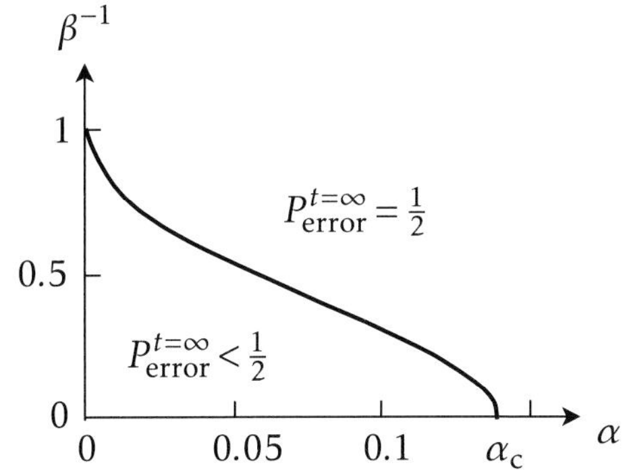

Figure 3.5 Phase diagram of the Hopfield network in the limit of large N (schematic). The region with $P_{\text{error}}^{t=\infty} < \frac{1}{2}$ is the ordered phase; the region with $P_{\text{error}}^{t=\infty} = \frac{1}{2}$ is the disordered phase. After Figure 2 in Ref. [34]

for the critical storage capacity α_c. When $\alpha > \alpha_c$, the network produces just noise. When $\alpha < \alpha_c$, by contrast, the network works well. The smaller the storage capacity, the better the network performs.

Figure 3.4 shows that the steady-state error probability changes very abruptly near α_c. Suppose we store 137 patterns with 1000 bits in a Hopfield network. In this case the network can recognise the patterns with a comparatively small error probability. However, if we try to store one or two more patterns, the network fails to produce output meaningfully related to the stored patterns. This rapid change is an example of a *phase transition*. In many physical systems, one observes similar transitions between ordered and disordered phases [32].

What happens at higher noise levels? The numerical solution of Equations (3.34), (3.26), and (3.38) shows that the critical storage capacity α_c decreases as the noise level increases (smaller values of β). This is shown schematically in Figure 3.5. Below the solid line, the error probability is smaller than $\frac{1}{2}$, so that the network operates reliably (although less so as one approaches the phase-transition boundary). Outside this region, the error probability equals $\frac{1}{2}$. In this region the network fails. In the limit of small α, the critical noise level is $\beta_c = 1$. In this regime, the network is described by the theory explained in Section 3.3, Equation (3.14).

Alternatively, these two different phases of the Hopfield network are characterised in terms of the order parameter m_1. We see that $m_1 \neq 0$ below the solid line, while $m_1 = 0$ above this line, in the limit of large N.

3.5 Beyond Mean-Field Theory

The theory summarised in this chapter rests on a mean-field approximation for the local field, Equation (3.6). The main result is the phase diagram shown in Figure 3.5, derived in the limit $N \to \infty$. For smaller values of N, one expects the transition to be less sharp, so that m_1 is non-zero also for values of α larger than the critical storage capacity α_c.

But even for large values of N, the question remains of how reliable the mean-field theory really is. To answer this question, one uses a more accurate theory, based on the so-called *replica trick*. One starts from the steady-state distribution of $\boldsymbol{s}$ for fixed patterns $\boldsymbol{x}^{(\mu)}$. In Chapter 4, we will see that the steady-state distribution for the McCulloch-Pitts dynamics is the *Boltzmann distribution*,

$$P_{\rm B}(\boldsymbol{s}) = Z^{-1}{\rm e}^{-\beta H(\boldsymbol{S})} \tag{3.43}$$

(the proof in Chapter 4 assumes that the diagonal weights are set to zero). The normalisation factor Z is called the *partition function*,

$$Z = \sum_{\boldsymbol{S}} {\rm e}^{-\beta H(\boldsymbol{S})} \,. \tag{3.44}$$

In order to compute the order parameter, one adds a threshold term to the energy function (2.45):

$$H = -\frac{1}{2}\sum_{ij} w_{ij} s_i s_j + \sum_{\mu} \lambda_\mu \sum_i x_i^{(\mu)} s_i \,. \tag{3.45}$$

Then the order parameter m_μ is obtained by taking a derivative w.r.t λ_μ:

$$m_\mu = \Big\langle \frac{1}{N}\sum_i x_i^{(\mu)} \langle n_i \rangle \Big\rangle = -\frac{1}{N\beta}\frac{\partial}{\partial \lambda_\mu} \big\langle \log Z \big\rangle \,. \tag{3.46}$$

The outer average is over different realisations of random patterns. The logarithm of Z is averaged using the replica trick. The idea is to represent the average of the logarithm as

$$\langle \log Z \rangle = \lim_{n\to 0} \tfrac{1}{n}(\langle Z^n \rangle - 1) \,. \tag{3.47}$$

The function Z^n looks like the partition function of n copies of the system, hence the name *replica* trick. If one assumes that all copies yield the same order parameter, one obtains the mean-field solution described in Section 3.4. If one allows different copies to have different order parameters (*replica-symmetry breaking*), one obtains a more accurate solution for the critical storage capacity [39],

$$\alpha_c = 0.138187 \,. \tag{3.48}$$

The mean-field result (3.42) differs only slightly from Equation (3.48). The most precise Monte-Carlo simulations (Section 4.2) for finite values of N [40] yield upon extrapolation to $N = \infty$

$$\alpha_c = 0.143 \pm 0.002 \,. \tag{3.49}$$

This is close to yet significantly different from the best theoretical estimate, Equation (3.48), and also different from the mean-field result (3.42).

To put these results into context, note that for other systems, mean-field theories tend to give results much worse than here. Usually, mean-field theories yield at best a qualitative description of a phase transition. For the Hopfield network, by contrast, the mean-field theory works very well because every neuron is connected with every other neuron. This helps to average out the fluctuations in Equation (3.6). In physical systems with local interactions, mean-field theories tend to work better in higher dimensions, because there are more neighbours to average over (Exercise 3.5).

3.6 Correlated and Non-Random Patterns

In the two previous sections, we assumed that the stored patterns are random with independently identically distributed bits. This allowed us to calculate the storage capacity of the Hopfield network using the central-limit theorem. The hope is that the result describes what happens for typical, non-random patterns, or for random patterns with correlated bits. Correlations affect the distribution of the cross-talk term and thus the storage capacity of the Hopfield network. It has been argued that the storage capacity increases when the patterns are more strongly correlated, while others have claimed that the capacity decreases in this limit (see Ref. [41] for a discussion).

For a set of definite patterns (no randomness to average over), the situation seems to be even more challenging. Yet there is a way of modifying Hebb's rule to deal with this problem, at least when the patterns are linearly independent. The recipe is explained by Hertz, Krogh, and Palmer [1]. One simply incorporates the overlaps

$$Q_{\mu\nu} = \frac{1}{N}\boldsymbol{x}^{(\mu)} \cdot \boldsymbol{x}^{(\nu)} \tag{3.50}$$

into Hebb's rule. To this end, one defines the $p \times p$ *overlap matrix* $\mathbb{Q}$ with elements $Q_{\mu\nu}$ and writes:

$$w_{ij} = \frac{1}{N}\sum_{\mu\nu} x_i^{(\mu)} \left(\mathbb{Q}^{-1}\right)_{\mu\nu} x_j^{(\nu)} \,. \tag{3.51}$$

For orthogonal patterns ($Q_{\mu\nu} = \delta_{\mu\nu}$), this modified Hebb's rule is identical to Equation (2.25). For non-orthogonal patterns, the rule (3.51) ensures that all patterns are recognised. Equation (3.51) requires that the matrix $\mathbb{Q}$ is invertible: its columns must be linearly independent (and this implies that the rows are linearly independent too). This limits the number of patterns one can store with the rule (3.51), because $p > N$ implies linear dependence.

For linearly independent patterns one can find the weights w_{ij} iteratively, by successive improvement from an arbitrary starting point. We can say that the network learns the task through a sequence of weight changes. This is the idea used to solve classification tasks with perceptrons (Part II).

3.7 Summary

In this chapter, we analysed the stochastic dynamics of Hopfield networks. We asked under which circumstances the network dynamics can reliably recognise stored patterns. If the stored patterns are random, the performance of the Hopfield network depends on their number, on the number N of bits per pattern, and upon the noise level. The storage capacity α equals the ratio of the number of stored patterns to the number of bits per pattern. The network operates reliably when this ratio is small, and provided the noise level is not too large. A mean-field analysis of the $N \to \infty$-limit shows that there is a phase transition in the parameter plane of the Hopfield network (Figure 3.5): when α exceeds the critical storage capacity α_c, the network ceases to function.

Hopfield networks share many properties with the networks discussed later on in this book. The most important point is perhaps that introducing noise in the dynamics allows one to study the convergence and performance of the network: in the presence of noise, there is a well-defined steady state that can be analysed. Without noise, in the deterministic limit, the network dynamics arrests in local minima of the energy function and may not reach the stored patterns. Naturally, the noise must be small enough for the network to function accurately. Finally, the building blocks of Hopfield networks are McCulloch-Pitts neurons and Hebb's rule for the weights. Many of the algorithms discussed in the coming chapters use these elements in some form.

3.8 Further Reading

The statistical mechanics of Hopfield networks is explained in *Introduction to the Theory of Neural Computation* by Hertz, Krogh, and Palmer [1]. Starting from the Boltzmann distribution, Chapter 10 in this book summarises how to compute the order parameters and how to evaluate the stability of the corresponding solutions. For more details on the replica trick, see the books by Müller, Reinhard, and Strickland [37] and by Engel and van den Broeck [42], as well as the review article [43].

3.9 Exercises

3.1 Mixed states. Write a computer program that implements the stochastic dynamics of a Hopfield model. Compute the order parameter for mixed states that are superpositions of the bits of three stored patterns. Determine how it depends on the noise level for $0.5 \leq \beta \leq 2.5$, small α, and large N. Solve the mean-field equation (3.21) numerically, and compare the results of this mean-field theory with those of your computer simulations. Repeat the exercise for mixed states that consist of superpositions of the bits of five stored patterns.

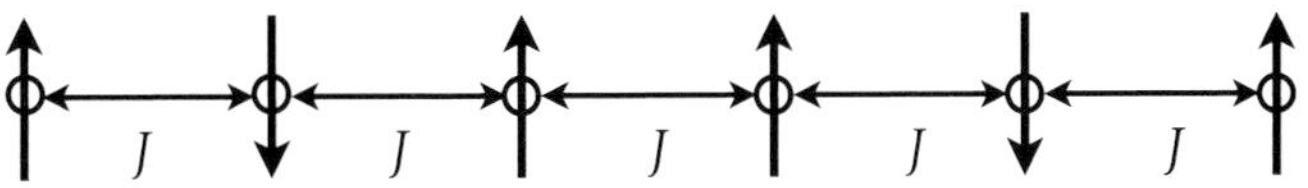

Figure 3.6 The Ising model is a model for ferromagnetism. It describes N spins that can either point up ($\uparrow$) or down ($\downarrow$), arranged on a lattice (here shown in one spatial dimension), interacting with their nearest neighbours with interaction strength J, and subject to an external magnetic field h. The state of spin i is described by the variable s_i, with $s_i = 1$ for $\uparrow$ and $s_i = -1$ for $\downarrow$

3.2 Deterministic limit. Derive the deterministic limit (3.39) of the three coupled Equations (3.34), (3.38), and (3.26) for m_1, q, and σ_z.

3.3 Phase diagram of the Hopfield network. Derive Equation (3.41) from Equation (3.39). Numerically solve (3.41) to find the critical storage capacity α_c in the deterministic limit. Quote your result with three-digit accuracy. To determine how the critical storage capacity depends on the noise level, numerically solve the three coupled Equations (3.26), (3.33), and (3.38). Compare your result with the schematic Figure 3.5.

3.4 Non-orthogonal patterns. Show that the rule (3.51) ensures that all patterns are recognised, for any set of non-orthogonal patterns that gives rise to an invertible matrix $\mathbb{Q}$. Demonstrate this by showing that the cross-talk term evaluates to zero, assuming that $\mathbb{Q}^{-1}$ exists.

3.5 Ising model. The Ising model is a model for ferromagnetism, N spins $s_i = \pm 1$ are arranged on a d-dimensional hypercubic lattice as shown in Figure 3.6 in one dimension. The energy function for the Ising model is $H = -J \sum_{i,j=\mathrm{nn}(i)} s_i s_j - h \sum_i s_i$. Here J is the ferromagnetic coupling between nearest-neighbour spins, h is an external magnetic field, and $\mathrm{nn}(i)$ denotes the nearest-neighbours of site i on the lattice. In equilibrium at temperature T, the states are distributed according to the Boltzmann distribution with $\beta = 1/(k_B T)$, where k_B is the Boltzmann constant. Derive a mean-field approximation for the magnetisation of the system, $m = \langle \frac{1}{N} \sum_i s_i \rangle$, assuming that N is large enough that the contribution of the boundary spins can be neglected. Derive an expression for the critical temperature below which mean-field theory predicts ferromagnetism, $m \neq 0$. Discuss how the critical temperature depends on the dimension d. Note: mean-field theory fails for the one-dimensional Ising model, but its predictions become more accurate as d increases.

3.6 Error probability. Show that Equations (2.39) and (3.40) agree if the condition (3.11) holds.

4
The Boltzmann Distribution

In Chapter 2, we saw that the deterministic dynamics (2.5) of Hopfield networks admits the Lyapunov function,

$$H = -\frac{1}{2}\sum_{ij} w_{ij} s_i s_j + \sum_i \theta_i s_i \,, \tag{4.1}$$

if the weights w_{ij} are symmetric and $w_{ii} \geq 0$. In this chapter,[1] we show that the asynchronous stochastic McCulloch-Pitts dynamics (3.1) converges to a steady state where the state vector $\boldsymbol{s}$ follows the Boltzmann distribution,

$$P_{\mathrm{B}}(\boldsymbol{s}) = Z^{-1} \mathrm{e}^{-\beta H(\boldsymbol{s})} \quad \text{with normalisation} \quad Z = \sum_{\boldsymbol{s}} \mathrm{e}^{-\beta H(\boldsymbol{s})} \,. \tag{4.2}$$

The stochastic dynamics (3.1) is closely related to that of *Markov-chain Monte-Carlo* algorithms, designed to efficiently sample from the Boltzmann distribution.

We also discuss how to solve optimisation tasks by Monte-Carlo simulation: one assigns a suitable energy H to each configuration $\boldsymbol{s}$, so that the function $H(\boldsymbol{s})$ has a global minimum for the optimal configuration $\boldsymbol{s}_{\min}$. The stochastic dynamics finds low-energy configurations (but not necessarily $\boldsymbol{s}_{\min}$), in particular if one iteratively decreases the noise level by increasing β (*simulated annealing* [44]).

Last but not least, we look at *Boltzmann machines* [14, 15, 45–47]. Boltzmann machines are stochastic neural networks that can be *trained* to learn the properties of a distribution $P_{\text{data}}(\boldsymbol{x})$ of binary input patterns $\boldsymbol{x}$. The idea is to iteratively change the weights in Equation (4.1) until the Boltzmann distribution represents the input distribution. This idea, to iterate the weights until the network learns the input distribution P_{data}, is used in a slightly different form in *supervised learning* (Part II). Boltzmann machines are closely related to Hopfield networks. Without so-called *hidden* neurons, both models learn to represent two-point correlations $\langle x_i^{(\mu)} x_j^{(\mu)} \rangle$ of pattern bits.

[1] In this chapter, we set the diagonal weights to zero.

When important information about the inputs is encoded in higher-order correlations, one can use *hidden* neurons to represent these correlations, neurons that are neither used for input nor for output. Generally, Boltzmann machines are hard to train, particularly if they have many hidden neurons. *Restricted Boltzmann machines* are neural networks with hidden neurons, but with fewer connections: only those between *visible* and hidden neurons are allowed. These neural networks can be fairly efficiently trained and can solve a number of different tasks. Apart from learning a distribution of input patterns, they can for instance be trained to recognise incomplete input patterns, and to classify inputs [25].

4.1 Convergence of the Stochastic Dynamics

We begin by showing that the stochastic dynamics (3.1) has a steady state where $\boldsymbol{s}$ is distributed according to the Boltzmann distribution (4.2). To this end, we consider an alternative yet equivalent formulation of the network dynamics. It consists of two steps. First, choose a neuron randomly, number m say. Second, update s_m to $s'_m \neq s_m$ with probability

$$\text{Prob}(s_m \to s'_m) = \frac{1}{1+e^{\beta \Delta H_m}}\,, \tag{4.3a}$$

with

$$\Delta H_m = H(\ldots, s'_m, \ldots) - H(\ldots, s_m, \ldots)\,. \tag{4.3b}$$

To explore the relation between the stochastic rules (4.3) and (3.1), we use

$$\Delta H_m = -b_m (s'_m - s_m) \tag{4.4}$$

with local field $b_m = \sum_j w_{mj} s_j - \theta_m$. To derive Equation (4.4), we assumed that the weights are symmetric and that the diagonal weights vanish. The result is obtained with a calculation similar to the one leading to Equation (2.48), except that we have non-zero thresholds here. To proceed, we break the rule (4.3) up into different cases. The state of neuron m changes with probability

$$\text{if } s_m = -1 \quad \text{obtain } s'_m = 1 \text{ with prob.} \quad \frac{1}{1+e^{-2\beta b_m}} = p(b_m)\,, \tag{4.5a}$$

$$\text{if } s_m = 1 \quad \text{obtain } s'_m = -1 \text{ with prob.} \quad \frac{1}{1+e^{2\beta b_m}} = 1 - p(b_m)\,. \tag{4.5b}$$

In the second row, we used $1 - p(b) = 1 - \frac{1}{1+e^{-2\beta b}} = \frac{1+e^{-2\beta b}-1}{1+e^{-2\beta b}} = \frac{1}{1+e^{2\beta b}}$. The state remains unchanged with probability:

$$\text{if } s_m = -1 \quad \text{obtain } s'_m = -1 \text{ with prob.} \quad \frac{1}{1+e^{-\beta b_m}} = 1 - p(b_m)\,, \tag{4.5c}$$

$$\text{if } s_m = 1 \qquad \text{obtain } s'_m = 1 \text{ with prob.} \qquad \frac{1}{1 + e^{\beta b_m}} = p(b_m)\,. \tag{4.5d}$$

Comparing with Equation (3.1), we conclude that the two schemes (3.1) and (4.3) are equivalent under the assumptions made ($w_{ij} = w_{ji}$ and $w_{ii} = 0$). Note that Equation (4.3) is more general than the stochastic Hopfield dynamics, because it does not require the energy function to be of the form (4.1). In particular, it is not necessary for the weights to be symmetric, nor is it necessary for the diagonal weights to vanish. Equations (3.1) and (4.3) are not equivalent if these conditions are not satisfied (Exercise 4.1).

The rule (4.3) defines a *Markov chain* of states:

$$\boldsymbol{s}_{t=0} \to \boldsymbol{s}_{t=1} \to \boldsymbol{s}_{t=2} \to \ldots \tag{4.6}$$

As before, the index t counts the iteration steps. A Markov chain is a *memoryless* random sequence of states defined by *transition probabilities* $p(\boldsymbol{s}'|\boldsymbol{s})$ from state $\boldsymbol{s}$ to $\boldsymbol{s}'$ [48]. The transition probability $p(\boldsymbol{s}'|\boldsymbol{s})$ connects arbitrary states. One distinguishes between *local moves* where only one neuron may change, as above, and *global moves* where many neurons may change their states in a single step.

In both cases, an update consists of two parts. First, a new state $\boldsymbol{s}'$ is suggested with probability $q(\boldsymbol{s}'|\boldsymbol{s})$. Second, the new state $\boldsymbol{s}'$ is accepted with *acceptance probability*

$$p_{\mathrm{a}}(\boldsymbol{s}'|\boldsymbol{s}) = \frac{1}{1 + \mathrm{e}^{\beta \Delta H}} \quad \text{with} \quad \Delta H = H(\boldsymbol{s}') - H(\boldsymbol{s})\,. \tag{4.7}$$

As result, the transition probability is given by a product of two factors

$$p(\boldsymbol{s}'|\boldsymbol{s}) = q(\boldsymbol{s}'|\boldsymbol{s})\, p_{\mathrm{a}}(\boldsymbol{s}'|\boldsymbol{s})\,. \tag{4.8}$$

These steps are repeated many times, creating the chain of states (4.6).

The Markov chain defined by the transition probability (4.8) has the Boltzmann distribution (4.2) as a steady-state distribution if the *detailed-balance* condition is satisfied:

$$p(\boldsymbol{s}'|\boldsymbol{s})\, P_{\mathrm{B}}(\boldsymbol{s}) = p(\boldsymbol{s}|\boldsymbol{s}')\, P_{\mathrm{B}}(\boldsymbol{s}')\,. \tag{4.9}$$

Note that this is a sufficient condition, not a necessary one [49]. There are Markov chains that do not satisfy detailed balance but still have a steady state (Exercise 4.4). Typically detailed balance implies not only that the Markov chain has $P_{\mathrm{B}}(\boldsymbol{s})$ [Equation (4.2)] as a steady-state distribution, but also that the distribution of states generated by the sequence (4.6) converges to $P_{\mathrm{B}}(\boldsymbol{s})$; see Ref. [48] for details.

Assume that a single neuron is picked randomly with uniform probability

$$q = N^{-1}\,, \tag{4.10}$$

where N is the number of neurons in the network. Since q does not depend on either s or s', the probability of suggesting a new state is clearly symmetric. Equations (4.2) and (4.7) then imply:

$$\frac{q e^{-\beta H(\boldsymbol{s})}}{1+e^{\beta[H(\boldsymbol{s}')-H(\boldsymbol{s})]}} = \frac{q}{e^{\beta H(\boldsymbol{s}')}+e^{\beta H(\boldsymbol{s})}} = \frac{q e^{-\beta H(\boldsymbol{s}')}}{1+e^{\beta[H(\boldsymbol{s})-H(\boldsymbol{s}')]}} . \tag{4.11}$$

This demonstrates that the detailed-balance condition (4.10) holds for the Markov chain defined by (4.7), (4.8), and (4.10). As a consequence, the Boltzmann distribution is a steady state of the Markov chain. If the simulation converges to the steady state (as it usually does), then states visited by the Markov chain are distributed according to the Boltzmann distribution. This also means that the steady-state distribution for the Hopfield model is the Boltzmann distribution, as stated in Section 3.5.

It is important to stress that Equation (4.9) is a condition for the transition probability $p(\boldsymbol{s}'|\boldsymbol{s}) = q(\boldsymbol{s}'|\boldsymbol{s})p_{\rm a}(\boldsymbol{s}'|\boldsymbol{s})$, not just for the acceptance probability $p_{\rm a}(\boldsymbol{s}'|\boldsymbol{s})$. For the local moves discussed above, q is a constant, so that $p(\boldsymbol{s}'|\boldsymbol{s}) \propto p_{\rm a}(\boldsymbol{s}'|\boldsymbol{s})$. In this case, it is sufficient to check the detailed-balance condition for the acceptance probability. In general, and in particular for global moves, it is necessary to include $q(\boldsymbol{s}'|\boldsymbol{s})$ in the detailed-balance check [50].

4.2 Monte-Carlo Simulation

The Markov chain described in the previous section is the basis for the *Markov-chain Monte-Carlo* algorithm. This method is widely used in statistical physics and in mathematical statistics. It is therefore important to understand the connections between the different formulations.

The Boltzmann distribution describes the probabilities of observing configurations of a large class of physical systems in their steady states [32]. The statistical mechanics of systems with energy function (also called *Hamiltonian*) H shows that their configurations are distributed according to the Boltzmann distribution in thermodynamic equilibrium at a given temperature T (in this context $\beta^{-1} = k_{\rm B}T$, where $k_{\rm B}$ is the Boltzmann constant), and free from any other constraints. If we denote the configuration of a system by the vector $\boldsymbol{s}$, then the Boltzmann distribution takes the form (4.2). The normalisation factor $Z = \sum_{\boldsymbol{s}} e^{-\beta H(\boldsymbol{s})}$ is also called *partition function*. For systems with a large number of interacting degrees of freedom, the partition function can be very expensive to compute, because the sum over $\boldsymbol{s}$ contains many terms. Therefore, instead of computing the distribution directly, one generates a Markov chain of states with a suitable transition probability, for instance (4.3).

In practice, one often uses a slightly different form of the transition probabilities (*Metropolis algorithm* [51]). Assuming that q does not depend on $\boldsymbol{s}$, one takes

$$p(\boldsymbol{s}'|\boldsymbol{s}) = q \begin{cases} \mathrm{e}^{-\beta \Delta H} & \text{when} \quad \Delta H > 0\,, \\ 1 & \text{when} \quad \Delta H \leq 0\,, \end{cases} \tag{4.12}$$

with $\Delta H = H(\boldsymbol{s}') - H(\boldsymbol{s})$ as before. That the Metropolis rates obey the detailed-balance condition (4.9) can be seen as follows:

$$\begin{aligned} p(\boldsymbol{s}'|\boldsymbol{s}) P_{\mathrm{B}}(\boldsymbol{s}) &= q Z^{-1} \mathrm{e}^{-\beta H(\boldsymbol{s})} \begin{cases} \mathrm{e}^{-\beta[H(\boldsymbol{s}')-H(\boldsymbol{s})]} & \text{if} \quad H(\boldsymbol{s}') > H(\boldsymbol{s}) \\ 1 & \text{otherwise} \end{cases} \\ &= q Z^{-1} \mathrm{e}^{-\beta \max\{H(\boldsymbol{s}), H(\boldsymbol{s}')\}} \\ &= q Z^{-1} \mathrm{e}^{-\beta H(\boldsymbol{s}')} \begin{cases} \mathrm{e}^{-\beta[H(\boldsymbol{s})-H(\boldsymbol{s}')]} & \text{if} \quad H(\boldsymbol{s}) > H(\boldsymbol{s}') \\ 1 & \text{otherwise} \end{cases} \\ &= p(\boldsymbol{s}|\boldsymbol{s}') P_{\mathrm{B}}(\boldsymbol{s}')\,. \end{aligned} \tag{4.13}$$

The Metropolis algorithm is summarised in Algorithm 2. It provides an elegant way of computing the average $\langle A \rangle$ of an observable $A(\boldsymbol{s})$ over the Boltzmann distribution of $\boldsymbol{s}$:

$$\langle A \rangle = Z^{-1} \sum_{\boldsymbol{s}} A(\boldsymbol{s})\, \mathrm{e}^{-\beta H(\boldsymbol{s})} \approx \frac{1}{T} \sum_{t=1}^{T} A(\boldsymbol{s}_t)\,. \tag{4.14}$$

This particular way of evaluating the average $\langle A \rangle$ is a special case of the more general method of *importance sampling* [52]. The central-limit theorem implies that the error of this estimate for $\langle A \rangle$ decreases $\propto T^{-1/2}$ as T increases. The prefactor is determined by the correlations between subsequent terms in the sum (4.14): the states in the sequence (4.6) are correlated, in particular when the moves are local, because then subsequent configurations are similar. Generating many quite strongly correlated samples from a distribution is not a very efficient way of sampling this distribution. Sometimes, it may be more efficient to suggest global moves instead, in order to avoid that subsequent states in the Markov chain are similar. But it is not guaranteed that global moves lead to weaker correlations. For global moves, ΔH may be more likely to assume large positive values, so that fewer suggested moves are accepted. As a consequence, the Markov chain may stay in certain states for a long time, increasing correlations in the sequence. Usually a compromise is most efficient – moves that are neither local nor global. In summary, the convergence of Monte-Carlo sampling is quite slow. This motivated Sokal to begin his lecture notes on Monte-Carlo simulation with the warning [49]

Monte Carlo is an extremely bad method; it should be used only when all alternative methods are worse.

Monte-Carlo algorithms are very widely used, and the original reference for the Metropolis algorithm [51] is considered one of the most significant scientific papers in computational physics. Sokal's point is of course that many problems cannot be solved in any other way, so that Monte-Carlo simulation is the only option. But we should be aware of the shortcomings of the method. The same caution applies to the topic of this book, machine-learning algorithms with neural networks.

Algorithm 2 Metropolis algorithm for symmetric $q(\boldsymbol{s}'|\boldsymbol{s})$

```
initialise s = s_0;
for t = 1, ..., T do
  suggest a new state s' with probability q(s'|s) ;
  compute ΔH = H(s') − H(s);
  if ΔH ≤ 0 then
    accept the new state: s = s';
  else
    draw a random number r uniformly distributed in [0, 1];
    if r < exp(−βΔH) then
      accept the new state: s = s';
    else
      reject s';
    end if
  end if
  sample s_t = s and A_t = A(s_t);
end for
```

4.3 Simulated Annealing

Combinatorial optimisation problems admit 2^k or $k!$ configurations – too many to find the optimal one by complete enumeration when k is large. An alternative strategy is to assign an energy $H(\boldsymbol{s})$ to each configuration $\boldsymbol{s}$, so that H is minimal at the optimal configuration $\boldsymbol{s}_{\min}$. One minimises $H(\boldsymbol{s})$ by Monte-Carlo simulation, using that the Monte-Carlo dynamics tends to decrease H when the temperature $k_{\mathrm{B}}T = \beta^{-1}$ is low, Figure 4.1. A common strategy is to lower the temperature on the fly. In the beginning of the simulation, the temperature is high, so that the dynamics first explores the rough features of the energy landscape. When the temperature is lowered, the dynamics perceives finer and finer features of $H(\boldsymbol{s})$. The hope is that it ends up in the global minimum $H_{\min} = H(\boldsymbol{s}_{\min})$ at zero temperature, where $P_{\mathrm{B}}(\boldsymbol{s}) = 0$ when $H(\boldsymbol{s}) > H_{\min}$ and $P_{\mathrm{B}}(\boldsymbol{s}) > 0$ only for $H(\boldsymbol{s}) = H_{\min}$. This method is called *simulated annealing* [44]; see also Section 10.9 in *Numerical Recipes* [53].

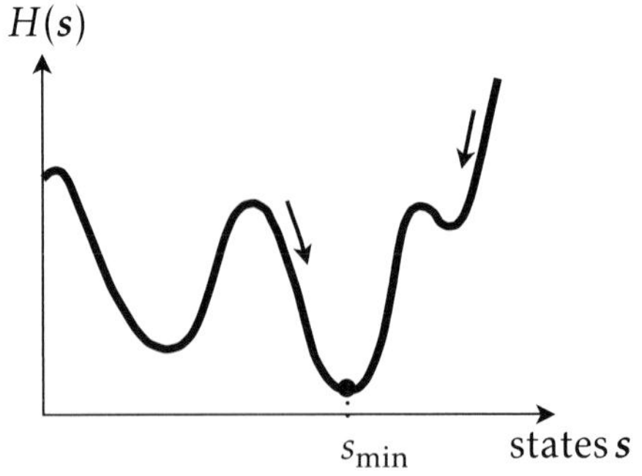

Figure 4.1 Schematic. Simulated annealing tends to reduce the energy function (arrows). Noise helps to prevent the dynamics from arresting in a local minimum

Slowly lowering the temperature during the simulation mimics the slow cooling of a physical system. It passes through a sequence of quasi-equilibrium Boltzmann distributions with lower and lower temperatures, until the system hopefully finds the global minimum $H_{\min}$.

For a number of combinatorial optimisation problems, one can write down energy functions that have the same form as Equation (2.50) with symmetric weights [54]. Since $s_{ij}^2 = 1$, one can always assume that the diagonal weights vanish, because they make only a constant contribution to H. In short, one can use the Hopfield dynamics (3.1) to minimise H. The *travelling-salesman* problem has been solved in this way [1, 54], gradually reducing the noise level as one iterates the stochastic dynamics.

It is by no means necessary to use a Hopfield model for this purpose. Instead, we can just use the stochastic dynamics (4.3) or the Metropolis algorithm (4.12) to solve combinatorial optimisation problems by simulated annealing. Nevertheless, a crucial step is to find a suitable energy function.

As an example, consider the *double-digest problem*. It arose when sequencing the human genome [55, 56]. The human genome sequence was first assembled by piecing together overlapping DNA segments in the right order by making sure that overlapping segments share the same DNA sequence. To this end, it is necessary to uniquely identify the DNA segments. The actual DNA sequence of a segment is a unique identifier. But it is sufficient and more efficient to identify a DNA segment by a *fingerprint*, for example the sequence of *restriction sites*. These are short subsequences (four or six base pairs long) that are recognised by enzymes that cut (*digest*) the DNA strand precisely at these sites. A DNA segment is identified by the types and locations of restriction sites that it contains, the so-called *restriction map*.

When a DNA segment is cut by two different enzymes, one can experimentally determine the lengths of the resulting fragments. Is it possible to infer how the cuts

Table 4.1 *Example configurations for the double-digest problem [55] for three different chromosome lengths L. For each example, three ordered fragment sets are given, corresponding to the result of digestion with enzyme A, with B, and with both A and B.*

$L = 10000$
$a =$ `[5976, 1543, 1319, 1120, 42]`
$b =$ `[4513, 2823, 2057, 607]`
$c =$ `[4513, 1543, 1319, 1120, 607, 514, 342, 42]`

$L = 20000$
$a =$ `[8479, 4868, 3696, 2646, 169, 142]`
$b =$ `[11968, 5026, 1081, 1050, 691, 184]`
$c =$ `[8479, 4167, 2646, 1081, 881, 859, 701, 691, 184, 169, 142]`

$L = 40000$
$a =$ `[9979, 9348, 8022, 4020, 2693, 1892, 1714, 1371, 510, 451]`
$b =$ `[9492, 8453, 7749, 7365, 2292, 2180, 1023, 959, 278, 124, 85]`
$c =$ `[7042, 5608, 5464, 4371, 3884, 3121, 1901, 1768, 1590, 959, 899, 707, 702, 510, 451, 412, 278, 124, 124, 85]`

were ordered in the DNA sequence of the segment from the fragment lengths, to find the restriction map? This is the double-digest problem [55]. In a double-digest experiment, a given DNA sequence is first digested by one enzyme (A say). Assume that this results in n fragments with lengths a_i ($i = 1, \ldots, n$). Second, the DNA sequence is digested by another enzyme, B. In this case, m fragments are found, with lengths $b_1, b_2, \ldots, b_m$. Third, the DNA sequence is digested with both enzymes A and B, yielding l fragments with lengths $c_1, \ldots, c_l$; see Table 4.1 for examples. The task is now to determine all possible orderings of the a- and b-cuts that result in l fragments with lengths $c_1, c_2, \ldots, c_l$. Since the solutions of the double-digest problem are degenerate, an important question is to determine how many distinct solutions there are (Exercise 4.5).

To write down an energy function, denote the ordered set of fragment lengths produced by digesting with enzyme A by $a = \{a_1, \ldots, a_n\}$, where $a_1 \geq a_2 \geq \ldots \geq a_n \geq 1$. Similarly $b = \{b_1, \ldots, b_m\}$ ($b_1 \geq b_2 \geq \ldots \geq b_m \geq 1$) for fragment lengths produced by enzyme B, and $c = \{c_1, \ldots, c_l\}$ ($c_1 \geq c_2 \geq \ldots \geq c_l \geq 1$) for fragment lengths produced by digesting first with A and then with B. Permutations σ and μ of the sets a and b result in a set of c-fragments that we call $\hat{c}(\sigma, \mu)$. Solutions of the double-digest problem correspond to permutations $[\sigma, \mu]$ that yield $\hat{c}(\sigma, \mu) = c$. A suitable energy function is therefore [55]

$$H(\sigma, \mu) = \sum_j c_j^{-1} [c_j - \hat{c}_j(\sigma, \mu)]^2 , \tag{4.15}$$

and configuration space is the space of all permutation pairs $s = [\sigma, \mu]$. Local moves in configuration space correspond to inversions of short subsequences of σ and/or μ. One can show that the corresponding $q(s'|s)$ is symmetric (Exercise 4.5). As noted previously, this is necessary for the stochastic dynamics to converge in its simplest form, Equation (4.12) and Algorithm 2.

For the simulation, one chooses a larger temperature $k_B T = \beta^{-1}$ to begin with, so that the stochastic dynamics explores the rough features of the energy landscape at first. As the simulation proceeds, the temperature is gradually reduced. This allows the dynamics to learn finer features of the landscape, as described above.

4.4 Boltzmann Machines

Boltzmann machines are generalised Hopfield networks that can learn to approximate data distributions of binary input patterns. Boltzmann machines differ from Hopfield networks in two essential ways. First, instead of using Hebb's rule, the weights are adjusted until the Boltzmann machine approximates the data distribution precisely. The weights are iteratively refined to minimise the difference between the data distribution and the model (the Boltzmann distribution). Nevertheless, this procedure is closely related to Hebb's rule, as we shall see. Second, to represent higher-order correlations between bits of input patterns, Boltzmann machines employ hidden neurons.

We begin with Boltzmann machines without hidden neurons (Figure 4.2), because they are simpler to analyse. Then we discuss why hidden neurons are necessary to learn the properties of general input distributions $P_{\text{data}}(\boldsymbol{x})$ of binary inputs $\boldsymbol{x}$. The training algorithm for Boltzmann machines with hidden neurons is described in Section 4.5.

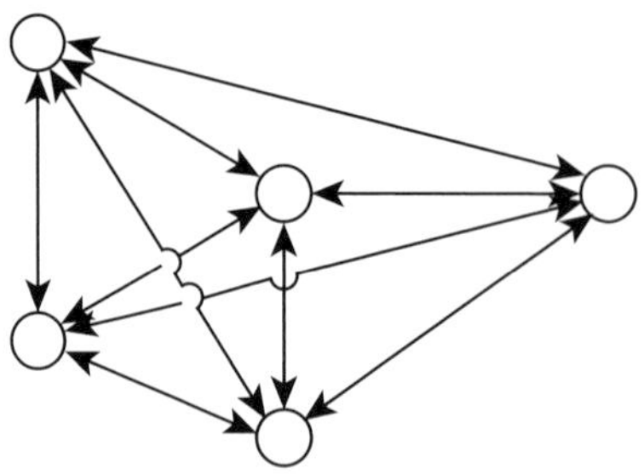

Figure 4.2 Boltzmann machine with five neurons. All weights are symmetric; the diagonal weights are set to zero. The states of the neurons are denoted by $s_i = \pm 1$. This neural network has no hidden neurons. It looks like a Hopfield network, but the weights are not given by Hebb's rule

The goal of the training algorithm is to find weights so that the Boltzmann distribution

$$P_{\mathrm{B}}(\boldsymbol{s} = \boldsymbol{x}) = Z^{-1}\exp\Big(\tfrac{1}{2}\sum_{i\neq j} w_{ij}x_i x_j\Big) \tag{4.16}$$

approximates the distribution $P_{\mathrm{data}}(\boldsymbol{x})$ as precisely as possible. Here and in the remainder of this chapter, we set $\beta = 1$. The input patterns have N binary bits [Equation (2.1)] with values ± 1. The weight matrix $\mathbb{W}$ is symmetric, $w_{ij} = w_{ji}$, and its diagonal elements are set to zero, $w_{ii} = 0$. In this section, we also set the thresholds to zero.

The Boltzmann machine is trained by iteratively adjusting the weights w_{ij}, using a sequence of input patterns $\boldsymbol{x}^{(\mu)}$ ($\mu = 1, \ldots, p$), independently sampled from the data distribution $P_{\mathrm{data}}(\boldsymbol{x})$. This is achieved by maximising the *likelihood* $\mathscr{L} = \prod_{\mu=1}^{p} P_{\mathrm{B}}\big(\boldsymbol{s} = \boldsymbol{x}^{(\mu)}\big)$ that the Boltzmann machine produces the sequence $\boldsymbol{x}^{(1)}, \ldots, \boldsymbol{x}^{(p)}$ of input patterns. Any pattern may appear more than once in the sequence, with frequency proportional to $P_{\mathrm{data}}(\boldsymbol{x})$. Maximising $\mathscr{L}$ therefore corresponds to approximating the data distribution as accurately as possible. Usually one maximises the logarithm of the likelihood, the *log-likelihood* function

$$\log \mathscr{L} = \log \prod_{\mu=1}^{p} P_{\mathrm{B}}\big(\boldsymbol{s} = \boldsymbol{x}^{(\mu)}\big) = \sum_{\mu=1}^{p} \log P_{\mathrm{B}}\big(\boldsymbol{s} = \boldsymbol{x}^{(\mu)}\big). \tag{4.17}$$

The logarithm is a monotonic function, so the log-likelihood has its maximum at the same weight values as the likelihood. Taking the logarithm simplifies the analysis of the learning algorithm, because $\log P_{\mathrm{B}}(\boldsymbol{s} = \boldsymbol{s}^{(\mu)})$ is simply a quadratic function of $x_j^{(\mu)}$. Also, a learning algorithm based on the log-likelihood is usually more stable numerically.

A different reasoning behind maximising the log-likelihood starts from the *Kullback-Leibler divergence*, defined as

$$D_{\mathrm{KL}} = \sum_{\mu=1}^{p} P_{\mathrm{data}}(\boldsymbol{x}^{(\mu)}) \log[P_{\mathrm{data}}(\boldsymbol{x}^{(\mu)})/P_{\mathrm{B}}(\boldsymbol{s} = \boldsymbol{x}^{(\mu)})]. \tag{4.18}$$

Terms in the sum with $P_{\mathrm{data}}(\boldsymbol{x}^{(\mu)}) = 0$ are set to zero, and D_{KL} is defined to equal infinity when there are patterns for which $P_{\mathrm{B}} = 0$ but $P_{\mathrm{data}} \neq 0$. The Kullback-Leibler divergence is a measure of the difference between the two distributions: D_{KL} is non-negative, and it assumes its global minimum $D_{\mathrm{KL}} = 0$ for $P_{\mathrm{data}}(\boldsymbol{x}^{(\mu)}) = P_{\mathrm{B}}(\boldsymbol{s} = \boldsymbol{x}^{(\mu)})$; see Exercise 4.6. We infer from Equation (4.18) that minimising D_{KL} corresponds to maximising $\log \mathscr{L}$.

To find the global maximum of the log-likelihood, we use *gradient ascent*: we repeatedly change the weights by adding increments

$$w'_{mn} = w_{mn} + \delta w_{mn} \quad \text{with} \quad \delta w_{mn} = \eta \frac{\partial \log \mathscr{L}}{\partial w_{mn}} \,. \tag{4.19}$$

The small parameter $\eta > 0$ is the *learning rate*. The gradient points in the steepest uphill direction of $\mathscr{L}$. The idea is to take many small uphill steps until one hopefully (but not necessarily) reaches the global maximum. Since the likelihood is a product of many possibly quite small factors, $\mathscr{L}$ can become very small. This can lead to numerical instabilities. Maximising $\log \mathscr{L}$ instead of $\mathscr{L}$ can be more stable because it yields an additional factor $\mathscr{L}^{-1}$ in the gradient: $\partial \log \mathscr{L} / \partial w_{mn} = \mathscr{L}^{-1} \partial \mathscr{L} / \partial w_{mn}$.

To evaluate the gradient of $\mathscr{L}$, we start from Equation (4.17)

$$\log \mathscr{L} = \sum_{\mu=1}^{p} \Big[-\log Z + \tfrac{1}{2} \sum_{i \neq j} w_{ij} x_i^{(\mu)} x_j^{(\mu)} \Big] \,. \tag{4.20}$$

This expression assumes that the diagonal weights vanish, just like Equation (4.16). The first step is to evaluate the derivative of

$$\log Z = \log \sum_{s_1=\pm 1, \ldots, s_N = \pm 1} \exp\Big(\tfrac{1}{2} \sum_{i \neq j} w_{ij} s_i s_j \Big) \,. \tag{4.21}$$

To compute $\partial \log Z / \partial w_{mn}$, we use the chain rule together with

$$\frac{\partial w_{ij}}{\partial w_{mn}} = \delta_{im} \delta_{jn} + \delta_{jm} \delta_{in} \,. \tag{4.22}$$

This relation is valid for symmetric weights and provided that $i \neq j$ and $m \neq n$. In Equation (4.22), δ_{kl} is the Kronecker delta, $\delta_{kl} = 1$ if $k = l$ and zero otherwise (Chapter 2). In particular, $\delta_{im} \delta_{jn} = 1$ only if $i = m$ and $j = n$. Otherwise, the product of Kronecker deltas equals zero. Equation (4.22) is illustrated by the following story (a modification of a well-known maths joke):

The linear function, x, and the constant function are going for a walk. When they suddenly see the derivative approaching, the constant function gets worried. 'I'm not worried' says the function x confidently, 'I'm not put to zero by the derivative.' When the derivative comes closer, it says 'Hi! I'm $\partial/\partial y$. How are you?'

The moral is: since x and y are independent variables, $\partial x / \partial y = 0$. Equation (4.22) reflects the same principle: the weights w_{ij} and w_{mn} are independent variables unless their indices agree. Equation (4.22) is valid for off-diagonal weights, and there are two terms on the r.h.s. because the weights are symmetric.

Returning to the derivative of $\log Z$ with respect to w_{mn}, one finds using Equation (4.22):

$$\frac{\partial \log Z}{\partial w_{mn}} = \sum_{s_1=\pm 1, \ldots, s_N = \pm 1} s_m s_n P_{\rm B}(\boldsymbol{s}) \equiv \langle s_m s_n \rangle_{\rm model} \,. \tag{4.23}$$

The last equality defines the two-point correlations of the model, $\langle s_m s_n \rangle_{\text{model}}$, computed using the steady-state distribution (4.16) of the Boltzmann machine. Evaluating the derivative of the second term in Equation (4.20) gives

$$\frac{\partial}{\partial w_{mn}} \frac{1}{2} \sum_{i \neq j} w_{ij} x_i^{(\mu)} x_j^{(\mu)} = x_m^{(\mu)} x_n^{(\mu)} . \tag{4.24}$$

In summary,

$$\frac{\partial \log \mathscr{L}}{\partial w_{mn}} = \sum_{\mu=1}^{p} \big(x_m^{(\mu)} x_n^{(\mu)} - \langle s_m s_n \rangle_{\text{model}} \big) = p \big(\langle x_m x_n \rangle_{\text{data}} - \langle s_m s_n \rangle_{\text{model}} \big) . \tag{4.25}$$

Here $\langle x_m x_n \rangle_{\text{data}} = p^{-1} \sum_{\mu=1}^{p} x_m^{(\mu)} x_n^{(\mu)}$ is the two-point correlation of the input data. Using (4.19), the *learning rule* becomes

$$\delta w_{mn} = \eta \big(\langle x_m x_n \rangle_{\text{data}} - \langle s_m s_n \rangle_{\text{model}} \big) , \tag{4.26}$$

where we dropped a factor of p that only affects the numerical value of the learning rate η. The weight increments are determined by the two-point pattern correlations, just like Hebb's rule (2.25). The first term on the r.h.s. of Equation (4.26) has precisely the same form as Equation (2.25), a sum over two-point correlations of the input patterns. The second average is over the steady-state distribution (4.16) of the Boltzmann machine. The learning rule takes the form of the difference between two-point correlations because the task is to minimise the difference between two distributions. It is plausible that the learning rule may converge because the weight increments vanish when the model correlations equal the data correlations.

The average $\langle s_m s_n \rangle_{\text{model}}$ can be approximated by numerical simulation of the McCulloch-Pitts dynamics

$$s_i' = \begin{cases} 1 & \text{with probability} \quad p(b_i) , \\ -1 & \text{with probability} \quad 1 - p(b_i) , \end{cases} \tag{4.27}$$

with $b_i = \sum_j w_{ij} s_j$ and $p(b_i) = \frac{1}{1+e^{-2b_i}}$. One must iterate Equation (4.27) until the system has reached its steady state, long enough so that any initial transient becomes negligible.

The training algorithm can be summarised as follows. One initialises all weights and computes $\langle x_m x_n \rangle_{\text{data}}$ from the given sequence of input patterns. One estimates $\langle s_m s_n \rangle_{\text{model}}$ by numerical simulation of the dynamics of the Boltzmann machine, and changes the weights using (4.26). This step is iterated, either with a sequence of new inputs or with the same inputs but in permuted sequence. In each iteration, one must compute $\langle s_m s_n \rangle_{\text{model}}$ again, because the weights changed. This procedure is quite slow, because it usually takes long simulations to estimate $\langle x_m x_n \rangle_{\rm model}$ accurately, in each iteration of the learning algorithm.

There is a more fundamental problem [47]. Like Hebb's rule, the learning rule (4.26) relies entirely upon two-point correlations of the input bits. This means that the Boltzmann machine cannot learn higher-order correlations between inputs. However, two-point correlations may not be sufficient to represent the information encoded in the input data. To illustrate this point, consider the Boolean XOR function (Exercise 2.13 and Chapter 5). It can be encoded in the four patterns $[-1,-1,-1]$, $[1,1,-1]$, $[-1,1,1]$, and $[1,-1,1]$. The first two components represent the input to the XOR function. The third component represents the output, which depends on both input variables as prescribed by the XOR function. Let us define an input distribution that reflects these three-point correlations by assigning $P_{\rm data} = \frac{1}{4}$ to the four patterns, and setting $P_{\rm data} = 0$ otherwise. A Boltzmann machine with three neurons cannot represent this input distribution, because there is no energy function of the form (4.1) that has four minima at these patterns. So the three-point correlations encoded in the four patterns cannot be represented in terms of a Boltzmann machine in its simplest form.

Also, Hopfield networks fail for the XOR function: the four states are not attractors of a Hopfield network with three neurons (Exercise 2.13). One could consider neural networks with third- or higher-order interactions [47],

$$H = -\tfrac{1}{2}\sum_{ij} w_{ij}^{(2)} s_i s_j - \tfrac{1}{6}\sum_{ijk} w_{ijk}^{(3)} s_i s_j s_k + \ldots \tag{4.28}$$

(Exercise 2.7). But the number of weights proliferates as the order increases, rendering the training very slow.

An alternative is to use Boltzmann machines with *hidden* neurons, that are neither input nor output units. The idea is that the hidden neurons can learn to represent such correlations [47]. The learning rule for the Boltzmann machines with hidden neurons is very similar to Equation (4.26), but when the number of hidden neurons is large, the Boltzmann machine is very slow to train. It is more efficient to remove all weights between visible neurons, and between hidden neurons. This is described in the next section.

4.5 Restricted Boltzmann Machines

Restricted Boltzmann machines [57] consist of visible and hidden neurons arranged in an undirected bipartite graph (Figure 4.3): the only connections are between neurons of different kinds, there are no connections between visible neurons, and no

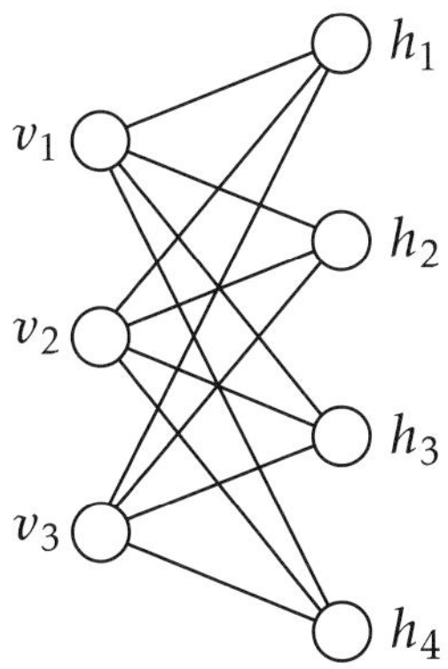

Figure 4.3 Restricted Boltzmann machine with three visible neurons, v_j, and four hidden neurons, h_i

connections between hidden neurons either. So the energy function for a restricted Boltzmann machine for N visible neurons v_j and M hidden neurons h_i takes the form

$$H = -\sum_{i=1}^{M}\sum_{j=1}^{N} w_{ij} h_i v_j + \sum_{j=1}^{N} \theta_j^{(\mathrm{v})} v_j + \sum_{i=1}^{M} \theta_i^{(\mathrm{h})} h_i \,, \tag{4.29}$$

with weights w_{ij} and thresholds $\theta_j^{(\mathrm{v})}$ and $\theta_i^{(\mathrm{h})}$. The McCulloch-Pitts dynamics reads

$$h_i' = \begin{cases} 1 & \text{with probability} \quad p(b_i^{(\mathrm{h})}) \\ -1 & \text{with probability} \quad 1 - p(b_i^{(\mathrm{h})}) \end{cases} \quad \text{with} \quad b_i^{(\mathrm{h})} = \sum_{j=1}^{N} w_{ij} v_j - \theta_i^{(\mathrm{h})} \,, \tag{4.30a}$$

and

$$v_j' = \begin{cases} 1 & \text{with probability} \quad p(b_j^{(\mathrm{v})}) \\ -1 & \text{with probability} \quad 1 - p(b_j^{(\mathrm{v})}) \end{cases} \quad \text{with} \quad b_j^{(\mathrm{v})} = \sum_{i=1}^{M} h_i w_{ij} - \theta_j^{(\mathrm{v})} \,. \tag{4.30b}$$

The diagonal weights are assumed to vanish, but the weight matrix is not required to be symmetric (Exercise 4.9). Since most often $M \gg N$, it is usually not even a square matrix.

The learning rule for the weights of the restricted Boltzmann machine is derived using gradient ascent on the log-likelihood for a single pattern $\boldsymbol{x}^{(\mu)}$:

$$\log P(\boldsymbol{x}^{(\mu)}) = \log \sum_{h_1=\pm 1,\ldots,h_M=\pm 1} P_{\mathrm{B}}(\boldsymbol{v} = \boldsymbol{x}^{(\mu)}, \boldsymbol{h}) \,. \tag{4.31}$$

Proceeding as in the previous section, one finds

$$\delta w_{mn}^{(\mu)} = \eta \big(\langle h_m x_n^{(\mu)} \rangle_{\text{data}} - \langle h_m v_n \rangle_{\text{model}} \big) \,. \tag{4.32}$$

The first average,

$$\langle h_m x_n^{(\mu)} \rangle_{\text{data}} = \sum_{h_1=\pm1,\ldots,h_M=\pm1} h_m x_n^{(\mu)} \Big[\prod_{i=1}^{M} P(h_i | \boldsymbol{v} = \boldsymbol{x}^{(\mu)}) \Big], \tag{4.33}$$

can be evaluated further, using the fact that there are no connections between the hidden neurons. Making use of the update rule (4.30a), we find

$$\sum_{h_m=\pm1} h_m P(h_m | \boldsymbol{v} = \boldsymbol{x}^{(\mu)}) = p(b_m^{(\text{h})}) - [1 - p(b_m^{(\text{h})})] = \tanh(b_m^{(\text{h})}), \tag{4.34}$$

just like Equation (3.7). For the other sums in Equation (4.33), we use the normalisation condition $1 = \sum_{h_k=\pm1} P(h_k | \boldsymbol{v} = \boldsymbol{x}^{(\mu)})$ to obtain

$$\langle h_m x_n^{(\mu)} \rangle_{\text{data}} = \tanh(b_m^{(\text{h})})\, x_n^{(\mu)} = \tanh\Big(\sum_{j=1}^{N} w_{mj} x_j^{(\mu)} - \theta_m^{(\text{h})} \Big) x_n^{(\mu)}.$$

The second average on the r.h.s. of Equation (4.32) simplifies to

$$\langle h_m v_n \rangle_{\text{model}} = \Big\langle \tanh\Big(\sum_{j=1}^{N} w_{mj} v_j - \theta_m^{(\text{h})} \Big) v_n \Big\rangle_{\text{model}}. \tag{4.35}$$

The average $\langle \cdots \rangle_{\text{model}}$ is computed by Monte-Carlo sampling, using the McCulloch-Pitts dynamics (4.30) to generate the Markov chain

$$\boldsymbol{v}_{t=0} \to \boldsymbol{h}_{t=0} \to \boldsymbol{v}_{t=1} \to \boldsymbol{h}_{t=1} \to \boldsymbol{v}_{t=2} \to \cdots. \tag{4.36}$$

In the limit $t \to \infty$, the steady state of this sequence is distributed according to the model distribution, the Boltzmann distribution with energy function (4.29). In general, only the asynchronous McCulloch-Pitts dynamics can be proven to converge (Sections 2.5 and 4.1). Here, however, the Markov chain can be generated more efficiently by updating all hidden neurons $\boldsymbol{h}_t$ at the same time, given $\boldsymbol{v}_t$, because the components of $\boldsymbol{h}_t$ are independent from each other since there are no connections between them. In the same way, the visible neurons $\boldsymbol{v}_t$ are updated in parallel. To speed up the computation further, one usually only iterates for a finite number of steps, up to $t = k$ say, and initialises the chain with $\boldsymbol{v}_{t=0} = \boldsymbol{x}^{(\mu)}$. After k steps, one approximates

$$\Big\langle \tanh\Big(\sum_{j=1}^{N} w_{mj} v_j - \theta_m^{(\text{h})} \Big) v_n \Big\rangle_{\text{model}} \approx \tanh\Big(\sum_{j=1}^{N} w_{mj} v_{j,t=k} - \theta_m^{(\text{h})} \Big) v_{n,t=k}. \tag{4.37}$$

This algorithm is called a *contrastive-divergence* or CD-k algorithm (Algorithm 3). Since the average over the model distribution is approximated

Algorithm 3 Contrastive divergence algorithm CD-k for ± 1 neurons

initialise weights and thresholds;
for $\nu = 1, \ldots, \nu_{\max}$ **do**
  sample p_0 patterns from the data distribution ($p_0 \leq p$);
  for $\mu = 1, \ldots, p_0$ **do**
    initialise $\boldsymbol{v}(0) \leftarrow \boldsymbol{x}^{(\mu)}$;
    update all hidden neurons: $\boldsymbol{b}^{(\mathrm{h})}(0) \leftarrow \mathbb{W}\boldsymbol{v}(0) - \boldsymbol{\theta}^{(\mathrm{h})}$;
    for $i = 1, \ldots, M$ **do**
      $h_i(0) \leftarrow +1$ with probability $p\big(b_i^{(\mathrm{h})}(0)\big)$ otherwise $h_i(0) \leftarrow -1$;
    end for
    for $t = 1, \ldots, k$ **do**
      update all visible neurons: $\boldsymbol{b}^{(\mathrm{v})}(t-1) \leftarrow \boldsymbol{h}(t-1) \cdot \mathbb{W} - \boldsymbol{\theta}^{(\mathrm{v})}$;
      for $j = 1, \ldots, N$ **do**
        $v_j(t) \leftarrow +1$ with probability $p\big(b_j^{(\mathrm{v})}(t-1)\big)$ otherwise $v_j(t) \leftarrow -1$;
      end for
      update all hidden neurons: $\boldsymbol{b}^{(\mathrm{h})}(t) \leftarrow \mathbb{W}\boldsymbol{v}(t) - \boldsymbol{\theta}^{(\mathrm{h})}$;
      for $i = 1, \ldots, M$ **do**
        $h_i(t) \leftarrow +1$ with probability $p\big(b_i^{(\mathrm{h})}(t)\big)$ otherwise $h_i(t) \leftarrow -1$;
      end for
    end for
    compute weight and threshold increments:
    $\delta w_{mn} \leftarrow \eta\Big[\tanh\big(b_m^{(\mathrm{h})}(0)\big)v_n(0) - \tanh\big(b_m^{(\mathrm{h})}(k)\big)v_n(k)\Big]$;
    $\delta\theta_n^{(\mathrm{v})} \leftarrow -\eta[v_n(0) - v_n(k)]$;
    $\delta\theta_m^{(\mathrm{h})} \leftarrow -\eta\big[\tanh\big(b_m^{(\mathrm{h})}(0)\big) - \tanh\big(b_m^{(\mathrm{h})}(k)\big)\big]$;
  end for
  for $\mu = 1, \ldots, p_0$ **do**
    adjust weights and thresholds;
  end for
end for

[Equation (4.37)], the CD-k algorithm does not precisely correspond to gradient ascent. In summary,

$$\delta w_{mn} = \eta\Big[\tanh\Big(\sum_j w_{mj} v_{j,t=0} - \theta_m^{(\mathrm{h})}\Big) v_{n,t=0} - \tanh\Big(\sum_j w_{mj} v_{j,t=k} - \theta_m^{(\mathrm{h})}\Big) v_{n,t=k}\Big]. \tag{4.38}$$

The analogous learning rules for the thresholds read

$$\delta\theta_n^{(\mathrm{v})} = -\eta\big(v_{n,t=0} - v_{n,t=k}\big), \tag{4.39a}$$

$$\delta\theta_m^{(\mathrm{h})} = -\eta\Big[\tanh\Big(\sum_j w_{mj} v_{j,t=0} - \theta_m^{(\mathrm{h})}\Big) - \tanh\Big(\sum_j w_{mj} v_{j,t=k} - \theta_m^{(\mathrm{h})}\Big)\Big]. \tag{4.39b}$$

The derivation of Equation (4.39) is left as an exercise (Exercise 4.10). Restricted Boltzmann machines may have 0/1 neurons with state values 0 and 1 instead of -1 and 1. For 0/1 neurons, the CD-k algorithm is slightly different (Exercise 4.11).

Figure 4.4 illustrates how a restricted Boltzmann machine can learn to complete patterns, using the bars-and-stripes data set [25, 47] as an example. To begin with, the restricted Boltzmann machine is trained using the CD-k algorithm. Then consider a partially obscured pattern. Assume, for instance, that only the upper row of its bits is known: $v_1 = +1$ (■), $v_2 = -1$ (□), and $v_3 = +1$ (■). The remaining bits $v_4, \ldots, v_9$ are obscured, their states are set to zero, as shown in Figure 4.4(**b**). To complete the pattern, one samples from the Boltzmann distribution $P_{\mathrm{B}}(v_4, \ldots, v_9 | v_1 = +1, v_2 = -1, v_3 = +1)$, keeping $v_1 = +1, v_2 = -1, v_3 = +1$ fixed (*clamping* these neurons), and iterates the McCulloch-Pitts dynamics for the remaining ones. Panel (**b**) shows how the machine outputs the correct completed pattern.

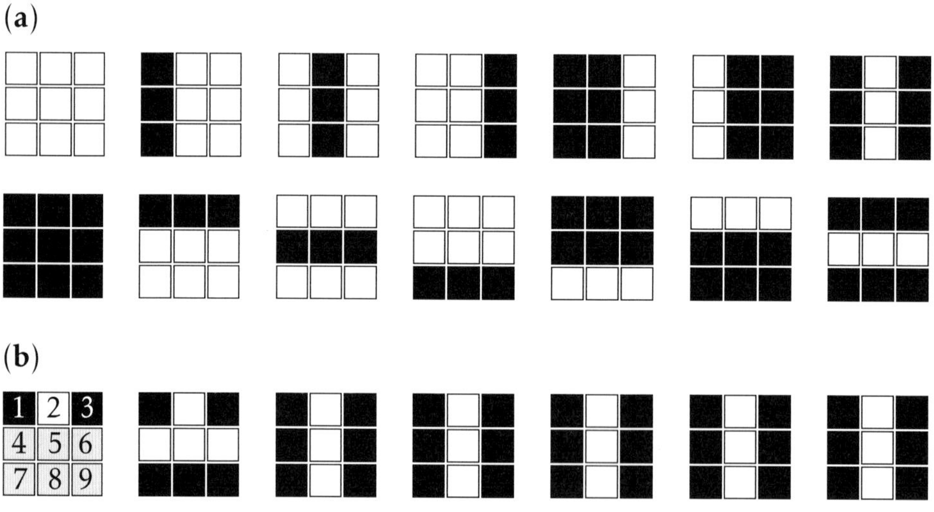

Figure 4.4 Pattern completion for bars-and-stripes data set [47]. (**a**) All patterns in the 3×3 bars-and-stripes data set, □ corresponds to -1, ■ to $+1$. (**b**) The three visible units $[v_1, v_2, v_3]$ in the first row are clamped to $[+1, -1, +1]$ and remain fixed to these values. The remaining units are initially set to 0 (gray bits); their states are allowed to change while sampling from the restricted Boltzmann machine. After a short transient of the McCulloch-Pitts dynamics, the pattern is correctly completed. After Figure 7 in Ref. [25]

This requires hidden neurons, because three-point correlations are needed to discriminate between bar and stripe patterns [47]. In general, a restricted Boltzmann machine can approximate a distribution $P_{\rm data}$ of binary input data better with more hidden neurons. How many are needed [58, 59]? The answer is not known in general, but it is plausible $M \sim 2^N$ hidden neurons are sufficient, because each hidden neuron can encode one of the binary input patterns (*winning neuron*, Section 7.1). More precisely, it can be shown that $M = 2^{N-1}$ hidden neurons are sufficient to reach arbitrarily small Kullback-Leibler divergence [60]. For binary data, an upper bound for the Kullback-Leibler divergence was derived in Refs. [61, 62]:

$$D_{\rm KL} \leq \log 2 \begin{cases} N - \lfloor \log_2(M+1) \rfloor - \frac{M+1}{2^{\lfloor \log_2(M+1) \rfloor}} & M < 2^{N-1} - 1\,, \\ 0 & M \geq 2^{N-1} - 1\,. \end{cases} \tag{4.40}$$

Here $\lfloor \cdots \rfloor$ denotes the integer part. Figure 4.5 illustrates this result. It demonstrates how well a restricted Boltzmann machine approximates the XOR distribution introduced in Section 4.4. The figure shows how the Kullback-Leibler divergence depends on the number of hidden neurons (Exercise 4.7). In this example, there are $N = 3$ inputs. We see that three hidden neurons are sufficient to allow the restricted Boltzmann machine to approximate the data distribution very precisely, consistent with Equation (4.40). In general, however, the CD-k algorithm is not guaranteed to converge to the optimal solution corresponding to the estimate (4.40).

Restricted Boltzmann machines are *generative models*, they can be used to sample from a distribution the machine has learned [25]. In this way, the machine can complete missing information, as illustrated in Figure 4.4. Restricted Boltzmann machines can also learn to classify patterns, by learning a distribution of

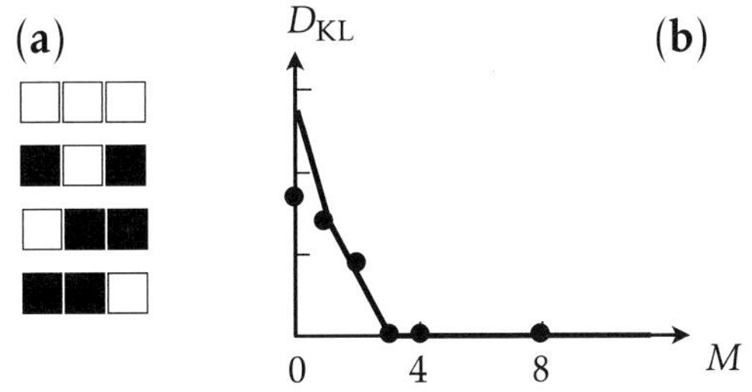

Figure 4.5 Restricted-Boltzmann-machine learning for the XOR problem [panel (**a**)]; see Section 4.4. Panel (**b**) shows numerical estimates of the Kullback-Leibler divergence $D_{\rm KL}$ versus the number M of hidden neurons, in comparison with the upper bound (4.40). Schematic, based on simulations performed by Arvid Wenzel Wartenberg using the CD-k algorithm for $k = 100$, with learning rate $\eta = 0.1$, averaging over 500 realisations

binary inputs together with their labels. To this end, one splits the visible neurons into input neurons and output neurons with labels or targets. This is a supervised-learning task, the subject of Part II. Recently, restricted Boltzmann machines were used to represent and analyse ground-state wave functions of quantum many-body systems [63].

4.6 Summary

This chapter dealt with the Boltzmann distribution. Two main points are, first, that the stochastic McCulloch-Pitts dynamics (3.1) has the Boltzmann distribution as a steady state. Second, the update rule (3.1) is a special case of the Markov-chain Monte-Carlo algorithm, for Hopfield models with the energy function (2.45). Since this algorithm tends to decrease the energy function, it can be used to solve complex optimisation problems. In simulated annealing one gradually reduces the noise level as the simulation proceeds. This mimics the slow cooling of a physical system, usually an efficient way of bringing the system into its global optimum.

Boltzmann machines are generalisations of Hopfield networks that can learn distributions of binary data by iteratively changing the weights and thresholds until the corresponding Boltzmann distribution approximates the data distribution. The learning rule is derived using gradient ascent on a target function, in this case the log-likelihood. A related idea is used for training deep neural networks with stochastic gradient descent (Part II). To learn general input distributions of binary patterns requires hidden neurons, also this is a central topic of Part II. Since Boltzmann machines with many hidden neurons are hard to train, one removes connections that are not needed. Restricted Boltzmann machines have connections only between visible and hidden neurons.

4.7 Further Reading

Older but still helpful references for Monte-Carlo methods in statistical physics are the book *Monte-Carlo Methods in Statistical Physics* edited by Binder [52], and Sokal's lecture notes [49]. Some historical notes are found in Ref. [64].

For a concise introduction to Boltzmann machines, refer to *Information Theory, Inference and Learning Algorithms* by MacKay [47] or *Machine Learning: A Probabilistic Perspective* by Murphy [65]. Ref. [66] is a more mathematical review of restricted Boltzmann machines.

How many hidden neurons should one allow for in a restricted Boltzmann machine? Little is known apart from the upper bound (4.40) for the Kullback-Leibler divergence, and simulations [60] show that one can get very precise approximations of P_{data} with less hidden neurons than stipulated by Equation (4.40).

Deep-belief networks consist of layers of restricted Boltzmann machines [2]. Contrastive-divergence training for such deep architectures (networks with many layers) is one of the first examples of deep-learning algorithms [67] (Chapter 7).

Helmholtz machines [68, 69] are generalisations of Boltzmann machines designed as more efficient generative models. They consist of two networks, encoder and decoder, just like variational autoencoders (Section 10.6). The encoder (called recognition model in Ref. [69]) generates a compressed representation of the data distribution, and the decoder (generative model) generates patterns from the compressed representation.

4.8 Exercises

4.1 Asymmetric weights. Show that Equations (3.1) and (4.3) are not equivalent for the network shown in Figure 2.9. The reason is that the weights are not symmetric.

4.2 Stochastic dynamics with 0/1-neurons. Derive the equivalent of the stochastic dynamics (3.1) for 0/1-neurons with states $n_j = 0$ or 1. Derive the equivalent of Equation (4.3) for 0/1 neurons, with energy function $H = -\frac{1}{2}\sum_{ij} w_{ij} n_i n_j + \sum_i \mu_i n_i$. Why are the two formulations *not* equivalent if the weights are asymmetric or if some w_{ii} are positive?

4.3 Metropolis algorithm. Use the Metropolis algorithm to generate a Markov chain that samples the exponential distribution $P(x) = \exp(-x)$.

4.4 Markov chain. Figure 4.6 illustrates the transition probabilities $p_{l \to k}$ for a Markov chain on a state space with three states. Find the steady state of this Markov chain. Does this chain satisfy detailed balance?

4.5 Double-digest problem. Implement simulated annealing for the double-digest problems given in Table 4.1. Use the energy function (4.15). Configuration space is the space of all permutation pairs $[\sigma, \mu]$. Local moves correspond to inversions

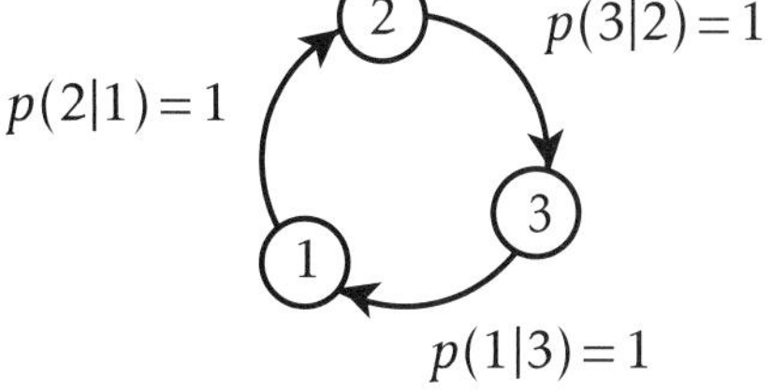

Figure 4.6 Markov chain with three states. The transition probability from state l to state k is denoted by $p(k|l)$ (Exercise 4.4)

of short subsequences of σ and/or μ. Check that the scheme of suggesting new states is symmetric. Using simulated annealing, determine the degeneracy of the solutions for the fragment sets shown in Table 4.1.

4.6 Kullback-Leibler divergence. Show that the Kullback-Leibler divergence $D_{\rm KL}$, Equation (4.18), is non-negative, and that it assumes its global minimum $D_{\rm KL} = 0$ at $P_{\rm data}(\boldsymbol{x}^{(\mu)}) = P_{\rm B}(\boldsymbol{s} = \boldsymbol{x}^{(\mu)})$. Show that minimising $D_{\rm KL}$ is equivalent to maximising the log-likelihood (4.17).

4.7 XOR function. Program a restricted Boltzmann machine to learn the XOR function, by approximating the following data distribution over three-bit binary patterns: $P_{\rm data} = \frac{1}{4}$ for $[-1, -1, -1]$, $[1, 1, -1]$, $[-1, 1, 1]$, and $[1, -1, 1]$, and $P_{\rm data} = 0$ otherwise. Plot the Kullback-Leibler divergence as a function of iteration number for different numbers of hidden neurons: 0, 2, 4, and 8.

4.8 Shifter ensemble. Explain why the shifter ensemble [15, 47] cannot be approximated by a Boltzmann machine without hidden neurons.

4.9 McCulloch-Pitts dynamics for restricted Boltzmann machine. Write down the deterministic analogue of the update rule (4.30) and show that the energy function of the restricted Boltzmann machine cannot increase under this rule. Note that it is not required that the weight matrix is symmetric, or that the diagonal elements are non-positive.

4.10 Thresholds in restricted Boltzmann machines. Derive the learning rule (4.39) for the thresholds for a restricted Boltzmann machine.

4.11 Restricted Boltzmann machine with 0/1 neurons. Derive the contrastive-divergence algorithm for training a restricted Boltzmann machine with 0/1 neurons.

4.12 Bars-and-stripes data set. Train a restricted Boltzmann machine with Algorithm 3 to learn the bars-and-stripes data set (Figure 4.4). After training, sample the model distribution for $M = 2, 4, 8, 16$, and 32 hidden neurons and use the numerical results to estimate the Kullback-Leibler divergence $D_{\rm KL}$ as a function of k. Compare with the theoretical upper bound (4.40).

Part II

Supervised Learning

The Hopfield network described in Part I recognises patterns stored in the network using Hebb's rule. Its neurons act as inputs and outputs. After feeding a distorted pattern into the network, the network dynamics runs until it reaches a steady state which hopefully corresponds to the stored pattern closest to the distorted one. In this case, the network classifies the distorted pattern by associating it with the closest one amongst the stored patterns.

Part II describes *supervised learning*, a different way of solving *classification tasks* with neural networks using labelled data sets. The *machine-learning repository* [70] at the University of California Irvine contains a number of such data sets. An example is the *iris* data set, which lists certain properties of 150 iris plants. For each plant, four attributes are given (Figure 5.1): its sepal length, sepal width, petal length, and petal width. Each entry in the data set contains a *label* (or *target*) that says which class the plant belongs to: *iris setosa*, *iris versicolor*, or *iris virginica*. This data set was described by the geneticist R. A. Fisher [71].

The machine-learning task is to adjust weights and thresholds of a neural network so that it correctly determines the class of each plant from its attributes. To this end, one uses a *training* set of labelled data. Each set of attributes is an input pattern to the network. The neural network is supposed to output the correct label (or target), in this case whether the plant is an *iris setosa*, *iris versicolor*, or *iris virginica*. One compares the network output with the corresponding target, for all input patterns in the training set, and changes the weights and thresholds until the network computes the correct output for each input pattern. The crucial question is whether the trained network can *generalise*: does it find the correct labels for an input pattern not contained in the training set?

The networks used for supervised learning are called *perceptrons* [10]. They consist of layers of McCulloch-Pitts neurons: a number of layers of *hidden* neurons, and an output layer. We briefly discussed the idea of hidden neurons in connection with restricted Boltzmann machines (Section 4.5), but perceptrons have different

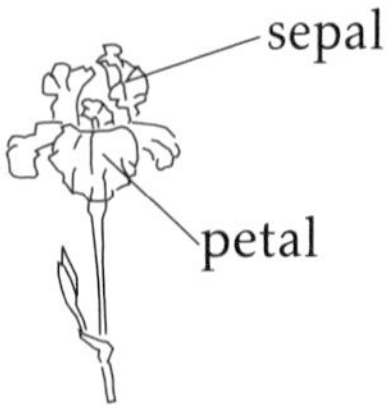

sepal length	width	petal length	width	classification
6.3	2.5	5.0	1.9	virginica
5.1	3.5	1.4	0.2	setosa
5.5	2.6	4.4	1.2	versicolor
4.9	3.0	1.4	0.2	setosa
6.1	3.0	4.6	1.4	versicolor
6.5	3.0	5.2	2.0	virginica

Figure 5.1 Left: petals and sepals of the iris flower. Right: six entries of the iris data set [70]. All lengths in cm. The whole data set contains 150 entries.

layouts, and they are trained in a different way. The layers are usually arranged from the left (input) to the right (output). All connections are one-way, from neurons in one layer to neurons in the layer immediately to the right. There are no connections between neurons in a given layer, or back to layers on the left. This arrangement ensures convergence of the training algorithm (*stochastic gradient descent*). During training with this algorithm, the network parameters are changed iteratively. In each step, an input is applied, and weights and thresholds of the network are updated to reduce the output error. Loosely speaking, each step corresponds to adding a little bit of Hebb's rule to the weights. This is repeated until the network classifies the training set correctly.

Stochastic gradient descent for multilayer perceptrons has received much attention recently, after it was realised that networks with many hidden layers can be trained to reliably recognise and classify image data (*deep learning*).

5
Perceptrons

In 1958, Rosenblatt [10] suggested to connect McCulloch-Pitts neurons into layered *feed-forward* networks to process information. He referred to these networks as *perceptrons*. The layout is illustrated in Figure 5.2. The leftmost layer consists of input terminals, drawn in black in Figure 5.2. To the right are two layers of McCulloch-Pitts neurons. The rightmost layer consists of output neurons. The intermediate layer is a *hidden* layer. The states of its neurons are not read out. All connections are one-way: every neuron feeds *forward*, only to neurons in the layer immediately to the right. There are no connections within layers, no back connections, no connections that skip a layer. There are N input terminals. As in Part I, we denote the input patterns by

$$\boldsymbol{x}^{(\mu)} = \begin{bmatrix} x_1^{(\mu)} \\ x_2^{(\mu)} \\ \vdots \\ x_N^{(\mu)} \end{bmatrix} . \tag{5.1}$$

The index $\mu = 1, \ldots, p$ labels the different input patterns. The hidden neurons compute

$$V_j = g\left(b_j\right) \quad \text{with} \quad b_j = \sum_k w_{jk} x_k - \theta_j \,, \tag{5.2}$$

with weights w_{jk} and thresholds θ_j. The function $g(b)$ is an activation function, and its argument is called local field (Section 1.2). The output neurons of the network shown in Figure 5.2 perform the computation

$$O_i = g\left(B_i\right) \quad \text{with} \quad B_i = \sum_j W_{ij} V_j - \Theta_i \,. \tag{5.3}$$

The index $i = 1, \ldots, M$ labels the output neurons with weights W_{ij}, and with thresholds Θ_i.

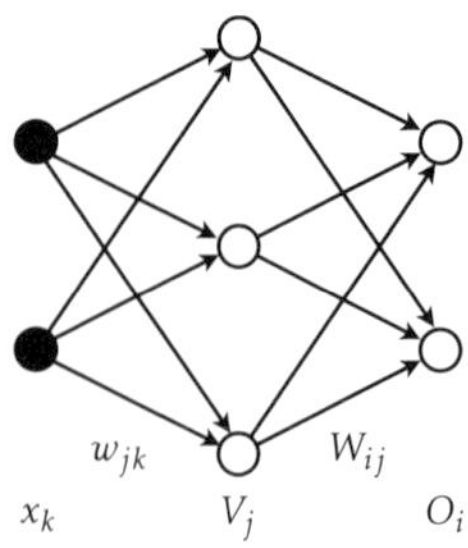

Figure 5.2 Feed-forward network with one hidden layer. The input terminals are coloured black. We use the notation of Ref. [1]: W_{ij} for the weights connecting to the output neuron O_i (with threshold Θ_i), and w_{jk} for the weights connecting to the hidden neuron V_j (with threshold θ_j)

A classification problem is given by a training set of input patterns $\boldsymbol{x}^{(\mu)}$ and the corresponding *target* vectors

$$\boldsymbol{t}^{(\mu)} = \begin{bmatrix} t_1^{(\mu)} \\ t_2^{(\mu)} \\ \vdots \\ t_M^{(\mu)} \end{bmatrix} . \tag{5.4}$$

The idea is to choose all weights and thresholds so that the network produces the desired output:

$$O_i^{(\mu)} = t_i^{(\mu)} \quad \text{for all} \quad i \quad \text{and} \quad \mu . \tag{5.5}$$

In the Hopfield networks described in Part I, the weights were assigned using Hebb's rule (2.26). Perceptrons, by contrast, are trained by iteratively updating their weights and thresholds until Equation (5.5) is satisfied. This is achieved by repeatedly adding small multiples of Hebb's rule to the weights (Section 5.2). An alternative approach is to define an energy function, a function of the weights of the network, that has a global minimum when Equation (5.5) is satisfied. The network is trained by taking small steps in weight space that reduce the energy function (gradient descent, Section 5.3).

5.1 A Classification Problem

To illustrate how perceptrons can solve classification problems, consider the simple example shown in Figure 5.3. There are 10 patterns, each has two real-valued components:

$$\boldsymbol{x}^{(\mu)} = \begin{bmatrix} x_1^{(\mu)} \\ x_2^{(\mu)} \end{bmatrix} . \tag{5.6}$$

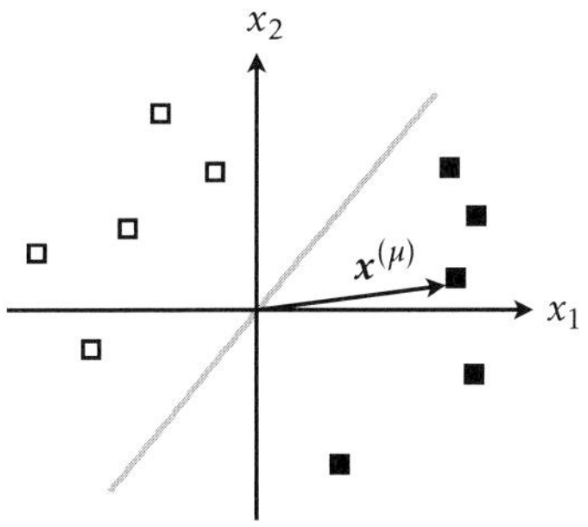

Figure 5.3 Classification problem with two-dimensional real-valued inputs and targets equal to ± 1. The gray solid line is the decision boundary. Legend: ■ corresponds to $t^{(\mu)} = 1$, and □ to $t^{(\mu)} = -1$

In Figure 5.3 the patterns are drawn as points in the x_1-x_2 plane, the *input plane*. There are two classes of patterns, with targets ± 1:

$$t^{(\mu)} = 1 \quad \text{for} \quad \blacksquare \quad \text{and} \quad t^{(\mu)} = -1 \quad \text{for} \quad \square\,. \tag{5.7}$$

A single neuron suffices to classify these patterns, a binary threshold unit with activation function $g(b) = \text{sgn}(b)$, consistent with the possible target values. Since there is only one neuron, we can arrange the weights into a weight vector

$$\boldsymbol{w} = \begin{bmatrix} w_1 \\ w_2 \end{bmatrix}. \tag{5.8}$$

The network performs the computation

$$O = \text{sgn}(w_1 x_1 + w_2 x_2 - \theta) = \text{sgn}(\boldsymbol{w} \cdot \boldsymbol{x} - \theta)\,. \tag{5.9}$$

Here $\boldsymbol{w} \cdot \boldsymbol{x} = w_1 x_1 + w_2 x_2$ is the scalar product between the vectors $\boldsymbol{w}$ and $\boldsymbol{x}$ (Chapter 2).

This example allows us to find a geometrical interpretation of the classification problem. We see in Figure 5.3 that the patterns fall into two clusters: □ to the left and ■ to the right. We can classify the patterns by drawing a line that separates the two clusters, so that everything to the right of the line has $t = 1$, while the patterns to the left of the line have $t = -1$. This line is called the *decision boundary*. To find the geometrical significance of Equation (5.9), let us put the threshold to zero for a moment, so that

$$O = \text{sgn}(\boldsymbol{w} \cdot \boldsymbol{x})\,. \tag{5.10}$$

Then the classification problem takes the form

$$\text{sgn}\left(\boldsymbol{w} \cdot \boldsymbol{x}^{(\mu)}\right) = t^{(\mu)}\,. \tag{5.11}$$

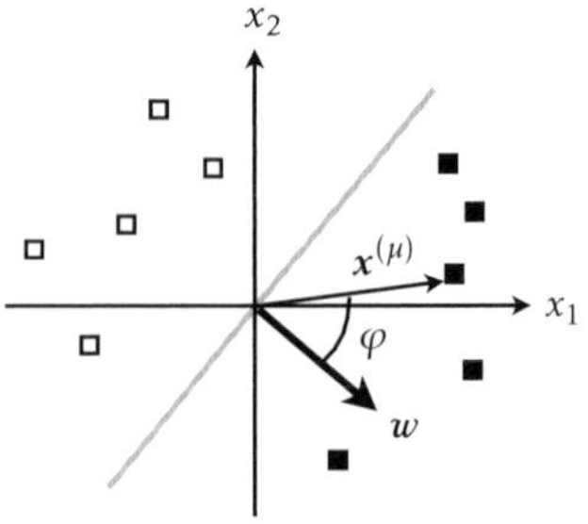

Figure 5.4 The perceptron classifies the patterns correctly for the weight vector $\boldsymbol{w}$ shown, orthogonal to the decision boundary (gray solid line). Legend: ■ corresponds to $t^{(\mu)} = 1$, and □ to $t^{(\mu)} = -1$

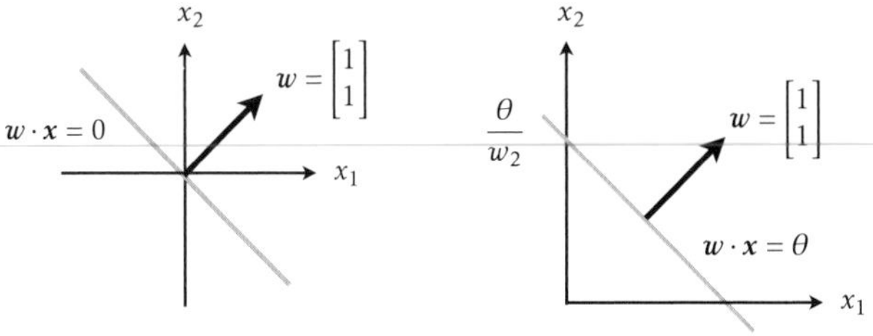

Figure 5.5 Decision boundaries without and with threshold

To evaluate the scalar product, we write the vectors as

$$\boldsymbol{w} = |\boldsymbol{w}| \begin{bmatrix} \cos\alpha \\ \sin\alpha \end{bmatrix} \quad \text{and} \quad \boldsymbol{x} = |\boldsymbol{x}| \begin{bmatrix} \cos\beta \\ \sin\beta \end{bmatrix} . \tag{5.12}$$

Here $|\boldsymbol{w}| = \sqrt{w_1^2 + w_2^2}$ denotes the norm of the vector $\boldsymbol{w}$, and α and β are the angles of the vectors with the x_1-axis. Then $\boldsymbol{w} \cdot \boldsymbol{x} = |\boldsymbol{w}||\boldsymbol{x}|\cos(\alpha - \beta) = |\boldsymbol{w}||\boldsymbol{x}|\cos\varphi$, where φ is the angle between the two vectors. When φ is between $-\pi/2$ and $\pi/2$, the scalar product is positive; otherwise, it is negative. As a consequence, the network classifies the patterns in Figure 5.3 correctly if the weight vector is orthogonal to the decision boundary, as shown in Figure 5.4.

What is the role of the threshold θ? Equation (5.9) implies that the decision boundary is parameterised by $\boldsymbol{w} \cdot \boldsymbol{x} = \theta$, or

$$x_2 = -(w_1/w_2)\, x_1 + \theta/w_2 \, . \tag{5.13}$$

Therefore the threshold determines the intersection of the decision boundary with the x_2-axis (equal to θ/w_2). This is illustrated in Figure 5.5.

The decision boundary – the straight line orthogonal to $\boldsymbol{w}$ – should divide inputs with positive and negative targets. If such a line can be found, then the problem

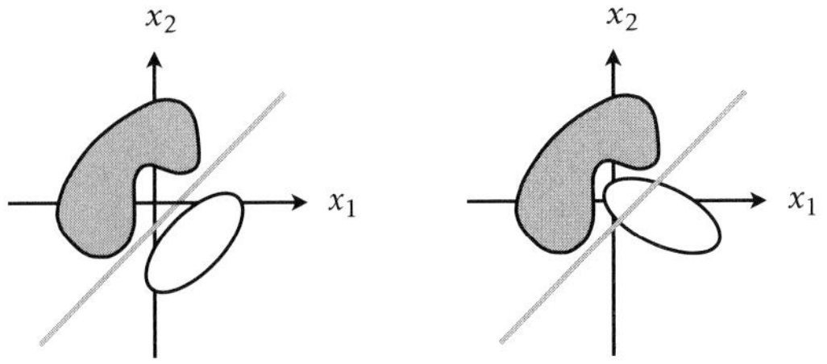

Figure 5.6 Linearly separable and non-separable data in two-dimensional input space

x_1	x_2	t
0	0	−1
0	1	−1
1	0	−1
1	1	+1

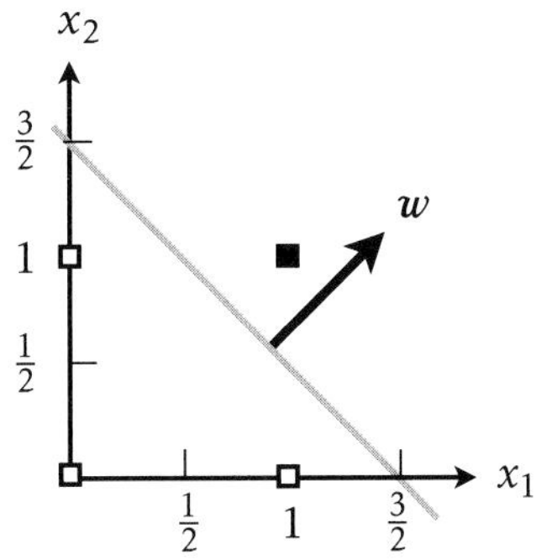

Figure 5.7 Boolean AND function: value table (left) and geometrical representation in the input plane (right). Legend: ■ corresponds to $t^{(\mu)} = 1$, and □ to $t^{(\mu)} = -1$

can be solved with a single neuron. We say that the problem is *linearly separable*. Conversely, if no such line exists, the problem not linearly separable. This can occur only when $p > N$. Figure 5.6 shows two problems. The left one is linearly separable, the right one is not.

Other examples are *Boolean functions*. A Boolean function takes N binary inputs and has one binary output. The Boolean AND function (two inputs) is illustrated in Figure 5.7. The value table of the function is shown on the left. The graphical representation is shown on the right of the Figure (□ corresponds to $t = -1$ and ■ to $t = +1$). Also shown is the decision boundary of a binary threshold unit and its weight vector $\boldsymbol{w}$. It is important to note that the decision boundary is not unique, neither are the weight vector and threshold value that solve the problem. The norm of the weight vector, in particular, is arbitrary. Figure 5.8 illustrates that the Boolean XOR function is not linearly separable [11]. There are 16 different Boolean functions of two variables. Only two are not linearly separable (Exercise 5.2): XOR (Figure 5.8) and XNOR.

Up until this point, we have discussed only one single neuron. If the classification problem requires several output neurons, each has its own weight vector $\boldsymbol{w}_i$

and threshold θ_i. We can group the weight vectors into a weight matrix $\mathbb{W}$ as in Part I, so that the row vectors $\boldsymbol{w}_i^{\mathsf{T}}$ are the rows of the weight matrix $\mathbb{W}$.

5.2 Iterative Learning Algorithm

In the previous section, we determined the weights and threshold for the Boolean AND function by inspection (Figure 5.7). Now we discuss an algorithm that finds the weights iteratively. This is illustrated in Figure 5.9. In panel (**a**), the pattern $\boldsymbol{x}^{(8)}$ ($t^{(8)} = 1$) is on the wrong side of the decision boundary. In order to correct this error, one turns the decision boundary anti-clockwise. To this end, one *adds* a small multiple of the pattern vector $\boldsymbol{x}^{(8)}$ to the weight vector:

$$\boldsymbol{w}' = \boldsymbol{w} + \delta\boldsymbol{w} \quad \text{with} \quad \delta\boldsymbol{w} = \eta\boldsymbol{x}^{(8)} . \tag{5.14}$$

The parameter $\eta > 0$ is called the *learning rate*. It must be small, so that the decision boundary is not rotated too far. The result is shown in panel (**b**). Panel (**c**) shows another case, where pattern $\boldsymbol{x}^{(4)}$ ($t^{(4)} = -1$) is on the wrong side of the decision boundary. In order to turn the decision boundary in the right way, anti-clockwise, one *subtracts* a small multiple of $\boldsymbol{x}^{(4)}$:

$$\boldsymbol{w}' = \boldsymbol{w} + \delta\boldsymbol{w} \quad \text{with} \quad \delta\boldsymbol{w} = -\eta\boldsymbol{x}^{(4)} . \tag{5.15}$$

These two *learning rules* combine to the learning rule of Rosenblatt [10]:

$$\boldsymbol{w}' = \boldsymbol{w} + \delta\boldsymbol{w}^{(\mu)} \quad \text{with} \quad \delta\boldsymbol{w}^{(\mu)} = \eta t^{(\mu)}\boldsymbol{x}^{(\mu)} . \tag{5.16}$$

For more than one neuron, the rule reads

$$w'_{ij} = w_{ij} + \delta w_{ij}^{(\mu)} \quad \text{with} \quad \delta w_{ij}^{(\mu)} = \eta t_i^{(\mu)} x_j^{(\mu)} . \tag{5.17}$$

This rule is reminiscent of Hebb's rule (2.9), except that here inputs and outputs are associated with distinct units. Therefore we have $t_i^{(\mu)} x_j^{(\mu)}$ instead of $x_i^{(\mu)} x_j^{(\mu)}$.

x_1	x_2	t
0	0	−1
0	1	+1
1	0	+1
1	1	−1

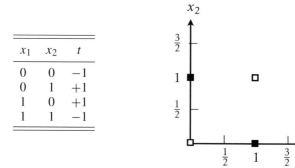

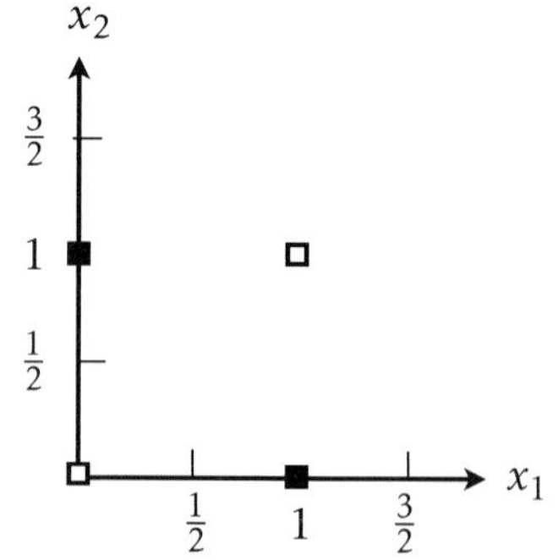

Figure 5.8 The Boolean XOR function is not linearly separable. Legend: ■ corresponds to $t^{(\mu)} = 1$, and □ to $t^{(\mu)} = -1$

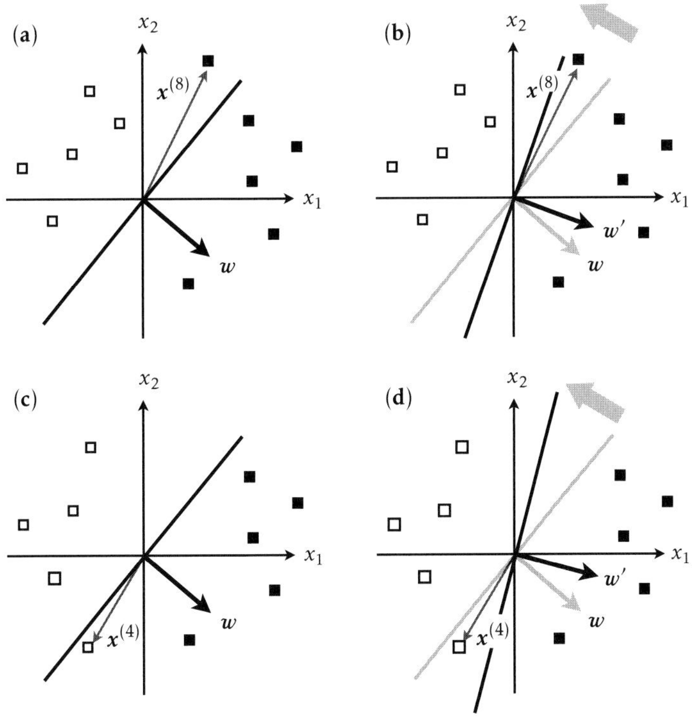

Figure 5.9 Illustration of the learning algorithm. In panel (**a**) the $t = 1$ pattern $x^{(8)}$ is on the wrong side of the decision boundary (solid black line). To correct the error, the weight must be rotated anti-clockwise [panel (**b**)]. In panel (**c**), the $t = -1$ pattern $x^{(4)}$ is on the wrong side of the decision boundary. To correct the error, the weight must also be rotated anti-clockwise [panel (**d**)]

One applies (5.17) iteratively for a sequence of randomly chosen patterns μ, until the problem is solved. This corresponds to adding a little bit of Hebb's rule in each iteration. To ensure that the algorithm stops when the problem is solved, one can use the learning rule [1]

$$\delta w_{ij}^{(\mu)} = \eta\big(t_i^{(\mu)} - O_i^{(\mu)}\big)x_j^{(\mu)} . \tag{5.18}$$

5.3 Gradient Descent for Linear Units

In this section, the learning algorithm (5.18) is derived in a different way, by minimising an energy function using gradient descent. This requires differentiation, therefore we must choose a differentiable activation function. The simplest choice is a linear activation function, $g(b) = b$. We set $\theta = 0$, so that the network computes

$$O_i^{(\mu)} = \sum_k w_{ik} x_k^{(\mu)} . \tag{5.19}$$

A neuron with a linear activation function is called a *linear unit*. The outputs $O_i^{(\mu)}$ assume continuous values, but not necessarily the targets $t_i^{(\mu)}$. For linear units, the classification problem

$$O_i^{(\mu)} = t_i^{(\mu)} \quad \text{for} \quad i = 1, \ldots, N \quad \text{and} \quad \mu = 1, \ldots, p \tag{5.20}$$

has the formal solution

$$w_{ik} = \frac{1}{N} \sum_{\mu\nu} t_i^{(\mu)} \left(\mathbb{Q}^{-1}\right)_{\mu\nu} x_k^{(\nu)} . \tag{5.21}$$

This can be verified by inserting Equation (5.21) into (5.19). Here $\mathbb{Q}$ is the overlap matrix with elements

$$Q_{\mu\nu} = \tfrac{1}{N} \boldsymbol{x}^{(\mu)} \cdot \boldsymbol{x}^{(\nu)} \tag{5.22}$$

(Section 3.6). For the solution (5.21) to exist, the matrix $\mathbb{Q}$ must be invertible. As mentioned in Section 3.6, this requires that $p \leq N$, because otherwise the input-pattern vectors are *linearly dependent*, along with the columns (and rows) of $\mathbb{Q}$. If the matrix $\mathbb{Q}$ has linearly dependent columns or rows, it cannot be inverted.

Let us assume that the input patterns are linearly independent, so that the solution (5.21) exists. In this case, we can find the solution iteratively. To this end, one defines the energy function

$$H = \frac{1}{2} \sum_{i\mu} \left(t_i^{(\mu)} - O_i^{(\mu)}\right)^2 . \tag{5.23}$$

This function is non-negative, and it vanishes when all outputs equal the corresponding targets, for all patterns.

The energy function (5.23) is regarded as a function of the weights w_{ij}, unlike the energy function in Part I, which is a function of the state-variables of the neurons. The goal is now to find weights that minimise H. If the input patterns are linearly independent, H vanishes at the global minimum, corresponding to the desired solution of the problem (Exercise 5.1). Let us use *gradient descent* to minimise H,

$$w'_{mn} = w_{mn} + \delta w_{mn} \quad \text{with weight increments} \quad \delta w_{mn} = -\eta \frac{\partial H}{\partial w_{mn}} , \tag{5.24}$$

with learning rate η. This is analogous to Equation (4.19), apart from the minus sign. In Section 4.4, the goal was to maximise the target function; here we want to minimise H by taking many downhill steps in search of the global minimum. The derivatives in Equation (5.24) are evaluated with the chain rule, together with Equation (4.22), which takes the form

$$\frac{\partial w_{ij}}{\partial w_{mn}} = \delta_{im} \delta_{jn} \tag{5.25}$$

for asymmetric weights. This yields the weight increments

$$\delta w_{mn} = \eta \sum_{\mu} \big(t_m^{(\mu)} - O_m^{(\mu)}\big)\, x_n^{(\mu)} \,. \tag{5.26}$$

This learning rule is very similar to Equation (5.18). One difference is that Equation (5.26) contains a sum over all patterns. It is important to also keep in mind that the activation functions are different: while Equation (5.18) was derived for $g(b) = \text{sgn}(b)$, the learning rule (5.26) was derived for $g(b) = b$. An advantage of the rule (5.26) is that it is derived from an energy function. This helps analyse the convergence of the algorithm, as we have seen in Chapter 2.

Linear units [Equation (5.19)] are special. The Boolean AND problem (Figure 5.7) does not admit the solution (5.21), even though the problem is linearly separable. Since the pattern vectors $\boldsymbol{x}^{(\mu)}$ are linearly dependent, the solution (5.21) does not exist. Shifting the patterns or introducing a threshold does not change this fact.

In Section 5.5, we discuss how to solve problems that are not linearly separable using a hidden layer of neurons with non-linear activation functions. Note that introducing hidden layers with linear units does not help, because the resulting input-output mapping is still linear if all neurons have linear activation functions, so that only problems with $p \leq N$ can be solved. This is the main reason for using hidden layers with non-linear activation functions.

There are four points to keep in mind. First, if the patterns are linearly independent, then we can use gradient descent to determine suitable weights (and thresholds) of linear units. Second, in general hidden layers with non-linear units are required, because a single neuron with a continuous and monotonous activation function can only solve problems with linearly independent patterns (Exercise 5.11). Third, for gradient descent for non-linear units, we must require that the activation function $g(b)$ is differentiable, or at least piecewise differentiable. Fourth, in this case, we calculate the gradients using the chain rule, resulting in factors of derivatives $\frac{\mathrm{d}}{\mathrm{d}b} g(b)$. This is the origin of the *vanishing-gradient problem* (Chapter 7).

5.4 Classification Capacity

In Chapter 3, we analysed the storage capacity of Hopfield networks. The analogous question for the classification problem described in Section 5.1 is: how many patterns can a single neuron with activation function $g(b) = \text{sgn}(b)$ classify? As in the case of Hopfield networks, one can find a general answer for random binary classification problems.

Figure 5.10 Left: Five points in a general position in the plane. Right: these points are not in a general position because three points lie on a straight line

Consider p points with coordinate vectors $\boldsymbol{x}^{(\mu)}$ in N-dimensional input space, and assign random targets:

$$t^{(\mu)} = \begin{cases} +1 & \text{with probability } \frac{1}{2}\,, \\ -1 & \text{with probability } \frac{1}{2}\,. \end{cases} \tag{5.27}$$

This random classification problem is *homogeneously* linearly separable if we can find an N-dimensional weight vector $\boldsymbol{w}$, so that $\boldsymbol{w} \cdot \boldsymbol{x} = 0$ is a valid decision boundary that goes through the origin:

$$\boldsymbol{w} \cdot \boldsymbol{u}^{(\mu)} > 0 \quad \text{if} \quad t^{(\mu)} = 1 \quad \text{and} \quad \boldsymbol{w} \cdot \boldsymbol{u}^{(\mu)} < 0 \quad \text{if} \quad t^{(\mu)} = -1\,. \tag{5.28}$$

So *homogeneously linearly separable* problems are binary classification problems that are linearly separable by a hyperplane that contains the origin. Problems with this property can be solved by a binary threshold unit with threshold $\theta = 0$.

Now assume that the points (including the origin) are in *general position* (Figure 5.10). In this case, *Cover's theorem* [72] gives an expression for the probability that the random binary classification problem of p patterns in dimension N is homogeneously linearly separable:

$$P(p, N) = \begin{cases} \left(\frac{1}{2}\right)^{p-1} \sum_{k=0}^{N-1} \binom{p-1}{k} & \text{for } p > N\,, \\ 1 & \text{otherwise}\,. \end{cases} \tag{5.29}$$

Here $\binom{l}{k} = \frac{l!}{(l-k)!k!}$ are the binomial coefficients, for $l \geq k \geq 0$. Equation (5.29) is proven by recursion, starting from a set of $p-1$ points in general position. Assume that the number $C(p-1, N)$ of homogeneously linearly separable classification problems given these points is known. After adding one more point, one can compute the $C(p, N)$ in terms of $C(p-1, N)$, and recursion yields Equation (5.29). Figure 5.11 shows this result as a function of $\alpha = p/N$ for different values of N. For $p \leq N$, any random classification problem is homogeneously linearly separable. In this case, the pattern vectors are linearly independent, so that the problem can also be solved by a linear unit (Section 5.3). But a neuron with activation function $\text{sgn}(b)$ can classify problems with more than N patterns.

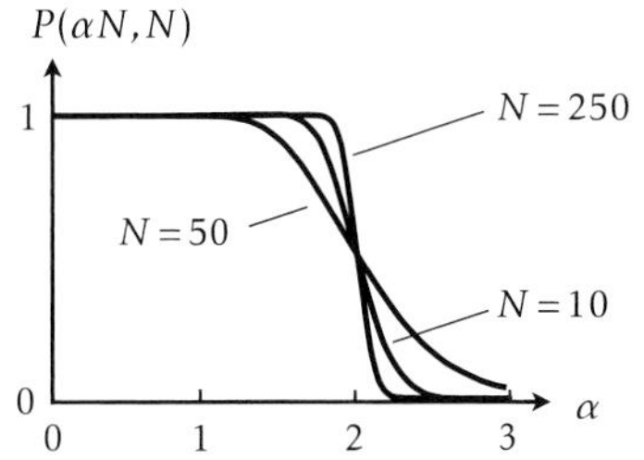

Figure 5.11 Probability (5.29) of separability as a function of $\alpha = p/N$ for three different values of the dimension N of input space

In the limit of $N \to \infty$, the function $P(\alpha N, N)$ approaches a step function $\theta_H(2-\alpha)$ (Exercise 5.12). In this limit, the maximal classification capacity is therefore $\alpha_{\max} = 2$.

What is the expected classification capacity for finite values of N? To answer this question, consider a random sequence of patterns $\boldsymbol{x}^{(1)}, \boldsymbol{x}^{(2)}, \ldots$ and targets $t^{(1)}, t^{(2)}, \ldots$ and ask [72]: what is the distribution of the largest integer so that the problem $\boldsymbol{x}^{(1)}, \boldsymbol{x}^{(2)}, \ldots, \boldsymbol{x}^{(n)}$ is separable in dimension N, but $\boldsymbol{x}^{(1)}, \boldsymbol{x}^{(2)}, \ldots, \boldsymbol{x}^{(n)}, \boldsymbol{x}^{(n+1)}$ is not? $P(n, N)$ is the probability that n patterns are linearly separable in N-dimensional input space. We can write $P(n+1, N) = q(n+1|n)P(n, N)$ where $q(n+1|n)$ is the conditional probability that $n+1$ patterns are linearly separable if the n patterns are. Then the probability that $n+1$ patterns are not separable (but n patterns are) reads $[(1-q(n+1|n)]P(n, N) = P(n, N) - P(n+1, N)$. We can interpret the right-hand side of this equation as a distribution p_n of the random variable n, the maximal number of separable patterns in dimension N:

$$p_n = P(n, N) - P(n+1, N) = \Big(\frac{1}{2}\Big)^n \binom{n-1}{N-1} \quad \text{for} \quad n = 0, 1, 2, \ldots .$$

It follows that the expected maximal number of separable patterns is

$$\langle n \rangle = \sum_{n=0}^{\infty} n p_n = 2N \,. \tag{5.30}$$

So the expected classification capacity is twice the input dimension:

$$\langle \alpha_{\max} \rangle = 2 \,. \tag{5.31}$$

This quantifies the notion that it is easier to separate patterns in higher-dimensional input space. As an illustration, consider the XOR problem which is not linearly separable in two-dimensional input space. The problem becomes separable when we *embed* the points in three-dimensional space, for instance by assigning $x_3 = 0$ to the $t = +1$ patterns and $x_3 = 1$ to the $t = -1$ patterns (Figure 5.12).

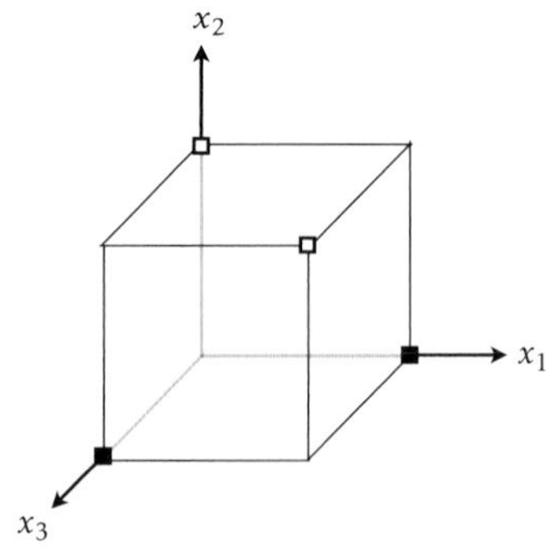

Figure 5.12 The XOR problem can be solved by embedding it into a three-dimensional input space

5.5 Multilayer Perceptrons

In Sections 5.1 and 5.2 we discussed how to solve linearly separable problems [Figure 5.13(**a**)]. The aim of this section is to show that non-separable problems like the one in Figure 5.13(**b**) can be solved by a perceptron with one hidden layer. A network that does the trick for the classification problem in Figure 5.13(**b**) is depicted in Figure 5.14. As in the previous section, all neurons have the signum function as activation function, with possible outputs ± 1:

$$V_j^{(\mu)} = \operatorname{sgn}\left(b_j^{(\mu)}\right) \quad \text{with} \quad b_j^{(\mu)} = \sum_k w_{jk} x_k^{(\mu)} - \theta_j \,,$$
$$O_1^{(\mu)} = \operatorname{sgn}\left(B_1^{(\mu)}\right) \quad \text{with} \quad B_1^{(\mu)} = \sum_j W_{1j} V_j^{(\mu)} - \Theta_1 \,. \tag{5.32}$$

Each of the three neurons in the hidden layer has its own decision boundary. The idea is to choose the weights w_{jk} and the thresholds θ_j in such a way that the three decision boundaries partition the input plane into distinct regions, so that each region contains either only $t = -1$ patterns or $t = +1$ patterns [3].

How this construction works is shown in Figure 5.15. The left part of the figure shows the three decision boundaries with their weight vectors, and how they divide the input plane into different regions which contain either only □ or only ■. Each region bears a three-digit code made out of the symbols + and −. The codes are determined by the states of the hidden neurons. A + sign in the j-th entry of the code means that $V_j = +1$. So the region in question is on the weight-vector side of the decision boundary j. A − sign, by contrast, corresponds to $V_j = -1$. In this case, the region is on the other side of the decision boundary, the one opposite the weight vector. The value table shows the targets associated with each region, together with the code of the region.

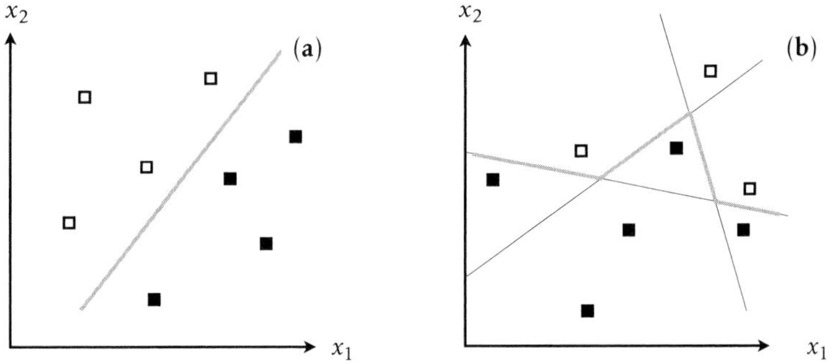

Figure 5.13 (**a**) Linearly separable problem. (**b**) Problems that are not linearly separable can be solved by a piecewise linear decision boundary (thick gray line). Legend: ■ corresponds to $t^{(\mu)} = 1$, and □ to $t^{(\mu)} = -1$

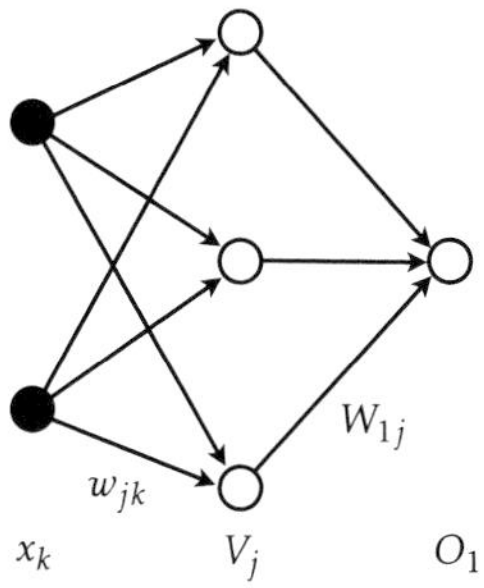

Figure 5.14 Hidden-layer perceptron to solve the problem shown in Figure 5.13(**b**)

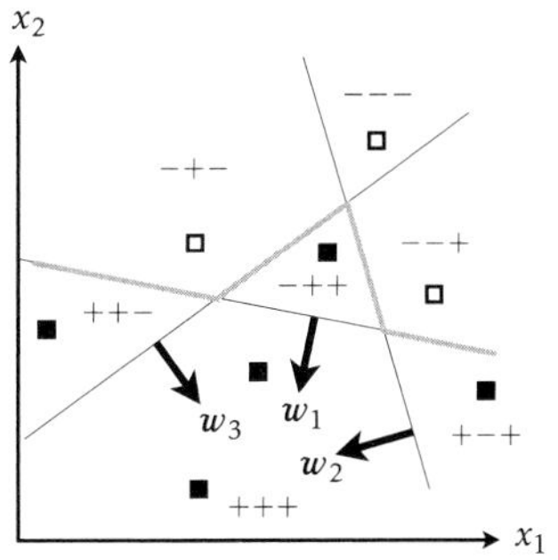

V_1	V_2	V_3	target
−	−	−	−1
+	−	−	
−	+	−	−1
−	−	+	−1
+	+	−	+1
+	−	+	+1
−	+	+	+1
+	+	+	+1

Figure 5.15 Left: Decision boundaries [Figure 5.13(**b**)], regions, and the corresponding binary codes determined by the states of the hidden neurons. Legend: ■ corresponds to $t^{(\mu)} = 1$, and □ to $t^{(\mu)} = -1$. Right: Encoding of the regions and corresponding targets. The region + − − does not exist

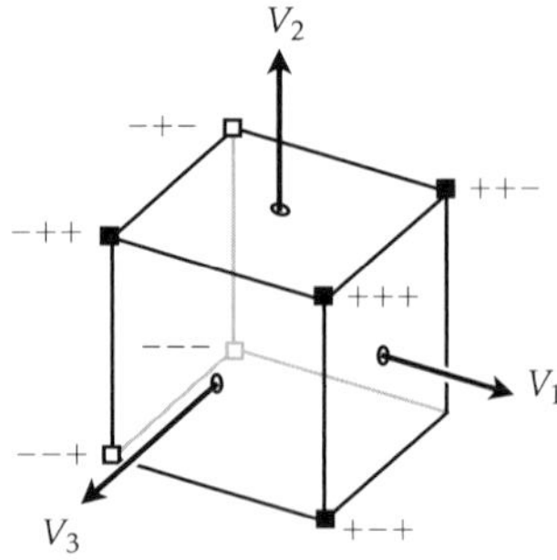

Figure 5.16 Graphical representation of the output problem for the classification problem shown in Figure 5.15

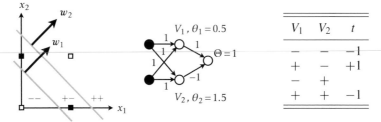

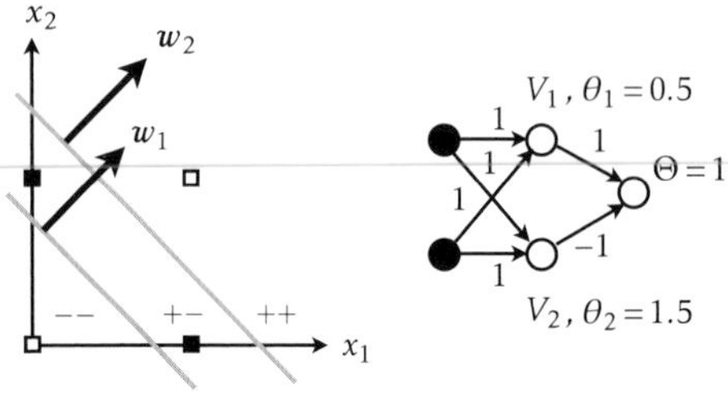

V_1	V_2	t
−	−	−1
+	−	+1
−	+	
+	+	−1

Figure 5.17 Boolean XOR function: geometrical representation, network layout, and value table for the output neuron. All neurons assume two possible states, +1 or −1. Legend for the geometrical representation: ■ corresponds to $t^{(\mu)} = 1$, and □ to $t^{(\mu)} = -1$

The weights W_{1j} and the threshold Θ_j of the output neuron are chosen so that it associates the correct target value with each region. A graphical representation of the output problem is shown in Figure 5.16. This problem is linearly separable (Exercise 5.3). The following function computes the correct output for each region:

$$O_1^{(\mu)} = \operatorname{sgn}\left(V_1^{(\mu)} + V_2^{(\mu)} + V_3^{(\mu)}\right) . \tag{5.33}$$

This solves the binary classification problem described in Figure 5.15, but note that the solution is not unique. There is a range of different weights and thresholds that solve the problem, and there are other solutions based on different network layouts. Nevertheless, the solution illustrates how non-linearly separable classification problems can be solved by adding a hidden layer to the network layout. The neurons in the hidden layer define segments of a piecewise linear decision boundary. More hidden neurons are needed if the decision boundary is very wiggly.

Figure 5.17 shows another example, how to solve the Boolean XOR problem with a perceptron that has two neurons in a hidden layer, with activation functions

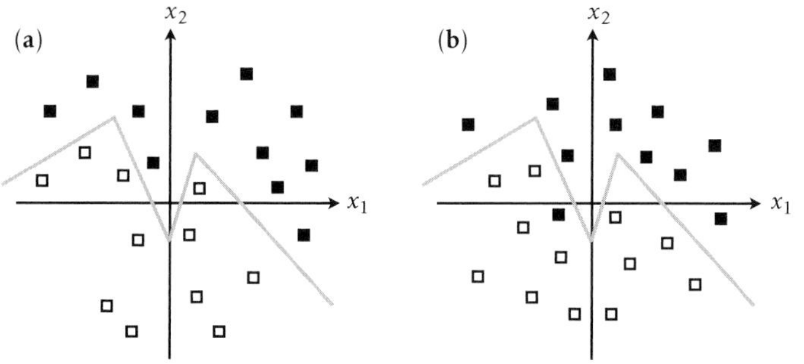

Figure 5.18 (**a**) Result of training the network on a training set. Legend: ■ corresponds to $t^{(\mu)} = 1$, and □ to $t^{(\mu)} = -1$. (**b**) Classification of a validation set. One pattern is incorrectly classified

sgn(b), thresholds $\frac{1}{2}$ and $\frac{3}{2}$, and all weights equal to unity. The output neuron has weights $+1$ and -1 and unit threshold:

$$O_1 = \text{sgn}(V_1 - V_2 - 1)\,. \tag{5.34}$$

Minsky and Papert [11] proved in 1969 that all Boolean functions can be represented by multilayer perceptrons, but that at least one hidden neuron must be connected to *all* input terminals. This means that not all neurons in the network are *locally* connected (the neurons have only a few incoming weights). Since fully connected networks are much harder to train than locally connected ones, this was considered a shortcoming at the time. Now, almost 50 years later, the perspective has changed. Convolutional networks (Chapter 8) have only local connections to the inputs and can be trained to recognise objects in images with high accuracy.

In summary, perceptrons are trained on a training set $[\boldsymbol{x}^{(\mu)}, \boldsymbol{t}^{(\mu)}]$, $\mu = 1, \ldots, p$, by moving the decision boundaries into the correct positions. This is achieved by repeatedly applying Hebb's rule to adjust all weights. A related learning rule is obtained by gradient descent on the energy function (5.23). We have not discussed how to update the *thresholds* yet, but it is clear that they too can be updated with gradient-descent learning.

Once all decision boundaries are in the right places, we must ask: what happens if we apply the trained network to a new dataset? Does it classify the new inputs correctly? In other words, can the network *generalise*? An example is shown in Figure 5.18. Panel (**a**) shows the result of training the network on a training set. The decision boundary separates $t = -1$ patterns from $t = 1$ patterns, so that the network classifies all patterns in the training set correctly. In panel (**b**) the trained network is applied to patterns in a *validation set*. We see that most patterns are correctly classified, save for one error. This means that the energy function (5.23) is not exactly zero for the validation set. Nevertheless, the network does quite a good job. Typically it is not a good idea to try to precisely classify all patterns near the decision boundary, because real-world data sets are subject to noise. It is a futile effort to try to learn and predict noise (Section 6.4).

5.6 Summary

Perceptrons are layered feed-forward networks that can learn to classify data in a training set $[\boldsymbol{x}^{(\mu)}, \boldsymbol{t}^{(\mu)}]$. For each input pattern $\boldsymbol{x}^{(\mu)}$, the network finds the correct target vector $\boldsymbol{t}^{(\mu)}$. We discussed the learning algorithm for a simple example: real-valued patterns with just two components and one binary target. This allowed us to represent the classification problem graphically, and to see how linearly separable classification problems can be solved by a simple perceptron. There are three different ways of understanding how the perceptron learns. First, geometrically, the perceptrons learn by moving decision boundaries into the correct locations. Second, this can be achieved by repeatedly adding a little bit of Hebb's rule. Third, these rules are similar to the learning rule derived by gradient descent on the energy function (5.23). Cover's theorem quantifies the capacity of a simple perceptron to separate patterns with binary targets. Finally, we discussed how to solve non-linearly separable classification problems with perceptrons with a hidden layer.

5.7 Further Reading

As mentioned in the Introduction, a short account of the history of perceptron research is the review by Kanal [23]. The remarkable book by Minsky and Papert explains the geometry of perceptron learning in great depth, and in a very elegant fashion. For a proof of Cover's theorem, see Ref. [73].

5.8 Exercises

5.1 Boolean AND problem. The Boolean AND problem (Figure 5.7) cannot be represented by a linear unit with weights $\boldsymbol{w}$ and threshold θ. To show this, solve $\partial H/\partial \boldsymbol{w} = 0$ and $\partial H/\partial \theta = 0$ for $\boldsymbol{w}$ and θ. Using the resulting weights and thresholds, demonstrate that $O^{(\mu)} \neq t^{(\mu)}$ for some μ. *Hint:* express the linear system to solve in terms of $\langle \boldsymbol{x}\boldsymbol{x}^{\mathsf{T}}\rangle$, $\langle \boldsymbol{x}\rangle$, $\langle t\boldsymbol{x}\rangle$, and $\langle t\rangle$, where $\langle\cdots\rangle$ is an average over patterns. Relate your findings to Equation (5.21) by computing the Moore-Penrose inverse of $\mathbb{Q}$ using its singular-value decomposition (see Section 2.6 in Ref. [53]).

5.2 Boolean functions. How many Boolean functions with two-dimensional inputs are there? How many of them are linearly separable? How many Boolean functions with three-dimensional inputs are there? How many of them are linearly

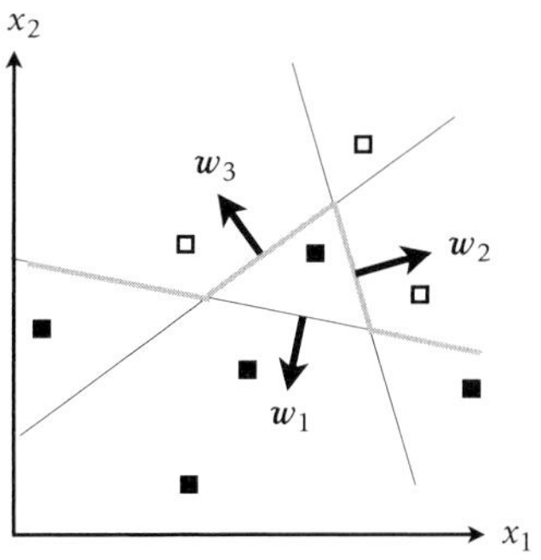

Figure 5.19 Alternative solution of the classification problem shown in Figure 5.15. See Exercise 5.4

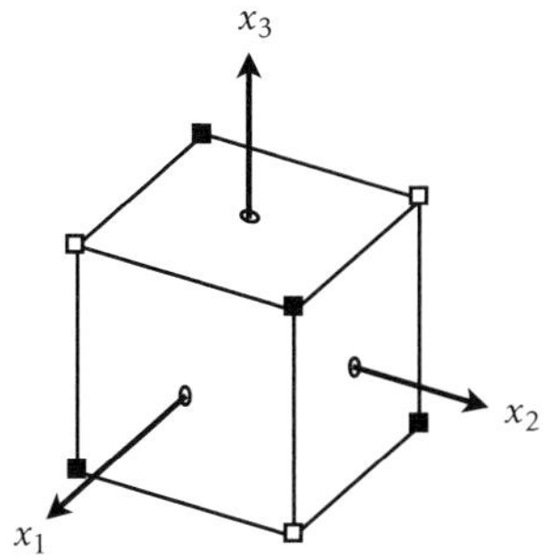

Figure 5.20 Three-dimensional parity function, with targets $t^{(\mu)} = 1$ (■), $t^{(\mu)} = -1$ (□). See Exercise 5.5

separable? Solve the problem graphically, considering its point-group symmetries. See also Exercise 5.13.

5.3 Output problem for binary classification. The binary classification problem shown in Figure 5.15 can be solved with a network with one hidden layer and one output neuron. Figure 5.16 shows the problem that the output neuron has to solve. Show that such output problems are linear separable if the decision boundaries corresponding to the hidden neurons allow one to partition the input plane into distinct regions that contain either only $t = +1$ or only $t = -1$ patterns.

5.4 Piecewise linear decision boundary. Find an alternative solution for the classification problem shown in Figure 5.15, where the weight vectors are chosen as depicted in Figure 5.19.

5.5 Three-dimensional parity function. The three-dimensional parity function is illustrated in Figure 5.20. The input bits $x_k^{(\mu)}$ for $k = 1, 2, 3$ are either $+1$ or -1. The output $O^{(\mu)}$ of the network is $+1$ if there is an odd number of positive bits in $\boldsymbol{x}^{(\mu)}$, and -1 if the number of positive bits are even. One representation of this

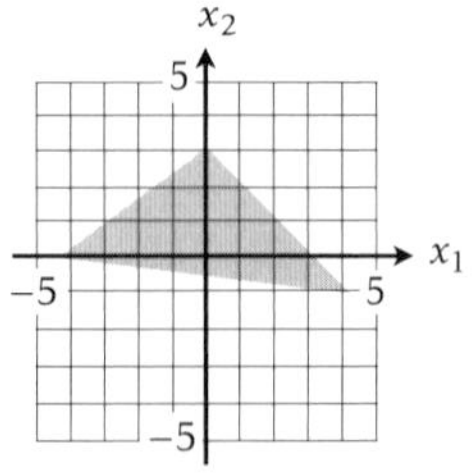

Figure 5.21 Classification problem. See Exercise 5.6

function uses a hidden layer with eight neurons. The state $V_j^{(\mu)}$ of hidden neuron j is computed as

$$V_j^{(\mu)} = \begin{cases} 1 & \text{if} \quad -\theta_j + \sum_k w_{jk} x_k^{(\mu)} > 0\,, \\ 0 & \text{if} \quad -\theta_j + \sum_k w_{jk} x_k^{(\mu)} \le 0\,. \end{cases} \tag{5.35}$$

Weights and thresholds are given by $w_{jk} = x_k^{(j)}$ and $\theta_j = 2$ $(j=1,\ldots,8)$. The network output is computed as $O^{(\mu)} = \sum_j W_j V_j^{(\mu)}$ (linear unit). Determine the weights W_j.

5.6 Linearly inseparable problem. A classification problem is given in Figure 5.21. Inputs $\boldsymbol{x}^{(\mu)}$ inside the gray triangle have targets $t^{(\mu)} = 1$, inputs outside the triangle have $t^{(\mu)} = 0$. The problem can be solved by a perceptron with one hidden layer with three neurons $V_j^{(\mu)} = \theta_{\mathrm{H}}\big(-\theta_j + \sum_{k=1}^{2} w_{jk} x_k^{(\mu)}\big)$, for $j = 1, 2, 3$. The network output is computed as $O^{(\mu)} = \theta_{\mathrm{H}}(-\Theta + \sum_{j=1}^{3} W_j V_j^{(\mu)})$. Here $\theta_{\mathrm{H}}(b)$ is the Heaviside function (Figure 2.10). Find weights w_{jk}, W_j and thresholds θ_j, Θ that solve the classification problem.

5.7 Perceptron with one hidden layer. A perceptron has one layer of hidden neurons, and a single output neuron. It receives two-dimensional input patterns $\boldsymbol{x}^{(\mu)} = [x_1^{(\mu)}, x_2^{(\mu)}]^{\mathsf{T}}$. They are mapped to four hidden neurons $V_i^{(\mu)}$ as

$$V_j^{(\mu)} = \begin{cases} 0 & \text{if} \quad -\theta_j + \sum_k w_{jk} x_k^{(\mu)} \le 0\,, \\ 1 & \text{if} \quad -\theta_j + \sum_k w_{jk} x_k^{(\mu)} > 0\,. \end{cases} \tag{5.36}$$

The network output is computed as

$$O^{(\mu)} = \begin{cases} 0 & \text{if} \quad -\Theta + \sum_j W_j V_j^{(\mu)} \le 0\,, \\ 1 & \text{if} \quad -\Theta + \sum_j W_j V_j^{(\mu)} > 0\,, \end{cases} \tag{5.37}$$

with $W_1 = W_3 = W_4 = 1$, $W_2 = -1$, and $\Theta = \frac{1}{2}$. Figure 5.22 (left) shows how input space is mapped to the hidden neurons. Draw the decision boundary of the network.

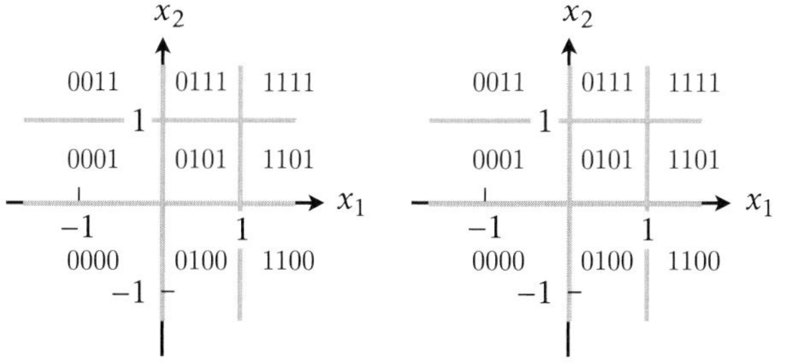

Figure 5.22 Left: Input plane with decision boundaries of hidden neurons V_j (gray lines). The boundaries partition input space into nine regions labelled by the binary code $V_1 V_2 V_3 V_4$. Right: The same, but for a different labelling. See Exercise 5.7

x_1	x_2	t
0.1	0.95	0
0.2	0.85	0
0.2	0.9	0
0.3	0.75	1
0.4	0.65	1
0.4	0.75	1
0.6	0.45	0
0.8	0.25	0
0.1	0.65	1
0.2	0.75	1
0.7	0.2	1

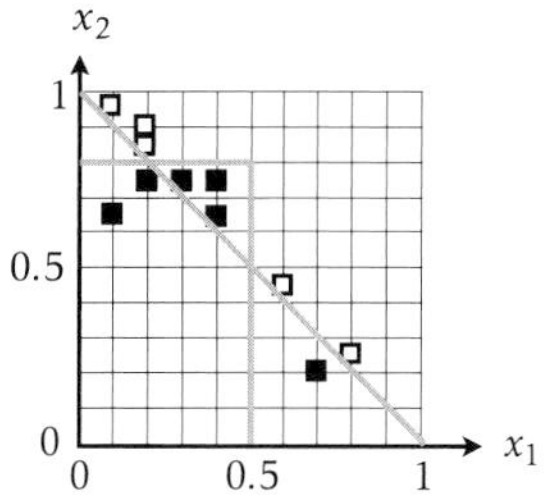

Figure 5.23 Inputs and targets for a classification problem. The targets are either $t = 0$ (□) or $t = 1$ (■). The three decision boundaries (gray lines) illustrate a solution to the problem using a multilayer perceptron. See Exercise 5.8

Give values for w_{ij} and θ_i that yield the pattern in Figure 5.22 (left). Show that one cannot map input space to the space of hidden neurons as in Figure 5.22 (right).

5.8 Multilayer perceptron. A classification problem is shown in Figure 5.23. It can be solved by a multilayer perceptron with two inputs, three hidden neurons $V_j^{(\mu)} = \theta_{\rm H}\Big(\sum_{i=1}^{2} w_{jk} x_k^{(\mu)} - \theta_j\Big)$, and one output $O^{(\mu)} = \theta_{\rm H}\Big(\sum_{j=1}^{3} W_j V_j^{(\mu)} - \Theta\Big)$, where $\theta_{\rm H}(b)$ is the Heaviside function (Figure 2.10). A possible solution is illustrated in Figure 5.23. Compute weights w_{jk} and thresholds θ_j of the hidden neurons that determine the three decision boundaries (gray lines). Draw a representation of the problem in the space with axes V_1, V_2, and V_3. Find output weights W_j and threshold Θ that solve the problem.

5.9 Expected maximal number of separable patterns. Show that the sum in Equation (5.30) sums to $2N$.

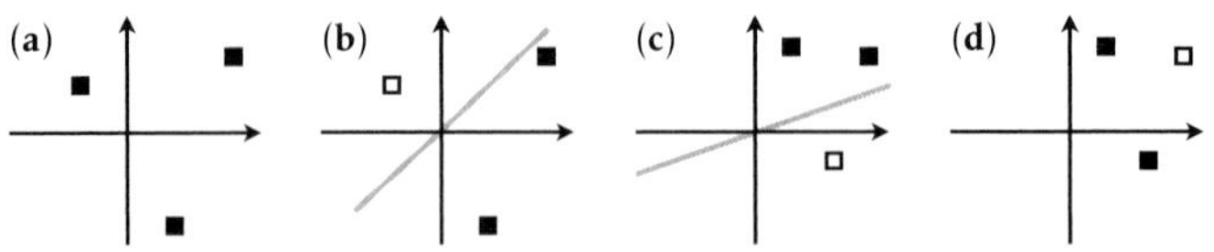

Figure 5.24 Cover's theorem for $p = 3$ and $m = 2$. Examples for problems that are homogeneously linearly separable, (**b**) and (**c**), and for problems that are not, (**a**) and (**d**). See Exercise 5.10

x_1	x_2	x_3	t
0	0	0	0
0	0	1	1
0	1	0	1
1	0	0	1
0	1	1	0
1	0	1	0
1	1	0	0
1	1	1	1

Figure 5.25 Value table for a three-dimensional Boolean function. See Exercise 5.14

5.10 Cover's theorem. Prove that $P(3,2) = \frac{3}{4}$ by complete enumeration of all cases. Some cases (but not all) are shown in Figure 5.24.

5.11 Non-linear activation function. Consider a single neuron with continuous, non-linear, and monotonically increasing activation function $g(b)$, and with N input components $x_1, \ldots, x_N$. Show that this neuron cannot solve binary classification problems $[\boldsymbol{x}^{(\mu)}, t^{(\mu)}]$ $(\mu = 1, \ldots, p)$ if $p > N$.

5.12 Random classification problem. The probability $P(p, N)$ that a random binary classification problem with p patterns in input dimension N is homogeneously linearly separable is given in Equation (5.29). Show that [1]

$$P(p,N) \sim \tfrac{1}{2}\Big\{1 + \operatorname{erf}\Big[\sqrt{\tfrac{\alpha N}{2}}\Big(\tfrac{2}{\alpha} - 1\Big)\Big]\Big\} \tag{5.38}$$

in the limit of $N \to \infty$ at fixed $\alpha = p/N$.

5.13 Boolean functions with n-dimensional inputs. What is the number $\mathcal{N}_n$ of linearly separable Boolean functions with n-dimensional inputs? Write a computer program that attempts to solve n-dimensional Boolean functions using the learning rule (5.18) for a McCulloch-Pitts neuron with $g(b) = \operatorname{sgn}(b)$. Try the program out on as many four- and five-dimensional Boolean functions as possible, to estimate $\mathcal{N}_4$ and $\mathcal{N}_5$. *Hint:* $\mathcal{N}_2 = 14$, $\mathcal{N}_3 = 104$, $\mathcal{N}_4 = 1882$, $\mathcal{N}_5 = 94572, \ldots$ (sequence A000609 in the online encyclopedia of integer sequences [74]).

5.14 Three-dimensional Boolean function. Figure 5.25 shows the value table for a three-dimensional Boolean function. Demonstrate that the function is not linearly separable by drawing it in three-dimensional input space. Construct a network with hidden layers that represents this function. *Hint:* one possibility is to wire together several two-dimensional XOR networks.

6
Stochastic Gradient Descent

In Chapter 5, we discussed how a hidden layer helps classify problems that are not linearly separable. We explained how the decision boundary in Figure 5.15 is represented in terms of the weights and thresholds of the hidden neurons and introduced a training algorithm based on gradient descent. In this section, the training algorithm is discussed in more detail.

Figure 5.2 shows the layout of the network to be trained. There are p input patterns $\boldsymbol{x}^{(\mu)}$ with N components each, as before. The output of the network has M components:

$$\boldsymbol{O}^{(\mu)} = \begin{bmatrix} O_1^{(\mu)} \\ O_2^{(\mu)} \\ \vdots \\ O_M^{(\mu)} \end{bmatrix}, \tag{6.1}$$

to be matched to the target vector $\boldsymbol{t}^{(\mu)}$. The network shown in Figure 5.2 computes

$$V_j^{(\mu)} = g\left(b_j^{(\mu)}\right) \quad \text{with} \quad b_j^{(\mu)} = \sum_{k=1}^{N} w_{jk} x_k^{(\mu)} - \theta_j\,, \tag{6.2a}$$

$$O_i^{(\mu)} = g\left(B_i^{(\mu)}\right) \quad \text{with} \quad B_i^{(\mu)} = \sum_j W_{ij} V_j^{(\mu)} - \Theta_i\,. \tag{6.2b}$$

Equation (6.2) shows that the outputs are obtained in terms of nested activation functions $g(b)$. They must be differentiable (or at least piecewise differentiable). Apart from that, there is no need to specify them further at this point.

6.1 Chain Rule and Error Backpropagation

The network in Figure 5.2 is trained by gradient-descent learning in the same way as in Section 5.3. The weight increments are given by

$$\delta W_{mn} = -\eta \frac{\partial H}{\partial W_{mn}} \quad \text{and} \quad \delta w_{mn} = -\eta \frac{\partial H}{\partial w_{mn}}, \tag{6.3}$$

with energy function

$$H = \frac{1}{2} \sum_{\mu i} \left(t_i^{(\mu)} - O_i^{(\mu)} \right)^2 . \tag{6.4}$$

The small parameter $\eta > 0$ in Equation (6.3) is the learning rate, as in Section 5.3. The derivatives of the energy function are evaluated with the *chain rule*. For the weights connecting to the output layer, we apply the chain rule once,

$$\frac{\partial H}{\partial W_{mn}} = -\sum_{\mu i} \left(t_i^{(\mu)} - O_i^{(\mu)} \right) \frac{\partial O_i^{(\mu)}}{\partial W_{mn}}, \tag{6.5a}$$

and then once more, using Equation (5.25):

$$\frac{\partial O_i^{(\mu)}}{\partial W_{mn}} = g'(B_i^{(\mu)}) \delta_{im} V_n^{(\mu)} . \tag{6.5b}$$

Here $g'(B) = \mathrm{d}g/\mathrm{d}B$ is the derivative of the activation function with respect to the local field B, and δ_{im} is the Kronecker delta: δ_{im}=1 if $i = m$ and zero otherwise.

An important point is that the states V_j of the neurons in the hidden layer do not depend on W_{mn} because these neurons do not have incoming connections with these weights, a consequence of the *feed-forward layout* of the network. In summary, we obtain for the increments of the weights connecting to the output layer:

$$\delta W_{mn} = -\eta \frac{\partial H}{\partial W_{mn}} = \eta \sum_{\mu=1}^{p} \left(t_m^{(\mu)} - O_m^{(\mu)} \right) g' \left(B_m^{(\mu)} \right) \equiv \eta \sum_{\mu=1}^{p} \Delta_m^{(\mu)} V_n^{(\mu)} . \tag{6.6a}$$

The quantity

$$\Delta_m^{(\mu)} = \left(t_m^{(\mu)} - O_m^{(\mu)} \right) g' \left(B_m^{(\mu)} \right) \tag{6.6b}$$

is a weighted output *error*: it vanishes when $O_m^{(\mu)} = t_m^{(\mu)}$. The weights connecting to the hidden layer are adjusted in a similar fashion, by applying the chain rule four times:

$$\frac{\partial H}{\partial w_{mn}} = -\sum_{\mu i} \left(t_i^{(\mu)} - O_i^{(\mu)} \right) \frac{\partial O_i^{(\mu)}}{\partial w_{mn}}, \tag{6.7a}$$

$$\frac{\partial O_i^{(\mu)}}{\partial w_{mn}} = \sum_{l} \frac{\partial O_i^{(\mu)}}{\partial V_l^{(\mu)}} \frac{\partial V_l^{(\mu)}}{\partial w_{mn}}, \tag{6.7b}$$

$$\frac{\partial O_i^{(\mu)}}{\partial V_l^{(\mu)}} = g' \left(B_i^{(\mu)} \right) W_{il}, \tag{6.7c}$$

$$\frac{\partial V_l^{(\mu)}}{\partial w_{mn}} = g'\big(b_l^{(\mu)}\big)\,\delta_{lm} x_n^{(\mu)}\,. \tag{6.7d}$$

Here we used Equation (5.25). With the definition of the output error, $\Delta_i^{(\mu)}$, Equation (6.3) yields

$$\delta w_{mn} = \eta \sum_\mu \sum_i \Delta_i^{(\mu)} W_{im} g'\big(b_m^{(\mu)}\big)\, x_n^{(\mu)} \equiv \eta \sum_\mu \delta_m^{(\mu)} x_n^{(\mu)}\,. \tag{6.8}$$

The last equality defines weighted errors,

$$\delta_m^{(\mu)} = \sum_i \Delta_i^{(\mu)} W_{im} g'\big(b_m^{(\mu)}\big)\,, \tag{6.9}$$

associated with the hidden layer. Note that the $\delta_m^{(\mu)}$ vanish when the output errors $\Delta_i^{(\mu)}$ are zero. Equation (6.9) shows that the errors are determined recursively. The neuron states are also updated recursively (Equation (6.2), but there is an important difference between Equations (6.9) and (6.2). The feed-forward structure of the layered network implies that the neurons are updated from left to right. Equation (6.9), by contrast, says that the errors are updated from right to left, from the output layer to the hidden layer. The term *backpropagation* refers to this difference: the neurons are updated forwards; the errors are updated backwards.

In terms of the errors $\Delta_m^{(\mu)}$ and $\delta_m^{(\mu)}$, the weight increments have the same form for both layers:

$$\delta W_{mn} = \eta \sum_{\mu=1}^{p} \Delta_m^{(\mu)} V_n^{(\mu)} \quad \text{and} \quad \delta w_{mn} = \eta \sum_{\mu=1}^{p} \delta_m^{(\mu)} x_n^{(\mu)}\,. \tag{6.10}$$

The rule (6.10) is also called δ-rule [1]. The thresholds are adjusted in a similar way:

$$\delta\Theta_m = -\eta \frac{\partial H}{\partial \Theta_m} = \eta \sum_{\mu=1}^{p} \big(t_m^{(\mu)} - O_m^{(\mu)}\big)\big[-g'\big(B_m^{(\mu)}\big)\big] = -\eta \sum_{\mu=1}^{p} \Delta_m^{(\mu)}\,, \tag{6.11a}$$

$$\delta\theta_m = -\eta \frac{\partial H}{\partial \theta_m} = \eta \sum_{\mu=1}^{p} \sum_i \Delta_i^{(\mu)} W_{im}\big[-g'\big(b_m^{(\mu)}\big)\big] = -\eta \sum_{\mu=1}^{p} \delta_m^{(\mu)}\,. \tag{6.11b}$$

So, the general form for the threshold increments is analogous to Equation (6.10),

$$\delta\Theta_m = -\eta \sum_{\mu=1}^{p} \Delta_m^{(\mu)} \quad \text{and} \quad \delta\theta_m = -\eta \sum_{\mu=1}^{p} \delta_m^{(\mu)}\,, \tag{6.12}$$

but without the state variables of the neurons (or the inputs). A way to remember the difference between Equations (6.10) and (6.12) is to note that the formula for the threshold increments looks like the one for the weight increments if one sets the state values of the neurons to -1. This follows from Equation (6.2).

The backpropagation rules (6.10) and (6.12) contain sums over patterns. This corresponds to feeding all patterns at the same time to compute the increments of weights and thresholds (*batch* training). Alternatively, one may choose a single pattern, update the weights by backpropagation, and then continue to iterate these training steps many times (*sequential* training). One iteration corresponds to feeding a single pattern, and p iterations are called one *epoch* (in batch training, one iteration corresponds to one epoch). If one chooses the patterns randomly, then sequential training results in *stochastic gradient descent*:

$$\delta W_{mn} = \eta \Delta_m^{(\mu)} V_n^{(\mu)} \quad \text{and} \quad \delta w_{mn} = \eta \delta_m^{(\mu)} x_n^{(\mu)}\,, \tag{6.13a}$$

$$\delta \Theta_m = -\eta \Delta_m^{(\mu)} \quad \text{and} \quad \delta \theta_m = -\eta \delta_m^{(\mu)}\,. \tag{6.13b}$$

Since the sum over pattern is absent, the steps do not necessarily decrease the energy function. Their directions fluctuate, but the average weight increment (averaged over all patterns) points downhill. The result is a *stochastic path* through parameter space, less prone to getting stuck in local minima (but see Section 7.8).

6.2 Stochastic Gradient-Descent Algorithm

The stochastic gradient-descent formulae derived in the previous section were derived for a network with one hidden layer. This section describes the details of the stochastic-gradient algorithm for deep networks with many hidden layers. To this end, we need to adapt our notation, as described in Figure 6.1. We label the layers by the index ℓ. The layer of input terminals has label $\ell = 0$, while layer $\ell = L$ denotes the layer of output neurons. The state variables for the neurons in layer ℓ are $V_j^{(\ell)}$, the weights connecting into these neurons from the left are $w_{jk}^{(\ell)}$, and the errors associated with layer ℓ are denoted by $\delta_k^{(\ell)}$. In this notation, Equation (6.2) reads

$$V_j^{(\ell)} = g\Big(\sum_k w_{jk}^{(\ell)} V_k^{(\ell-1)} - \theta_j^{(\ell)}\Big)\,. \tag{6.14}$$

Repeating the steps outlined in the previous section, we arrive at the update formulae

$$\delta w_{mn}^{(\ell)} = \eta \delta_m^{(\ell)} V_n^{(\ell-1)} \quad \text{and} \quad \delta \theta_m^{(\ell)} = -\eta \delta_m^{(\ell)}\,, \tag{6.15}$$

with errors

$$\delta_j^{(\ell-1)} = \sum_i \big(t_i - V_i^{(L)}\big) \frac{\partial V_i^{(L)}}{\partial V_j^{(\ell-1)}} g'\big(b_j^{(\ell-1)}\big)\,, \tag{6.16}$$

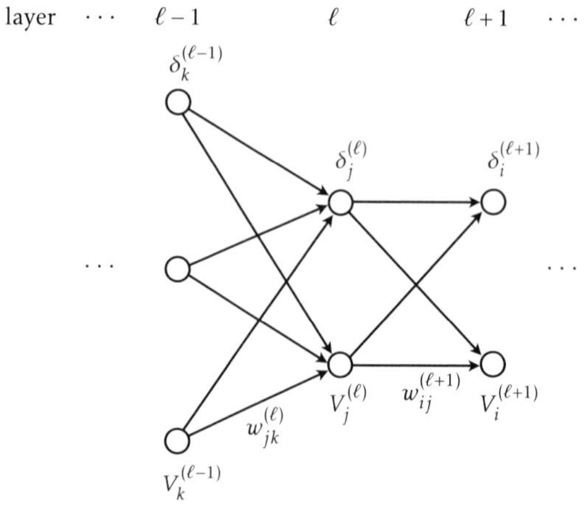

Figure 6.1 Illustration of the notation used in Algorithm 4

where $b_j^{(\ell)} = \sum_k w_{jk}^{(\ell)} V_k^{(\ell-1)} - \theta_j^{(\ell)}$ is the local field of $V_j^{(\ell)}$. It involves the matrix-vector product between the weight matrix $\mathbb{W}^{(\ell)}$ and the vector $\boldsymbol{V}^{(\ell-1)}$. Evaluating the gradients $\partial V_i^{(L)}/\partial V_j^{(\ell-1)}$ with the chain rule, one obtains the recursion

$$\delta_j^{(\ell-1)} = \sum_i \delta_i^{(\ell)} w_{ij}^{(\ell)} g'\big(b_j^{(\ell-1)}\big), \tag{6.17}$$

with initial condition $\delta_i^{(L)} = \big(t_i - V_i^{(L)}\big) g'\big(b_i^{(L)}\big)$. For one hidden layer, Equation (6.17) is equivalent to (6.9). The result of the recursion (6.17) is a vector $\boldsymbol{\delta}^{(\ell-1)}$ with components $\delta_j^{(\ell-1)}$, obtained by component-wise multiplication of $[\mathbb{W}^{(\ell)\mathsf{T}}\boldsymbol{\delta}^{(\ell)}]_j$ with $g'\big(b_j^{(\ell-1)}\big)$. Component-wise multiplication of vectors is sometimes called Schur or Hadamard product [75], denoted by $\boldsymbol{a} \odot \boldsymbol{b} = [a_1 b_1, \ldots, a_N b_N]^{\mathsf{T}}$. It does not have a geometric meaning like the scalar product or the cross product of vectors, and therefore there is little point in using it. Also, note that the vector $\boldsymbol{\delta}^{(\ell)}$ is multiplied by the transpose of the weight matrix, $\mathbb{W}^{(\ell)\mathsf{T}}$, rather than by the weight matrix itself. We return to this point in Section 9.1.

The stochastic-gradient algorithm is summarised in Algorithm 4. One feeds an input $\boldsymbol{x}^{(\nu)}$, updates the weights using (6.15), and iterates these steps until the energy function (5.23) is deemed sufficiently small. Note that the resulting weights and thresholds are not unique. In Figure 5.17, all weights for the Boolean XOR function are equal to ± 1. But the training algorithm (6.10) corresponds to repeatedly adding weight increments. This may cause the weights to grow.

In practice, the stochastic gradient-descent dynamics may be too noisy. In this case, it is better to average over a small number of randomly chosen patterns. Such a set is called *mini batch*, of size m_B say. In *stochastic gradient descent with mini batches*, one replaces Equations (6.10) and (6.12) with

Algorithm 4 Stochastic gradient descent

initialise weights $w_{mn}^{(\ell)}$ to random numbers, thresholds to zero, $\theta_m^{(\ell)} = 0$;
for $\nu = 1, \ldots, \nu_{\max}$ **do**
  choose a value of μ and apply pattern $\boldsymbol{x}^{(\mu)}$ to input layer, $\boldsymbol{V}^{(0)} \leftarrow \boldsymbol{x}^{(\mu)}$;
  for $\ell = 1, \ldots, L$ **do**
    propagate forward: $V_j^{(\ell)} \leftarrow g\Big(\sum_k w_{jk}^{(\ell)} V_k^{(\ell-1)} - \theta_j^{(\ell)}\Big)$;
  end for
  compute errors for output layer: $\delta_i^{(L)} \leftarrow g'(b_i^{(L)})(t_i - V_i^{(L)})$;
  for $\ell = L, \ldots, 2$ **do**
    propagate backward: $\delta_j^{(\ell-1)} \leftarrow \sum_i \delta_i^{(\ell)} w_{ij}^{(\ell)} g'(b_j^{(\ell-1)})$;
  end for
  for $\ell = 1, \ldots, L$ **do**
    change weights and thresholds: $w_{mn}^{(\ell)} \leftarrow w_{mn}^{(\ell)} + \eta\delta_m^{(\ell)} V_n^{(\ell-1)}$ and $\theta_m^{(\ell)} \leftarrow \theta_m^{(\ell)} - \eta\delta_m^{(\ell)}$;
  end for
end for

$$\delta W_{mn} = \eta \sum_{\mu=1}^{m_B} \Delta_m^{(\mu)} V_n^{(\mu)} \quad \text{and} \quad \delta\Theta_m = -\eta \sum_{\mu=1}^{m_B} \Delta_m^{(\mu)}\,, \tag{6.18}$$

$$\delta w_{mn} = \eta \sum_{\mu=1}^{m_B} \delta_m^{(\mu)} x_n^{(\mu)} \quad \text{and} \quad \delta\theta_m = -\eta \sum_{\mu=1}^{m_B} \delta_m^{(\mu)}\,.$$

Sometimes the mini-batch rule is quoted with prefactors of m_B^{-1} before the sums. The factors m_B^{-1} can just be absorbed in the learning rate, but when comparing learning rates for different implementations, one needs to check whether or not there are factors of m_B^{-1} in front of the sums in Equation (6.18).

How does one select which inputs to include in a mini batch? This is discussed in Section 6.3: at the beginning of each epoch, one randomly *shuffles* the sequence of the input patterns in the training set. Then the first mini batch contains patterns $\mu = 1, \ldots, m_B$, and so forth.

Common choices for the activation functions $g(b)$ are the *sigmoid* function or tanh:

$$g(b) = \frac{1}{1+e^{-b}} \equiv \sigma(b)\,, \tag{6.19a}$$

$$g(b) = \tanh(b)\,. \tag{6.19b}$$

In both cases, the derivatives can be expressed in terms of the function itself:

$$\tfrac{\mathrm{d}}{\mathrm{d}b}\sigma(b) = \sigma(b)[1-\sigma(b)]\,, \quad \tfrac{\mathrm{d}}{\mathrm{d}b}\tanh(b) = \big[1-\tanh^2(b)\big]\,. \tag{6.20}$$

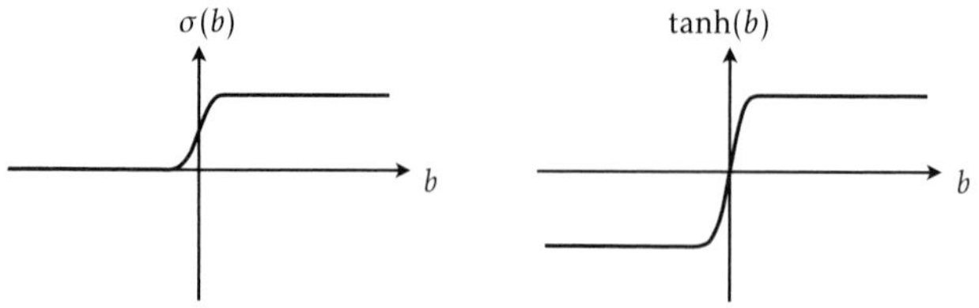

Figure 6.2 Saturation of the activation functions (6.19). The derivatives $g'(b) \equiv \frac{\mathrm{d}}{\mathrm{d}b} g(b)$ of both activation functions tend to zero for large values of $|b|$

The second equality was used in Section 3.4. The following shorthand notation for the derivative of the activation function $g(b)$ is common: $g'(b) \equiv \frac{\mathrm{d}}{\mathrm{d}b} g(b)$.

As illustrated in Figure 6.2, the activation functions (6.19) saturate at large values of $|b|$: their derivatives $g'(b)$ tend to zero. Since the backpropagation rule (6.16) contains factors of $g'(b)$, this implies that the algorithm slows down if $|b|$ becomes too large. For this reason, the initial weights and thresholds should be chosen so that the local fields b are not too large in magnitude, to avoid $g'(b)$ becoming too small. A standard procedure is to take all weights to be initially randomly distributed, for example Gaussian with zero mean, and with a suitable variance. The performance of networks with many hidden layers (*deep* networks) can be sensitive to the initialisation of the weights (Section 7.2).

It is sometimes argued that the initial values of the thresholds are not so critical. The idea is that they are learned more rapidly than the weights, at least initially, and a common choice is to initialise the thresholds to zero. Section 7.2 summarises a mean-field argument that comes to a different conclusion.

6.3 Preprocessing the Input Data

It can be useful to preprocess the input data, although any preprocessing may remove information from the data. Nevertheless, it is usually advisable to shift the data so the mean of each component over all p patterns vanishes:

$$\langle x_k \rangle = \frac{1}{p} \sum_{\mu=1}^{p} x_k^{(\mu)} = 0 \,. \tag{6.21}$$

There are several reasons for this. First, large mean values can cause steep gradients in the energy function (Exercise 6.9) that are difficult to navigate with gradient descent. Different input-data variances in different directions have a similar effect. Therefore one *scales* the inputs so that the input-data distribution has the same variance in all directions (Figure 6.3), equal to unity for instance:

$$\sigma_k^2 = \frac{1}{p} \sum_{\mu=1}^{p} \left(x_k^{(\mu)} - \langle x_k \rangle \right)^2 = 1 \,. \tag{6.22}$$

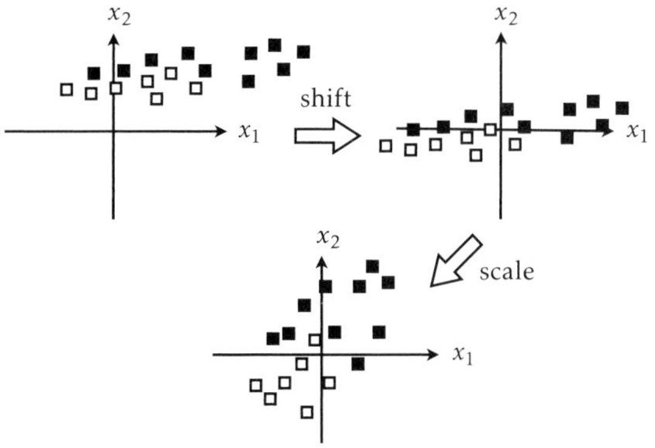

Figure 6.3 Shift and scale the input data to achieve zero mean and unit variance

Second, to avoid that the neurons connected to the inputs saturate, their local fields must not be too large (Section 6.2). If one initialises the weights to Gaussian random numbers with mean zero and unit variance, large activations are quite likely if the distribution of input patterns has a large mean or a large variance. Third, enforcing zero input mean by shifting the input data avoids that the weights of the neurons in the first hidden layer must decrease or increase together [76]. Equation (6.18) shows that the components of $\delta \boldsymbol{w}_m \propto \delta_m \boldsymbol{x}$ into hidden neuron m are likely to have the same signs if the input data has a large mean. This makes it difficult for the network to learn to differentiate. In summary, it is advisable to shift and scale the input-data distribution so that it has mean zero and unit variance, as illustrated in Figure 6.3. The same transformation (using the mean values and scaling factors determined for the training set) should be applied to any new data set that the network is supposed to classify after it has been trained on the training set.

Figure 6.4 shows a distribution of inputs that falls into two distinct clusters. The difference between the clusters is sometimes called *covariate shift*; here *covariate* is just another term for input. Imagine feeding just inputs from one of the clusters to the network. It will learn local properties of the decision boundary, instead of its global features. Such global properties are efficiently learned if the network is more frequently confronted with unfamiliar data. For sequential training (stochastic gradient descent) this is not a problem, because the sequence of input patterns presented to the network is random. However, if one trains with mini batches, the mini batches should contain randomly chosen patterns in order to avoid covariate shifts. To this end, one randomly *shuffles* the sequence of the input patterns in the training set, at the beginning of each epoch.

It is also recommended [76] to observe the output errors during training. If the errors are similar for a number of subsequent learning steps, the corresponding

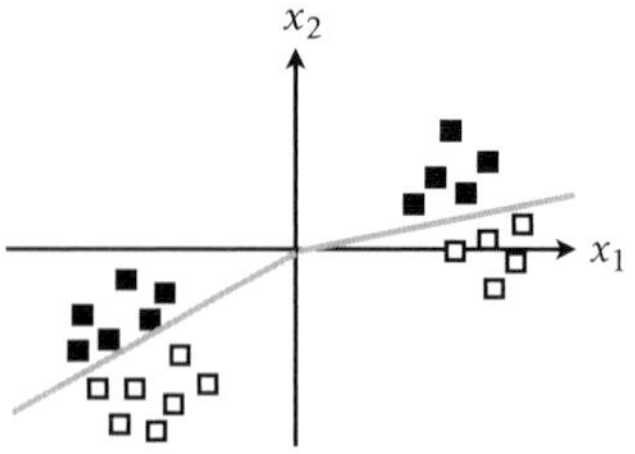

Figure 6.4 When the input data falls into clusters as shown in this figure, one should randomly pick data from either cluster. The decision boundary is shown as a solid gray line. It has different slopes for the two clusters

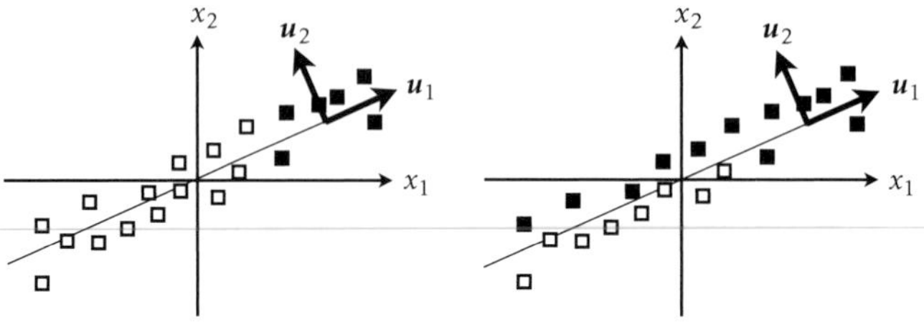

Figure 6.5 Principal-component analysis (schematic). The data set on the left can be classified keeping only the principal component $\boldsymbol{u}_1$ of the data. This is not true for the data set on the right

inputs appear familiar to the network. Larger errors correspond to unfamiliar inputs, and Ref. [76] suggests feeding such inputs more often.

When the input data is very high dimensional, many input terminals are needed. This usually means that one should use many neurons in the hidden layers. This can be problematic because it increases the risk of overfitting the input data. To avoid this as far as possible, one can reduce the dimensionality of the input data by *principal-component analysis*. This method allows one to project high-dimensional data to a lower-dimensional subspace (Figure 6.5).

The data shown on the left of Figure 6.5 falls approximately onto a straight line, the principal direction $\boldsymbol{u}_1$. We see that the coordinate orthogonal to the principal direction is not useful in classifying the data. Consequently, this coordinate can be disregarded, reducing the dimensionality of the data set. The idea of principal-component analysis is to rotate the basis in input space so that the variance of the data along the first axis of the new coordinate system, $\boldsymbol{u}_1$, is maximal. One keeps the input components corresponding to $\boldsymbol{u}_1$, discarding those corresponding to $\boldsymbol{u}_2$.

To determine the maximal-variance direction, consider the data variance along a unit direction $\boldsymbol{v}$ ($|\boldsymbol{v}| = 1$):

$$\sigma_v^2 = \left\langle (\boldsymbol{x} \cdot \boldsymbol{v})^2 \right\rangle - \langle \boldsymbol{x} \cdot \boldsymbol{v} \rangle^2 = \boldsymbol{v} \cdot \mathbb{C}\boldsymbol{v}\,. \tag{6.23}$$

Here

$$\mathbb{C} = \langle \delta\boldsymbol{x}\,\delta\boldsymbol{x}^{\mathsf{T}}\rangle \quad \text{with} \quad \delta\boldsymbol{x} = \boldsymbol{x} - \langle \boldsymbol{x}\rangle \tag{6.24}$$

is the data *covariance matrix*. The variance σ_v^2 is maximal when $\boldsymbol{v}$ points in the direction of the leading eigenvector of the covariance matrix $\mathbb{C}$. This can be seen as follows. The covariance matrix is symmetric; therefore its eigenvectors $\boldsymbol{u}_1, \ldots, \boldsymbol{u}_N$ form an orthonormal basis of input space. This allows us to express the matrix $\mathbb{C}$ as

$$\mathbb{C} = \sum_{\alpha=1}^{N} \lambda_\alpha \boldsymbol{u}_\alpha \boldsymbol{u}_\alpha^{\mathsf{T}}\,. \tag{6.25}$$

The eigenvalues λ_α are non-negative. This follows from Equation (6.24) and the eigenvalue equation $\mathbb{C}\boldsymbol{u}_\alpha = \lambda_\alpha \boldsymbol{u}_\alpha$. We arrange the eigenvalues by magnitude, $\lambda_1 \geq \lambda_2 \geq \ldots \geq \lambda_N \geq 0$. Using Equation (6.25), we can write for the variance

$$\sigma_v^2 = \sum_{\alpha=1}^{N} \lambda_\alpha v_\alpha^2 \tag{6.26}$$

with $v_\alpha = \boldsymbol{v} \cdot \boldsymbol{u}_\alpha$. We want to show that σ_v^2 is maximal for $\boldsymbol{v} = \pm \boldsymbol{u}_1$ subject to the constraint that $\boldsymbol{v}$ is normalised to unity,

$$\sum_{\alpha=1}^{N} v_\alpha^2 = 1\,. \tag{6.27}$$

To ensure that this constraint remains satisfied as the v_α are varied, one introduces a *Lagrange multiplier* λ (Exercises 6.10 and 6.11). The constraint (6.27) is multiplied with λ and added to the target function (6.26). The function to maximise reads

$$\mathscr{L} = \sum_\alpha \lambda_\alpha v_\alpha^2 - \lambda\big(1 - \sum_\alpha v_\alpha^2\big)\,. \tag{6.28}$$

To find the maximum of $\mathscr{L}$, we determine its singular points, defined by $\partial \mathscr{L}/\partial v_\beta = 0$. This yields $v_\beta(\lambda_\beta + \lambda) = 0$. The maximum of $\mathscr{L}$ is obtained for $\lambda = -\lambda_1$, where λ_1 is the maximal eigenvalue of $\mathbb{C}$ with eigenvector $\boldsymbol{u}_1$. We conclude that all components v_β must vanish, except for v_1 which must equal unity. This shows that the variance σ_v^2 is maximised by the principal direction.

In more than two dimensions, there is commonly more than one direction along which the data varies significantly. These k principal directions correspond to the k eigenvectors of $\mathbb{C}$ with the largest eigenvalues. This can be shown recursively. One projects the data to the subspace orthogonal to $\boldsymbol{u}_1$ by applying the projection matrix $\mathbb{P}_1 = \mathbb{1} - \boldsymbol{u}_1\boldsymbol{u}_1^{\mathsf{T}}$. Then one repeats the procedure outlined above, and finds that the data varies maximally along $\boldsymbol{u}_2$. Upon iteration, one obtains the k principal directions $\boldsymbol{u}_1, \ldots, \boldsymbol{u}_k$. Often there is a gap between the k largest eigenvalues and the small ones (all close to zero). Then one can safely project the data onto the subspace spanned by the k principal directions. If there is no gap, then it is less clear what should be done.

The data set shown on the right of Figure 6.5 illustrates another problem. This data set is much harder to classify if we use only the principal component alone. In this case, we lose important information by projecting the data on its principal component.

6.4 Overfitting and Cross Validation

The goal of supervised learning is to generalise from a training set to new data. Only general properties of the training set are of interest, not specific ones that are particular to the training set in question. A neural network with more neurons may classify the input data better, because it more accurately represents all specific features of the given data set. But a different set of patterns from the same input distribution could look quite different in detail, in which case the decision boundary may not classify the new data very well (Figure 6.6). In other words, the network may fit too fine details (for instance noise in the training set) that have no general meaning. This problem, illustrated in Figure 6.6, is called *overfitting*. The tendency to overfit is larger for networks with more neurons. In general, we should look for a compromise: between accurate classification of the training set and the risk of overfitting.

One way of avoiding overfitting is to use *cross validation* and *early stopping*. One splits the data into two sets: a *training set* and a *validation set*. The idea is that these sets share the general features to be learnt. But although training and validation sets are drawn from the same distribution, they may differ in details that are not of interest.

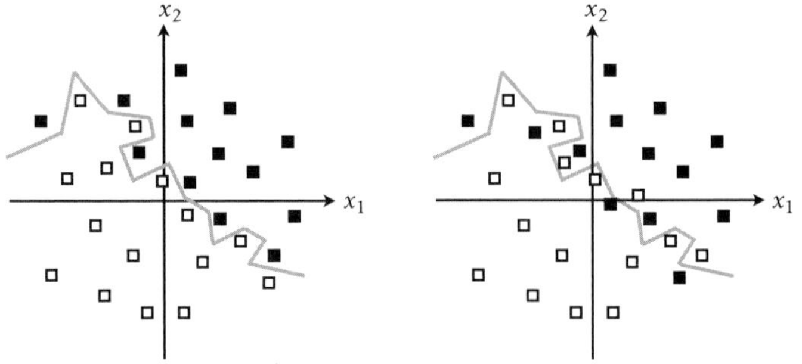

Figure 6.6 Overfitting. Left: Accurate representation of the decision boundary in the training set, for a network with a hidden layer with 15 neurons. Right: This new data set differs from the first one just by a little bit of noise. The points in the vicinity of the decision boundary are not correctly classified. Legend: ■ corresponds to $t^{(\mu)} = 1$, and □ to $t^{(\mu)} = -1$

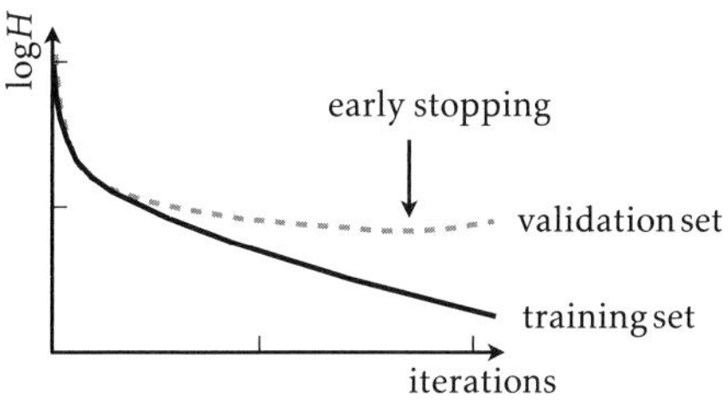

Figure 6.7 Progress of training and validation errors. The plot is schematic, and the data is smoothed. Based on simulations performed by Oleksandr Balabanov. Shown is the natural logarithm of the energy functions for the training set (solid line) and the validation set (dashed line) as a function of the number of training iterations. The training is stopped when the validation energy begins to increase

While the network is trained on the training set, one monitors not only the energy function for the training set, but also the energy function evaluated using the validation data. As long as the network learns general features of the input distribution, both training and validation energies decrease. But when the network starts to learn specific features of the training set, then the validation energy saturates, or may start to increase. At this point, the training is stopped. The scheme is illustrated in Figure 6.7.

Often the possible state values of the output neurons are continuous while the targets assume only discrete values. In this case, one may also monitor the *classification error* of the validation set. The definition of the classification error depends on the type of the classification problem. For one single output neuron with targets $t = 0/1$, the classification error is defined as

$$C = \frac{1}{p}\sum_{\mu=1}^{p}\left|t^{(\mu)} - \theta_{\rm H}(O^{(\mu)} - \tfrac{1}{2})\right|. \tag{6.29a}$$

If, by contrast, the targets take the values $t = \pm 1$, then the classification error reads

$$C = \frac{1}{2p}\sum_{\mu=1}^{p}\left|t^{(\mu)} - \operatorname{sgn}(O^{(\mu)})\right|. \tag{6.29b}$$

As a third example, consider a classification problem where inputs must be classified into M mutually exclusive classes, such as the MNIST data set of handwritten digits (Section 8.3) where $M = 10$. Another example is given in Table 6.1, with $M = 3$ classes. In both cases, one of the targets equals unity while all others equal zero. As a consequence, the targets sum to unity: $\sum_i^M t_i^{(\mu)} = 1$. Now assume that the network has sigmoid outputs, $O_i^{(\mu)} = \sigma\big(b_i^{(\mu)}\big)$. To classify input $\boldsymbol{x}^{(\mu)}$ from the network outputs $O_i^{(\mu)}$, we define

Table 6.1 *An illustration of the difference between energy function and classification error. The table shows network outputs for three different inputs from the iris data set, as well as the correct classifications. All inputs are classified correctly, but the difference between outputs and targets is substantial.*

μ	output $\boldsymbol{O}^{(\mu)}$	target $\boldsymbol{t}^{(\mu)}$	classification	correct?
1	[0.4, 0.5, 0.4]	[0,1,0]	versicolor	yes
2	[0.4, 0.3, 0.5]	[0,0,1]	setosa	yes
3	[0.6, 0.5, 0.4]	[1,0,0]	virginica	yes

$$y_i^{(\mu)} = \begin{cases} 1 & \text{if } O_i^{(\mu)} \text{ is the largest of all outputs } i = 1, \ldots, M, \\ 0 & \text{otherwise.} \end{cases} \tag{6.30a}$$

Then the classification error can be computed as

$$C = \frac{1}{2p} \sum_{\mu=1}^{p} \sum_{i=1}^{M} \left| t_i^{(\mu)} - y_i^{(\mu)} \right| . \tag{6.30b}$$

In all cases, the *classification accuracy* is defined as $(1 - C)\,100\%$; it is usually quoted as a percentage.

The classification error determines the fraction of inputs that are classified wrongly. However, it contains less information than the energy function, which is in fact a mean-squared error of the outputs. This is illustrated in Table 6.1. All three inputs are classified correctly, but there is a substantial mean-squared error. This indicates that the classification is not very reliable.

6.5 Adaptation of the Learning Rate

It is tempting to choose larger learning rates, because they enable the network to escape more efficiently from shallow minima. But this can lead to problems when the energy function varies rapidly, causing the training to fail. To avoid this, one uses an *adaptive* learning rule, such as

$$\delta w_{mn}^{(t)} = -\eta \frac{\partial H}{\partial w_{mn}}\Big|_{\{w_{ij}\}=\{w_{ij}^{(t)}\}} + \alpha \delta w_{mn}^{(t-1)} . \tag{6.31}$$

Here $t = 1, 2, \ldots, T$ labels the iteration number. We see that the increment at step t depends not only on the instantaneous gradient but also on the weight change $\delta w_{mn}^{(t-1)}$ of the previous iteration. We say that the dynamics becomes *inertial* and the weights gain *momentum*. The parameter $\alpha \geq 0$ is called momentum constant.

It determines how strong the inertial effect is. We see that $\alpha = 0$ corresponds to the usual backpropagation rule. When α is positive, then how does inertia change the learning process? Iterating Equation (6.31) yields

$$\delta w_{mn}^{(T)} = -\eta \sum_{t=0}^{T} \alpha^{T-t} \frac{\partial H}{\partial w_{mn}^{(t)}} . \tag{6.32}$$

Here and in the following, we use the short-hand notation

$$\frac{\partial H}{\partial w_{mn}^{(t)}} \equiv \frac{\partial H}{\partial w_{mn}}\Big|_{\{w_{ij}\}=\{w_{ij}^{(t)}\}} .$$

Equation (6.32) shows that $\delta w_{mn}^{(T)}$ is a weighted average of the gradients encountered during training. Now assume that the training is stuck in a shallow minimum. Then the gradient $\partial H/\partial w_{mn}^{(t)}$ remains roughly constant through many time steps. To illustrate what happens, let us assume that $\partial H/\partial w_{mn}^{(t)} = \partial H/\partial w_{mn}^{(0)}$ for $t = 1, \ldots, T$. In this case, we can write

$$\delta w_{mn}^{(T)} \approx -\eta \frac{\partial H}{\partial w_{mn}^{(0)}} \sum_{t=0}^{T} \alpha^{T-t} = -\eta \frac{\alpha^{T+1} - 1}{\alpha - 1} \frac{\partial H}{\partial w_{mn}^{(0)}} . \tag{6.33}$$

In this situation, convergence is accelerated when α is close to unity. We also see that it is necessary that $\alpha < 1$ for the sum in Equation (6.33) to converge.

The other limit to consider is that the gradient changes rapidly from iteration to iteration. How is the learning rule modified in this case? As an example, let us assume that the gradient remains of the same magnitude, but that its sign oscillates, $\partial H/\partial w_{mn}^{(t)} = (-1)^t \partial H/\partial w_{mn}^{(0)}$ for $t = 1, \ldots, T$. Inserting this into Equation (6.32), we obtain

$$\delta w_{mn}^{(T)} \approx -\eta \frac{\partial H}{\partial w_{mn}^{(0)}} \sum_{t=0}^{T} (-1)^t \alpha^{T-t} = -\eta \frac{\alpha^{T+1} + (-1)^T}{\alpha + 1} \frac{\partial H}{\partial w_{mn}^{(0)}} . \tag{6.34}$$

Here the increments are much smaller compared with those in Equation (6.33). This shows that introducing inertia can substantially accelerate convergence without sacrificing accuracy. The disadvantage is, of course, that there is yet another parameter to choose, namely the momentum constant α.

Nesterov's *accelerated gradient* method [77] is another means of implementing momentum. The algorithm was developed for smooth optimisation problems, but it has been suggested to use the method when training deep neural networks with gradient descent [78]:

$$\delta w_{mn}^{(t)} = -\eta \frac{\partial H}{\partial w_{mn}}\Big|_{\{w_{ij}^{(t)} + \alpha_{t-1} \delta w_{ij}^{(t-1)}\}} + \alpha_{t-1} \delta w_{mn}^{(t-1)} . \tag{6.35}$$

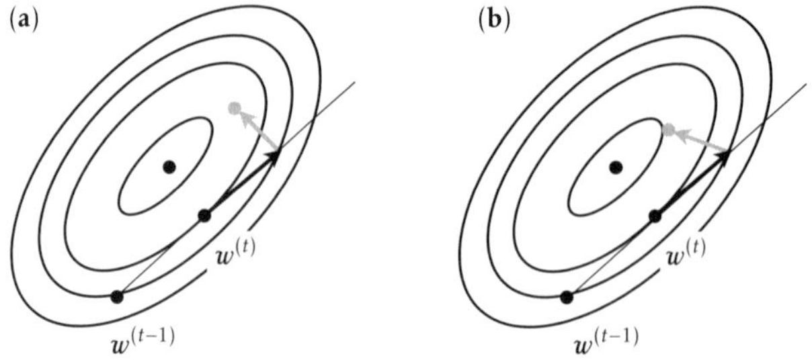

Figure 6.8 (**a**) Momentum method (6.31). The gray arrow represents the increment $-\eta(\partial H/\partial w_{mn})|_{\{w_{ij}^{(t)}\}}$. (**b**) Nesterov's accelerated gradient method (6.35). The gray arrow represents $-\eta(\partial H/\partial w_{mn})|_{\{w_{ij}^{(t)}+\alpha_{t-1}\delta w_{ij}^{(t-1)}\}}$. The location of $\boldsymbol{w}^{(t+1)}$ (gray point) is closer to the minimum (black centre point) than in panel (**a**)

A suitable sequence of coefficients α_t is defined by recursion [78]. The coefficients α_t approach unity from below as t increases.

Nesterov's accelerated-gradient method is more accurate than the simple momentum method, because the accelerated-gradient method evaluates the gradient at an extrapolated point, not at the initial point. Figure 6.8 illustrates a situation where Nesterov's method converges more rapidly. Nesterov's method is not much more difficult to implement than Equation (6.31), and it is not much more expensive in terms of computational cost.

There are other ways of adapting the learning rate during training, described in Section 4.10 in Haykin's book [2]. Finally, the learning rate need not be the same for all neurons. If the weights of neurons in different layers change at very different speeds (Section 7.2), one could define a layer-dependent learning rate η_ℓ that is larger for neurons with smaller gradients.

6.6 Summary

Backpropagation is an efficient algorithm for stochastic gradient descent on the energy function (6.4) in weight space, because it refers only to quantities that are local to the weight to be updated. Networks with many hidden neurons have many free parameters (their weights and thresholds). This increases the risk of overfitting, which reduces the power of the network to generalise. Deep networks with many hidden layers are particularly prone to overfitting (Chapter 7). The tendency of networks to overfit can be reduced by cross validation.

6.7 Further Reading

The backpropagation algorithm is explained in Section 6.1 of Hertz, Krogh, and Palmer [1], and in Chapter 4 of Haykin's book [2]. The paper [76] by LeCun et al.

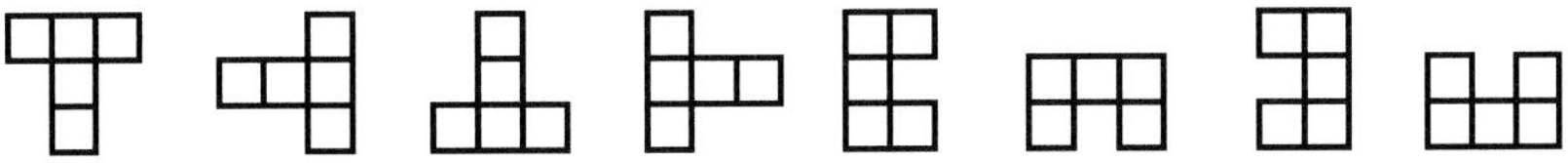

Figure 6.9 Patterns detected by the convolutional network of Ref. [12]. After Figure 13 in Ref. [12]

predates deep learning, but it is still a very nice collection of recipes for making backpropagation more efficient.

One of the first papers on error backpropagation is the one by Rumelhart et al. [12] from 1986. The authors provide an elegant explanation and summary of the backpropagation algorithm. They also describe the results of different numerical experiments, and one of them introduces convolutional networks (Chapter 8) to learn to tell the difference between the letters T and C (Figure 6.9).

6.8 Exercises

6.1 Covariance matrix. Show that the eigenvalues of the data covariance matrix $\mathbb{C}$ defined in Equation (6.24) are real and non-negative.

6.2 Principal-component analysis. Compute the data covariance matrix $\mathbb{C}$ for the example shown in Figure 6.10 and determine the principal direction. Determine the principal direction for the data shown in Figure 6.11.

6.3 Nesterov's accelerated-gradient method. The version (6.35) of Nesterov's algorithm is slightly different from the original formulation [77]. This point is discussed in Ref. [78]. Show that both versions are equivalent.

6.4 Momentum. Section 6.5 describes how to speed up gradient descent by introducing momentum. To explain how this works, consider the one-dimensional energy function shown in Figure 6.12. Iterate Equation (6.31) for $\alpha = \frac{1}{2}$, and determine how many iteration steps it takes to get from $w_{\rm A}$ to $w_{\rm B}$. Compare with the corresponding result for $\alpha = 0$. Then consider what happens after $w_{\rm B}$. How many steps does it take to reach $w_{\rm C}$?

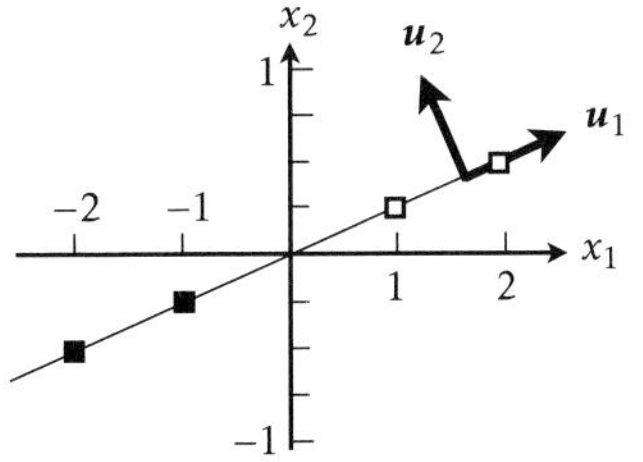

Figure 6.10 The principal direction of this data set is $\boldsymbol{u}_1$

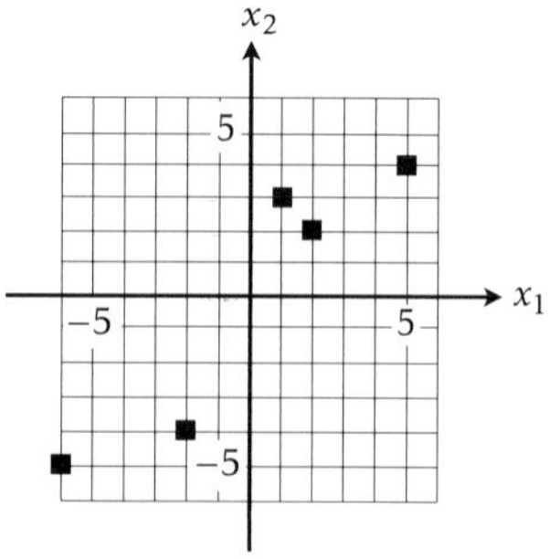

Figure 6.11 Calculate the principal component of this data set. See Exercise 6.2

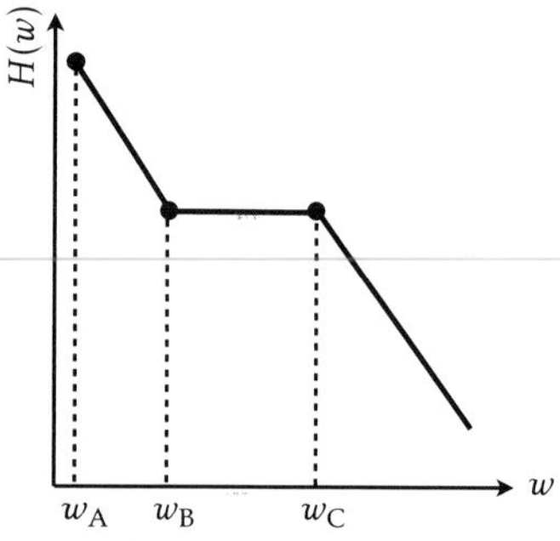

Figure 6.12 One-dimensional energy function used to illustrate how momentum accelerates gradient descent. See Exercise 6.4

6.5 Backpropagation. Derive stochastic gradient-descent learning rules for the weights of a network with the layout shown in Figure 5.2. Assume that all activation functions are of sigmoid form, $\sigma(b) = 1/(1 + \mathrm{e}^{-b})$, hidden thresholds are denoted by θ_j, and those of the output neurons by Θ_i. The energy function is $H = -\sum_{i,\mu}\big[t_i^{(\mu)} \log O_i^{(\mu)} + \big(1 - t_i^{(\mu)}\big) \log\big(1 - O_i^{(\mu)}\big)\big]$ (see Section 7.5), where log is the natural logarithm and $t_i^{(\mu)} = 0/1$ are the targets.

6.6 Stochastic gradient descent. To train a multilayer perceptron using stochastic gradient descent, one needs update formulae for the weights and thresholds in the network. Derive these update formulae for sequential training using backpropagation for the network shown in Figure 6.13. The weights for the first and second hidden layer, and for the output layer are denoted by $w_{jk}^{(1)}$, $w_{mj}^{(2)}$, and W_{im}. The corresponding thresholds are denoted by $\theta_j^{(1)}$, $\theta_m^{(2)}$, and Θ_i, and the activation function by $g(\cdots)$. The target value for input pattern $\boldsymbol{x}^{(\mu)}$ is $t_i^{(\mu)}$, and the pattern index μ ranges from 1 to p. The energy function is $H = \frac{1}{2}\sum_{i=1}^{M}\sum_{\mu=1}^{p}\big(t_i^{(\mu)} - O_i^{(\mu)}\big)^2$.

6.7 Multilayer perceptron. A perceptron has hidden layers $\ell = 1, \ldots, L - 1$ and output layer $l = L$. Neuron j in layer ℓ computes $V_j^{(\ell)} = g\big(b_j^{(\ell)}\big)$ with

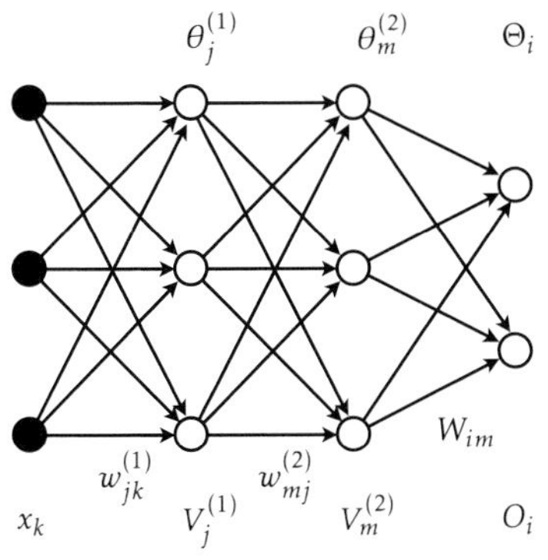

Figure 6.13 Multilayer perceptron with three input terminals, two hidden layers, and two output neurons. See Exercise 6.6

$b_j^{(\ell)} = -\theta_j^{(\ell)} + \sum_k w_{jk}^{(\ell)} V_k^{(\ell-1)}$, where $w_{jk}^{(\ell)}$ are weights, $\theta_j^{(\ell)}$ are thresholds, $g(b)$ is the activation function, $V_i^{(L)} = O_i = g\big(b_i^{(L)}\big)$, and $V_k^{(0)} = x_k$. Draw this network. Indicate where the elements x_k, O_i belong, as well as $b_j^{(\ell)}$, $V_j^{(\ell)}$, $w_{jk}^{(\ell)}$ and $\theta_j^{(\ell)}$ for $\ell = 0, \ldots, L$. Determine how the derivatives $\partial V_i^{(\ell)}/\partial w_{mn}^{(\ell')}$ depend upon the derivatives $\partial V_j^{(\ell-1)}/\partial w_{mn}^{(\ell')}$ for $\ell' < \ell$. Evaluate the derivative $\partial V_j^{(\ell)}/\partial w_{mn}^{(\ell')}$ for $\ell' = \ell$. Using gradient descent on the energy function $H = \frac{1}{2}\sum_{i\mu}\big(t_i^{(\mu)} - O_i^{(\mu)}\big)^2$, find the learning rule for the weight $w_{mn}^{(L-2)}$ with learning rate η.

6.8 Error backpropagation. Derive Equation (6.16) and use this result to deduce the recursion (6.17).

6.9 Preprocessing input data. Find an example that shows how large input means may generate steep gradients in the energy function (6.4).

6.10 Singular points of the Lagrangian $\mathscr{L}$. To use the method of Lagrange multipliers λ for constrained minimisation, one forms the Lagrangian $\mathscr{L}$ [Equation (6.28)] by adding λ times the constraint to the target function to be minimised. Then one searches for the *singular points* of $\mathscr{L}$, the points where the derivatives of $\mathscr{L}$ with respect to the parameters and λ vanish. The method builds on the assumption that the singular points of $\mathscr{L}$ correspond to extrema of the target function. Construct a counterexample.

6.11 Lagrange multiplier. Find the global minimum of the function $f(x_1, x_2) = x_1 - x_2^2/2$ subject to the constraint $x_1^2 + x_2^2 = 1$ using a Lagrange multiplier λ. Explain the geometric picture underlying this method. Note that it works equally well if one replaces λ with $-\lambda$.

7

Deep Learning

7.1 How Many Hidden Layers?

In Chapter 5, we saw why it is sometimes necessary to have a hidden layer: in order to solve problems that are not linearly separable. Under which circumstances is one hidden layer sufficient? Are there problems that require more than one hidden layer? Even if not necessary, may additional hidden layers improve the performance of the network?

To understand how many hidden layers suffice, it is useful to view the classification problem as an *approximation problem* [79]. Consider the classification problem $[\boldsymbol{x}^{(\mu)}, t^{(\mu)}]$ for $\mu = 1, \ldots, p$. This problem defines a *target function* $t(\boldsymbol{x})$. Training a network to solve this task corresponds to approximating the target function $t(\boldsymbol{x})$ by the output function $O(\boldsymbol{x})$ of the network, from N-dimensional input space to one-dimensional outputs.

How many hidden layers are necessary or sufficient to approximate a given set of functions to a certain accuracy, by choosing weights and thresholds? The answer depends on the nature of the target function. Is it real-valued, perhaps continuous, or does it assume only discrete values?

We start with real-valued inputs and a single output [1, 80]. Consider the network drawn in Figure 7.1. The neurons in the hidden layers have sigmoid activation functions $\sigma(b) = (1 + \mathrm{e}^{-b})^{-1}$. The output is a linear unit, with linear activation function $g(b) = b$. With two hidden layers one tries to approximate the function $t(\boldsymbol{x})$ by

$$O(\boldsymbol{x}) = \sum_m W_m g\Big(\sum_j w^{(2)}_{mj} g\Big(\sum_k w^{(1)}_{jk} x_k - \theta^{(1)}_j\Big) - \theta^{(2)}_m\Big) - \Theta\,. \tag{7.1}$$

In the simplest case, the inputs are one-dimensional (Figure 7.2). The training set consists of pairs $[x^{(\mu)}, t^{(\mu)}]$ that encode the target function $t(x)$. The task is then to approximate $t(x)$ by the network output $O(x)$:

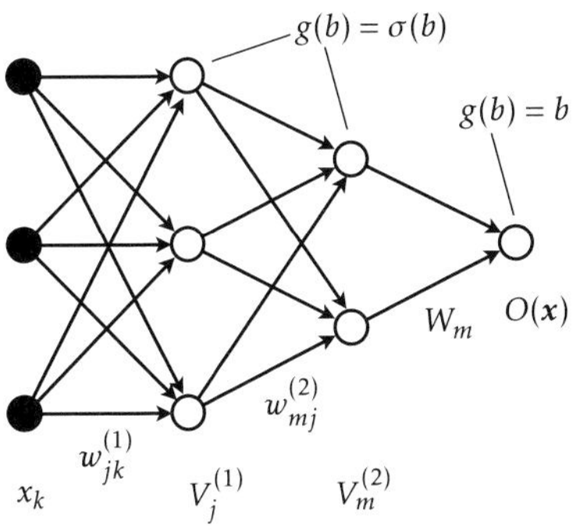

Figure 7.1 Multilayer perceptron for function approximation

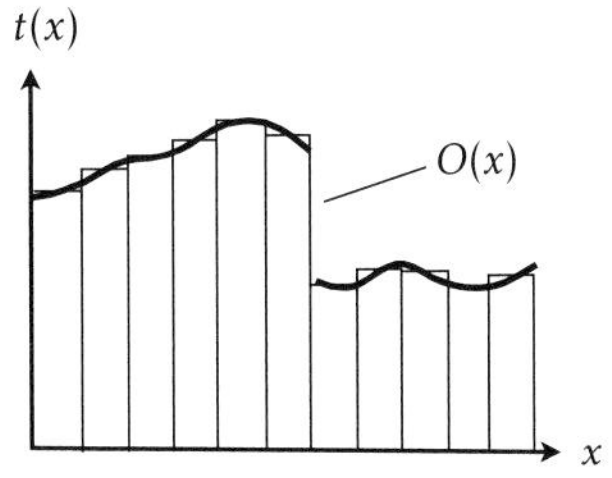

Figure 7.2 The neural-network output $O(x)$ approximates the target function $t(x)$ (thick black curve)

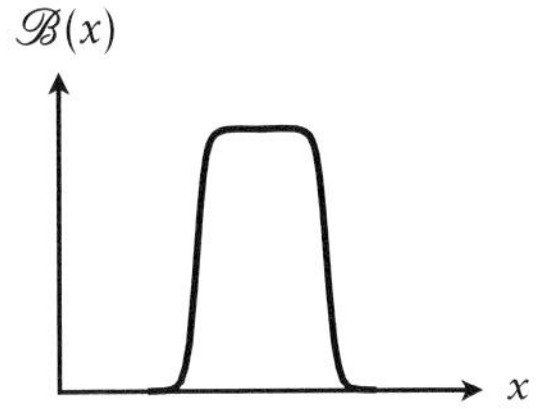

Figure 7.3 Basis function used to approximate a one-dimensional target function

$$O(x) \approx t(x) . \tag{7.2}$$

To this end, one uses linear combinations of the basis functions $\mathscr{B}(x)$ shown in Figure 7.3. Any reasonable real-valued function $t(x)$ can be approximated by sums of such basis functions, each suitably shifted and scaled. Furthermore, these basis functions can be expressed as scaled differences of sigmoid activation functions

$$\mathscr{B}(x) = W\big[\sigma\big(w^{(1)}x - \theta_1^{(1)}\big) - \sigma\big(w^{(1)}x - \theta_2^{(1)}\big)\big] . \tag{7.3}$$

Comparing with Equation (7.1) shows that one hidden layer is sufficient to construct the function $O(x)$ in this way. Now consider two-dimensional inputs. In this case, a suitable basis function is (Figure 7.4):

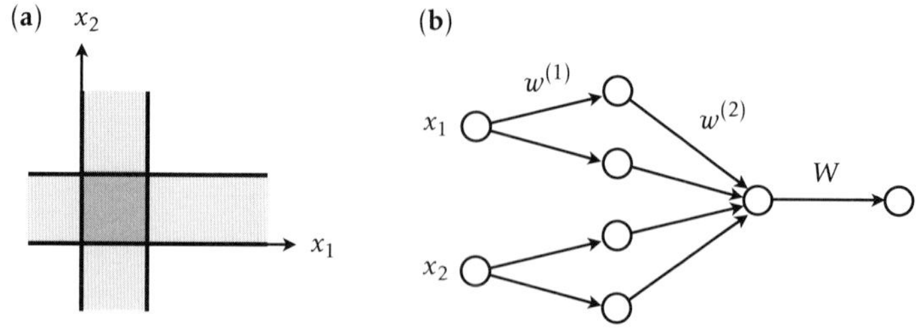

Figure 7.4 Two-dimensional basis functions. (**a**) To make a localised basis function with two inputs, one needs two hidden layers of neurons with sigmoid activation functions. One layer determines the lightly shaded cross in terms of a linear combination of four sigmoid outputs. The second layer localises the final output to the darker square [Equation (7.4)]. (**b**) Network layout for one basis function, after Figure 8 in Ref. [80]

$$\begin{aligned}\mathscr{B}(\boldsymbol{x}) = W\sigma\big\{w^{(2)}[\sigma(w^{(1)}x_1) - \sigma(w^{(1)}x_1 - \theta_2^{(1)}) + \sigma(w^{(1)}x_2) \\ - \sigma(w^{(1)}x_2 - \theta_4^{(1)})] - \theta^{(2)}\big\}. \end{aligned} \tag{7.4}$$

So two hidden layers are sufficient for two input dimensions. For each basis function, we require four neurons in the first hidden layer and one neuron in the second hidden layer. The construction is analogous for more than two input dimensions. Also, for each basis function, we need $2N$ neurons in the first layer and one neuron in the second layer. In conclusion, two hidden layers are sufficient to approximate a real-valued target function.

Yet it is not always necessary to use two layers for real-valued functions. For continuous functions, one hidden layer is sufficient. This is ensured by the *universal approximation theorem* [2]. It says any continuous function can be approximated to arbitrary accuracy by a network with a single hidden layer, for sufficiently many neurons in the hidden layer.

In Chapter 5, we considered discrete Boolean functions. Any Boolean function with N-dimensional inputs can be represented by a network with one hidden layer, using 2^N neurons in the hidden layer. An example for such a network is discussed in Ref. [1]:

$$\begin{array}{lrl} x_k \in \{+1,-1\} & k = 1,\ldots,N & \text{inputs} \\ V_j & j = 0,\ldots,2^N - 1 & \text{hidden neurons} \\ g(b) = \text{sgn}(b) & & \text{activation function of hidden neurons} \\ g(b) = \text{sgn}(b) & & \text{activation function of output neuron} \end{array} \tag{7.5}$$

A difference compared with the Boolean-function representations in Section 5.5 is that here the inputs take the values ± 1. The reason is that this simplifies the proof,

(a)

x_1	x_2	t
-1	-1	-1
$+1$	-1	$+1$
-1	$+1$	$+1$
$+1$	$+1$	-1

(b)

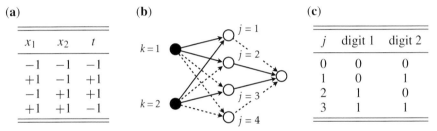

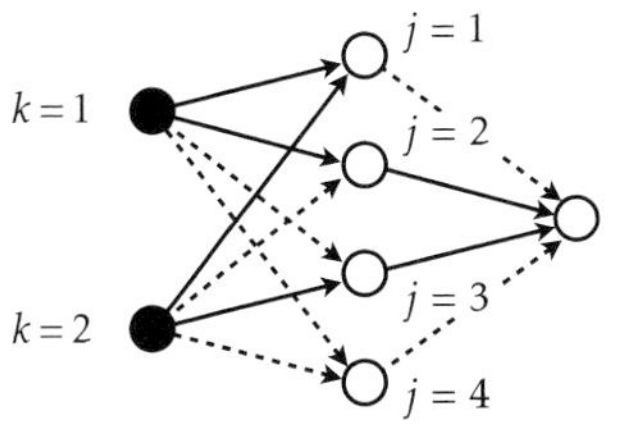

(c)

j	digit 1	digit 2
0	0	0
1	0	1
2	1	0
3	1	1

Figure 7.5 Boolean XOR function. (**a**) Value table, (**b**) Network layout. For the weights feeding into the hidden layer, dashed lines correspond to $w_{jk} = -\delta$ and solid lines to $w_{jk} = \delta$ (see the text). For the weights feeding into the output neuron, dashed lines correspond to $W_{1j} = -\gamma$, and solid lines to $W_{1j} = \gamma$. Panel (**c**) summarises the binary representation of j used to determine the weights w_{jk} of the hidden layer [Equation (7.6)]

which is by construction [1]. For each hidden neuron, one assigns the weights as follows

$$w_{jk} = \begin{cases} \delta & \text{if the } k^{\text{th}} \text{ digit of binary representation of } j \text{ is } 1, \\ -\delta & \text{otherwise,} \end{cases} \tag{7.6}$$

with $\delta > 1$ (see below). The thresholds θ_j of all hidden neurons are the same, equal to $N(\delta - 1)$. The idea is that each input pattern turns on exactly one neuron in the hidden layer, the *winning neuron*). The weights feeding into the output neuron are determined as follows. If the output for the pattern represented by neuron V_j is $+1$, let $W_{1j} = +1$, otherwise $W_{1j} = -1$. The threshold is set to $\Theta = \sum_j W_{1j}$.

To show how this construction works, consider the Boolean XOR function as an example (Figure 7.5). To confirm that only one winning neuron gives a positive signal, consider pattern $\boldsymbol{x}^{(1)} = [-1, -1]^{\mathsf{T}}$. It activates the first neuron in the hidden layer ($j = 0$). To see this, compute the local fields of the hidden neurons:

$$\begin{aligned} b_0^{(1)} &= 2\delta - 2(\delta - 1) = 2\,, \\ b_1^{(1)} &= -2(\delta - 1) = 2 - 2\delta\,, \\ b_2^{(1)} &= -2(\delta - 1) = 2 - 2\delta\,, \\ b_3^{(1)} &= -2\delta - 2(\delta - 1) = 2 - 4\delta\,. \end{aligned} \tag{7.7}$$

If we choose $\delta > 1$ then the output of the first hidden neuron gives a positive output ($V_0 > 0$), the other neurons produce negative outputs, $V_j < 0$ for $j = 1, 2, 3$. In conclusion, output neuron 1 is the winning neuron for this pattern.

Now consider $\boldsymbol{x}^{(3)} = [-1, +1]^{\mathsf{T}}$. In this case,

$$\begin{aligned} b_0^{(3)} &= -2(\delta - 1) = 2 - 2\delta\,, \\ b_1^{(3)} &= -2\delta - 2(\delta - 1) = 2 - 4\delta\,, \end{aligned} \tag{7.8}$$

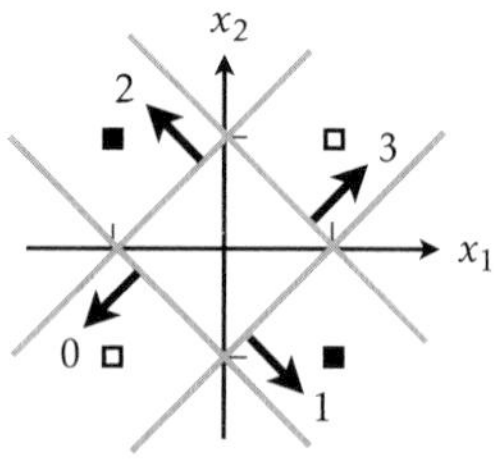

Figure 7.6 Shows how the XOR network depicted in Figure 7.5 partitions the input plane. Target values are encoded as in Figure 5.8: □ corresponds to $t = -1$ and ■ to $t = +1$

$$b_2^{(3)} = 2\delta - 2(\delta - 1) = 2\,,$$
$$b_3^{(3)} = -2(\delta - 1) = 2 - 2\delta\,.$$

Now the third hidden neuron gives a positive output, while the others yield negative values. It works in the same way for the other two patterns, $\boldsymbol{x}^{(2)}$ and $\boldsymbol{x}^{(4)}$. In summary, there is a unique winning neuron for each pattern.[1] Figure 7.6 shows how the four decision boundaries corresponding to V_j partition the input plane.

According to the scheme outlined above, the output neuron computes

$$O_1 = \text{sgn}(-V_1 + V_2 + V_3 - V_4) \tag{7.9}$$

with $\Theta = \sum_j W_{1j} = 0$. For $\boldsymbol{x}^{(1)}$ and $\boldsymbol{x}^{(4)}$, we find the correct result $O_1 = -1$. The same is true for $\boldsymbol{x}^{(2)}$ and $\boldsymbol{x}^{(3)}$, in this case we obtain $O_1 = 1$. In summary, this example illustrates how an N-dimensional Boolean function is represented by a network with one hidden layer with 2^N neurons. The problem is of course that this network is expensive to train for large values of N because the number of neurons becomes very large.

There are more efficient layouts if one uses more than one hidden layer. As an example, consider the *parity function* for N-dimensional binary inputs, with bits equal to 0 or 1. It measures the parity of the sequence of input bits. The function. It evaluates to unity if there is an odd number of ones in the input; otherwise, it evaluates to zero. A construction similar to the above yields a network layout with 2^N neurons in the hidden layer. If one instead wires together XOR networks, one can solve the parity problem with $O(N)$ neurons [81] (Figure 7.7). When N is a power of two, this network has $3(N-1)$ neurons. To see this, set the number of inputs to $N = 2^k$. Figure 7.7 shows that the number $\mathscr{N}_k$ of neurons satisfies the recursion $\mathscr{N}_{k+1} = 2\mathscr{N}_k + 3$ with $\mathscr{N}_1 = 3$. The solution of this recursion is $\mathscr{N}_k = 3(2^k - 1)$.

[1] That pattern $\mu = k$ gives the winning neuron $j = k - 1$ is of no importance; it is just a consequence of how the patterns are ordered in the value table in Figure 7.5.

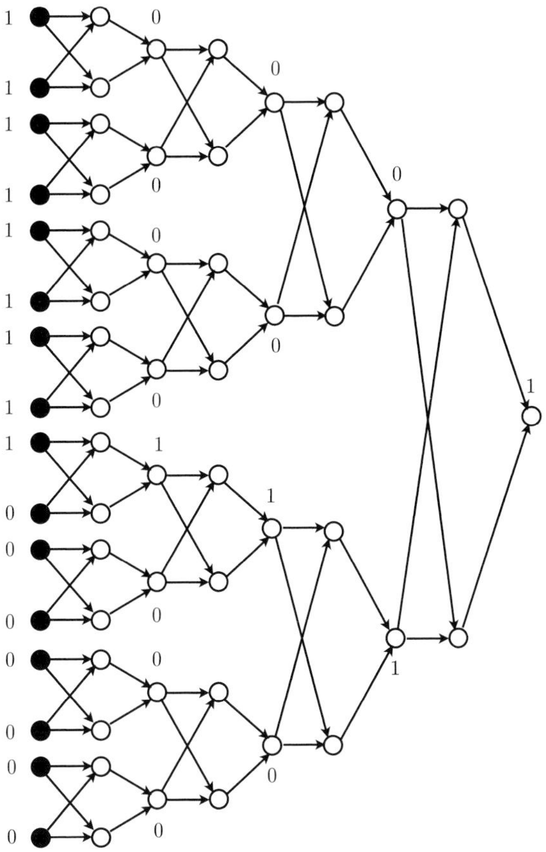

Figure 7.7 Solution of the parity problem for N-dimensional inputs. The network is built from XOR units, here with 0/1 neurons. Only the states of the inputs and outputs of the XOR units are shown, not those of the hidden neurons. In total, the whole network has only $O(N)$ neurons. After Figure 2 in Ref. [81]

This example also illustrates a second reason why it may be useful to have more than one hidden layer. To design a neural network for a certain task it is often convenient to build the network from well-studied building blocks. One wires them together, often in a hierarchical fashion. In Figure 7.7, there is only one building block, the XOR network from Figure 5.17. Other examples are convolutional networks for image analysis (Chapter 8). Here the fundamental building blocks are so-called feature maps, they recognise different geometrical features in the image, such as edges or corners.

7.2 Vanishing and Exploding Gradients

This section describes an inherent instability in the training of deep networks with stochastic gradient descent, the vanishing- or exploding-gradient problem.

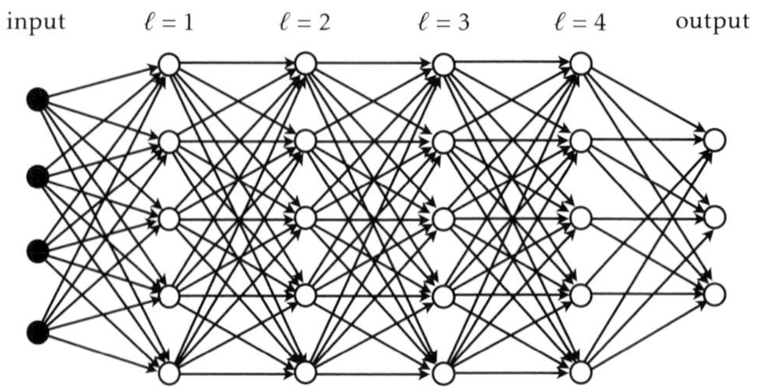

Figure 7.8 Fully connected deep network with four hidden layers

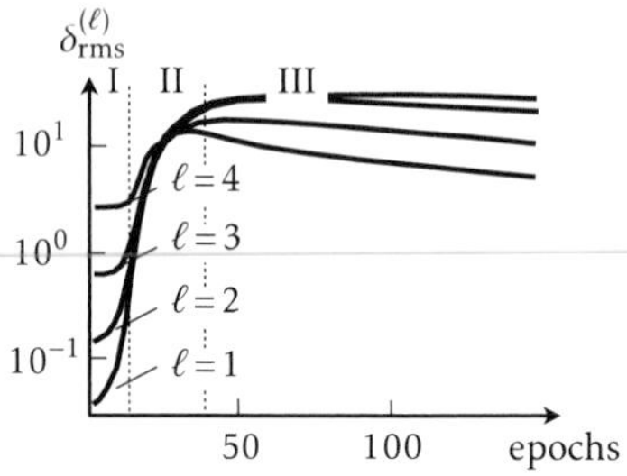

Figure 7.9 Vanishing-gradient problem for a network with four fully connected hidden layers. The figure illustrates schematically how the r.m.s. error $\delta_{\rm rms}^{(\ell)}$ in layer ℓ depends on the number of training epochs. During phase I, the vanishing-gradient problem is severe, during phase II the network starts to learn, phase III is the convergence phase where the errors decline. Schematic, based on simulations performed by Ludvig Storm, training a network with four hidden layers and $N = 30$ neurons per layer on the MNIST data set (Section 8.3).

In Chapter 6, we saw that learning slows down when the factors $g'(b)$ in the recursion (6.17) become small. When the network has several hidden layers, like the one shown in Figure 7.8, the potentially small factors of $g'(b)$ are multiplied, aggravating the problem. As a consequence, the weights of hidden neurons close to the input layer change only by small amounts, the smaller the more hidden layers the network has. This is the *vanishing-gradient problem.*

Figure 7.9 quantifies the problem. The figure shows that the r.m.s. errors averaged over different realisations of random initial weights, $\delta_{\rm rms}^{(\ell)} \equiv \big(\langle N^{-1}\sum_{j=1}^N [\delta_j^{(\ell)}]^2\rangle\big)^{1/2}$, tend to be very small during initial training. To explain this phenomenon, consider the very simple example discussed in Ref. [5]: a long chain of neurons with only one neuron per layer (Figure 7.10). The output $V^{(L)}$ is given by nested activation functions

$$V^{(L)}=g\Big(w^{(L)}g\Big(w^{(L-1)}\cdots g\big(w^{(2)}g(w^{(1)}x-\theta^{(1)})-\theta^{(2)}\big)\ldots-\theta^{(L-1)}\Big)-\theta^{(L)}\Big). \tag{7.10}$$

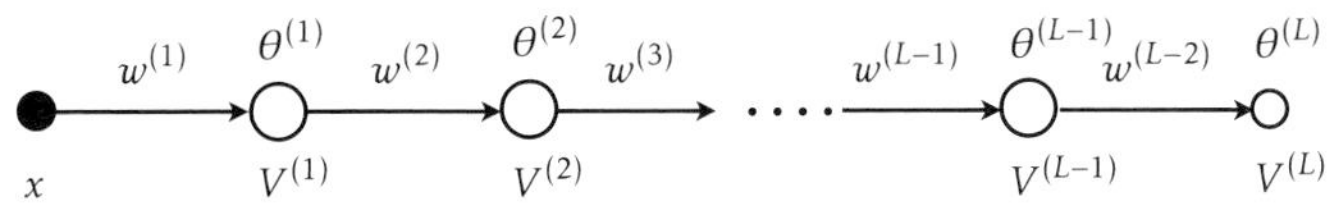

Figure 7.10 Chain of neurons illustrating the vanishing-gradient problem [5], with neurons $V^{(\ell)}$, weights $w^{(\ell)}$, and thresholds $\theta^{(\ell)}$

Let us compute the errors $\delta^{(\ell)}$ using Equation (6.16). The partial derivative in (6.16) is evaluated using the chain rule:

$$\begin{aligned}
\frac{\partial V^{(L)}}{\partial V^{(L-1)}} &= g'(b^{(L)})w^{(L)}\,, \\
\frac{\partial V^{(L)}}{\partial V^{(L-2)}} &= \frac{\partial V^{(L)}}{\partial V^{(L-1)}}\frac{\partial V^{(L-1)}}{\partial V^{(L-2)}} = g'(b^{(L)})w^{(L)}g'(b^{(L-1)})w^{(L-1)}\,, \\
&\vdots
\end{aligned} \tag{7.11}$$

where $b^{(k)} = w^{(k)}V^{(k-1)} - \theta^{(k)}$ is the local field for neuron k. This yields the following expression for $\partial V^{(L)}/\partial V^{(\ell)}$:

$$\frac{\partial V^{(L)}}{\partial V^{(\ell)}} = \prod_{k=L}^{\ell+1}[g'(b^{(k)})w^{(k)}]\,. \tag{7.12}$$

Inserting this expression into Equation (6.16), we find

$$\delta^{(\ell)} = [t - V^{(L)}(x)]g'(b^{(L)})\prod_{k=L}^{\ell+1}[w^{(k)}g'(b^{(k-1)})]\,. \tag{7.13}$$

One can also obtain this result by applying the recursion from Algorithm 4, $\delta^{(\ell)} = \delta^{(\ell+1)}w^{(\ell+1)}g'(b^{(\ell)})$.

Now consider the early stages of training [5]. For the activation functions (6.19), the maximum of $g'(b)$ is $\frac{1}{4}$ and 1, respectively, and $g'(b)$ becomes exponentially small if $|b|$ is large. If one initialises the weights as described in Section 6.2, to Gaussian random variables with mean zero and variance $\sigma_w^2 = 1$ say, then the factors $w^{(k)}g'(b^{(k-1)})$ tend to be smaller than unity. In this case, Equation (7.12) implies that the error or gradient $\delta^{(\ell)}$ vanishes quickly as ℓ decreases. The reason is simply that the number of small factors in the product (7.12) increases when ℓ becomes smaller, and multiplying many small numbers gives a very small product. As a result, the training slows down. As mentioned above, this is the *vanishing-gradient problem* (phase I in Figure 7.9).

What happens at later times? Figure 7.9 indicates that the network continues to learn slowly. For the particular example shown in Figure 7.9, the effect persists for about 20 epochs. Then the first layers begin to learn faster (phase II). There is to date no mathematical theory describing how this transition occurs. Much later in training, the errors decay as the learning converges (phase III in Figure 7.9).

There is a second, equivalent, point of view [5]: the learning is slow in a layer far from the output because the output is not very sensitive to the state of these neurons. The effect of a given neuron on the output is measured by Equation (7.12), which describes how the output of the network changes when changing the *state* of a neuron in a particular layer. At any rate, Equation (7.13) demonstrates that hidden layers far from the output learn slowly, at least initially when the weights are still random.

Suppose we try to combat the vanishing-gradient problem by increasing the weight variance σ_w^2. The problem is that this may cause the factors $w^{(k)} g'(b^{(k-1)})$ to become larger than unity. As a consequence, the gradients increase exponentially instead (*exploding gradients*). In conclusion, the training dynamics is fundamentally unstable. This is due to the multiplicative nature of the recursion for the errors. Taking the logarithm of the product in Equation (7.12) and assuming that the weights are independently distributed random numbers, the *central-limit theorem* (Chapter 2) implies that the distribution of $\log \delta^{(\ell)}$ is Gaussian. In other words, the distribution of the errors is lognormal, implying that very small and very large values of $\delta^{(\ell)}$ occur with high probability.

In networks like the one shown in Figure 7.8, the principle is the same. Assume that all layers $\ell = 1, \ldots, L$ have the same number N of neurons. When $N > 1$, one multiplies $N \times N$ matrices instead of numbers. The product (7.12) of random numbers becomes a product of random matrices. Using the chain rule, we find

$$\frac{\partial V_i^{(L)}}{\partial V_j^{(\ell)}} = \sum_{m=1}^{N} \sum_{n=1}^{N} \cdots \sum_{p=1}^{N} \frac{\partial V_i^{(L)}}{\partial V_m^{(L-1)}} \frac{\partial V_m^{(L-1)}}{\partial V_n^{(L-2)}} \cdots \frac{\partial V_p^{(\ell+1)}}{\partial V_j^{(\ell)}} . \tag{7.14}$$

With the update rule

$$V_m^{(k)} = g\Big(\sum_{j=1}^{N} w_{ij}^{(k)} V_j^{(k-1)} - \theta_i^{(k)} \Big) \tag{7.15}$$

we can evaluate each factor:

$$\frac{\partial V_m^{(k)}}{\partial V_n^{(k-1)}} = g'(b_m^{(k)}) w_{mn}^{(k)} . \tag{7.16}$$

Substituting this result into Equation (7.14), we see that the partial derivatives $\partial V_i^{(L)} / \partial V_j^{(\ell)}$ can be computed in the form of a matrix product. The matrix $\mathbb{J}_{L-\ell}$ with elements $[\mathbb{J}_{L-\ell}]_{ij} = \partial V_i^{(L)} / \partial V_j^{(\ell)}$ is given by

$$\mathbb{J}_{L-\ell} = \mathbb{D}^{(L)} \mathbb{W}^{(L)} \mathbb{D}^{(L-1)} \mathbb{W}^{(L-1)} \cdots \mathbb{D}^{(\ell+1)} \mathbb{W}^{(\ell+1)} . \tag{7.17}$$

Here $\mathbb{W}^{(k)}$ is the matrix of weights feeding into layer k, and

$$\mathbb{D}^{(k)} = \begin{bmatrix} g'\big(b_1^{(k)}\big) & & \\ & \ddots & \\ & & g'\big(b_N^{(k)}\big) \end{bmatrix} \tag{7.18}$$

is the diagonal matrix with entries $D_{jj}^{(k)} = g'(b_j^{(k)})$. The matrix product (7.17) determines the error dynamics, just like Equation (7.13):

$$\delta^{(\ell)} = \delta^{(L)} \mathbb{J}_{L-\ell}\,. \tag{7.19}$$

Does the error magnitude $|\delta^{(\ell)}|^2 = \delta^{(\ell)\mathsf{T}}\delta^{(\ell)}$ shrink or grow as they propagate through the layers? This is determined by the eigenvalues of the *left Cauchy-Green* matrix $\mathbb{J}_p\mathbb{J}_p^{\mathsf{T}}$, with $p = L - \ell$. This matrix is symmetric, and its eigenvalues are non-negative. Their square roots are the *singular values* of $\mathbb{J}_p$:

$$\Lambda_1^{(p)} \geq \Lambda_2^{(p)} \geq \cdots \geq \Lambda_N^{(p)} \geq 0\,. \tag{7.20}$$

It is customary to sort the singular values by their magnitudes, as in Equation (7.20). When there are many layers, the number $p = L - \ell$ of factors in Equation (7.17) is large. In this case, the maximal singular value either decreases or increases exponentially as a function of p [82]. The corresponding rate

$$\lambda_1 = \lim_{p\to\infty} \frac{1}{p} \log \Lambda_1^{(p)} \tag{7.21}$$

is called the maximal *Lyapunov exponent*. A negative maximal Lyapunov exponent indicates that the errors vanish exponentially. The eigenvectors of $\mathbb{J}_p\mathbb{J}_p^{\mathsf{T}}$ are called *backward Lyapunov vectors*. They describe how the errors change as they propagate through the network. How small differences between the inputs change, is determined by the *forward Lyapunov vectors*, the eigenvectors of $\mathbb{J}_p^{\mathsf{T}}\mathbb{J}_p$. Since $\mathbb{J}_p^{\mathsf{T}}\mathbb{J}_p$ and $\mathbb{J}_p\mathbb{J}_p^{\mathsf{T}}$ have the same eigenvalues, the rate of decay or increase of the magnitude of input differences is the same as that of the error magnitudes.

The concept of a maximal Lyapunov exponent is borrowed from chaos theory [83–85], where $\lambda_1 > 0$ implies that small perturbations of the initial conditions grow exponentially as a function of time. The iterated map (7.10) is a dynamical system.

The transition in Figure 7.9 is triggered by a change of the Lyapunov exponent from negative values to $\lambda_1 \approx 0$ [86]. In summary, the unstable-gradient problem in deep networks is due to the fact that the maximal singular value $\Lambda_1^{(p)}$ either increases or decreases exponentially as one moves away from the output layer, depending on whether the maximal Lyapunov exponent is negative or positive.

Pennington et al. [87] suggested to tackle the unstable-gradient problem by initialising weights and thresholds in such a way that the maximal Lyapunov exponent is close to zero, in order to make sure that the errors neither grow nor shrink exponentially. Consider the network shown in Figure 7.8, with N neurons per hidden layer, and initialise the weights to independent Gaussian random numbers with mean zero and variance σ_w^2. The thresholds are initialised in the same way, with variance σ_θ^2. In the limit of $N \to \infty$, one can use a mean-field theory [87], just as in Chapter 3, to estimate the maximal Lyapunov exponent.

Following Ref. [87], the first step is to compute how the errors propagate through the network. We assume uncorrelated random input patterns, Equation (2.29), and random weights with mean zero and variance $\langle w_{ij} w_{kl} \rangle = \sigma_w^2 \delta_{ij} \delta_{kl}$. When $N \to \infty$, the errors are sums of many random numbers [Equation (6.17)]. Invoking the central-limit theorem (Chapter 2), one concludes that the errors are approximately Gaussian distributed, with mean zero and with variance

$$\langle [\delta_j^{(\ell-1)}]^2 \rangle = \Big\langle \sum_{i,k=1}^{N} \delta_i^{(\ell)} \delta_k^{(\ell)} \, w_{ij}^{(\ell)} w_{kj}^{(\ell)} \, [g'(b_j^{(\ell-1)})]^2 \Big\rangle \approx \sigma_w^2 \sum_{i=1}^{N} \langle [\delta_i^{(\ell)}]^2 \rangle \langle [g'(b_j^{(\ell-1)})]^2 \rangle \,. \tag{7.22}$$

The last approximation neglects possible correlations with the local fields. The variance $\langle [\delta_i^{(\ell)}]^2 \rangle$ does not depend on i, so that the sum just gives a factor of N.

In the limit of large N, the central-limit theorem ensures that the local fields $b_j^{(\ell)}$ are Gaussian distributed too, with mean zero and variance

$$\sigma_\ell^2 = \frac{1}{N} \sum_{j=1}^{N} [b_j^{(\ell)}]^2 \,. \tag{7.23}$$

This allows us to estimate

$$\langle [g'(b_j^{(\ell)}]^2 \rangle \sim \int \mathrm{d}z_\ell \, \frac{\mathrm{e}^{-z^2/2\sigma_\ell^2}}{\sqrt{2\pi\sigma_\ell^2}} [g'(z_\ell)]^2 \equiv F(\sigma_\ell) \,. \tag{7.24}$$

Equations (7.23) describes how the distribution of local fields $b_j^{(\ell)}$ narrows or broadens as one iterates. For $g(b) = \tanh(b)$, it was shown in Ref. [87] that σ_ℓ approaches a fixed point, $\sigma^* = \lim_{\ell\to\infty} \sigma_\ell$, under certain conditions on the activation function and upon the variances of the weights and thresholds, σ_w^2 and σ_θ^2. If σ_ℓ is well approximated by σ^*, Equation (7.22) simplifies to $\delta_{\rm rms}^{(\ell-1)} \approx \delta_{\rm rms}^{(\ell)} \sqrt{\sigma_w^2 N F(\sigma^*)}$. This results in a mean-field estimate of the maximal Lyapunov exponent,

$$\lambda_1 \sim \log \left| \delta^{(\ell-1)} / \delta^{(\ell)} \right| \approx \tfrac{1}{2} \log[\sigma_w^2 N F(\sigma^*)]. \tag{7.25}$$

The network parameters should be adjusted so that this exponent is as close to zero as possible. This means, in particular, that one should take

$$\sigma_w^2 \propto N^{-1} \tag{7.26}$$

(see also Refs. [88, 89]). But we must keep in mind that Equation (7.25) relies on taking the limit $N \to \infty$. It is expected that the assumptions underlying Equation (7.25) break down when N is finite, causing the mean-field theory to fail. The tails of the error distribution, for example, are expected to become heavier as N decreases, as indicated by the results for $N = 1$ described above. Note also that $\mathbb{J}_p$ assumes rank zero with a small but non-zero probability when N is finite. In this case $\lambda_1 = -\infty$.

There are a number of other tricks that help to cope with unstable gradients in practice, to some extent at least. First, it is sometimes argued that activation functions which do not saturate at large b, such as the ReLU function, help against the vanishing-gradient problem. Second, batch normalisation (Section 7.6.5) may reduce the unstable-gradient problem. Third, introducing connections that skip layers (*residual networks*) can alleviate the unstable-gradient problem. This is discussed in Section 7.4.

Finally, there is an important aspect of the problem that we did not discuss: unstable gradients limit the extent to which information can propagate through the network in a meaningful way. This is explained in Ref. [90].

In this section, we assumed all along that the weights are random numbers. When the network starts to learn this is no longer the case. The question is how the singular values of $\mathbb{J}_p$ change when correlations between different factors in the product (7.17) develop.

7.3 Rectified Linear Units

Glorot et al. [91] suggested to use a piecewise activation function, the ReLU function[2] $\max\{0, b\}$ (Chapter 1). What is the point of using ReLU neurons? When training a deep network with ReLU activation functions, many of the hidden neurons produce output zero. This means that the network of active neurons (non-zero output) is *sparsely* connected. It is sometimes argued that sparse networks have desirable properties; at least sparse representations of a classification problem tend to be easier to learn because they are more likely to be linearly separable (Section 5.4). Figure 7.11 illustrates that for a given input pattern, only a certain fraction of hidden neurons is active. For these neurons the computation is linear, yet different input patterns give different sets of active neurons. The product in Equation (7.17) acquires a particularly simple structure: the matrices $\mathbb{D}^{(p)}$ are diagonal with 0/1 entries. But while the weight matrices are independent initially, they become correlated as training proceeds. Also the $\mathbb{D}^{(p)}$-matrices

[2] Since the derivative of the ReLU function is discontinuous at $b = 0$, a common convention is to set the derivative to zero at $b = 0$.

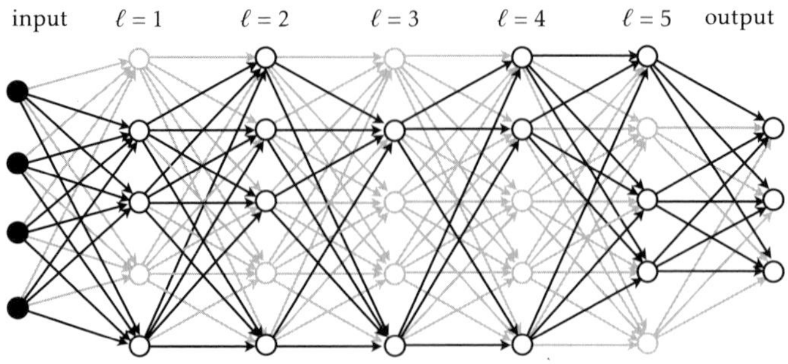

Figure 7.11 Sparse network of active neurons with ReLU activation functions. The black paths correspond to *active* neurons with positive local fields

develop correlations: which diagonal elements vanish depends on which pattern is clamped to the input terminals.

A hidden layer with only one or very few active neurons might act as a *bottleneck* preventing efficient backpropagation of output errors which could in principle slow down the training. For the examples given in Ref. [91], this does not occur. To describe information propagation through a network with ReLU neurons, one should compute the probability that a given number of singular values of the matrix $\mathbb{J}_p$ vanish (Section 7.2).

The ReLU function is unbounded for large positive local fields. Therefore, the vanishing-gradient problem (Section 7.2) is thought to be less severe in networks made of rectified linear units. However, since the ReLU function does not saturate, the weights tend to increase. Glorot et al. [91] suggested to employ L_1-regularisation (Section 7.6.1) to make sure that the weights do not grow.

Finally, using ReLU functions instead of sigmoid functions speeds up the training, because the ReLU function has piecewise constant derivatives. Such functions are faster to evaluate numerically than non-linear activation functions and their derivatives.

7.4 Residual Networks

One way of reducing the vanishing-gradient problem is to introduce short cuts, connections that skip layers [92]. Empirical evidence shows that networks with such short cuts are easier to train than standard multilayer perceptrons. The likely reason is that the vanishing-gradient problem is less severe in networks with short cuts, because error propagation is determined by the matrix product with the smallest number of factors.

This section explains how to train networks with short cuts [93]. The layout is illustrated schematically in Figure 7.12. Black arrows stand for usual feed-forward

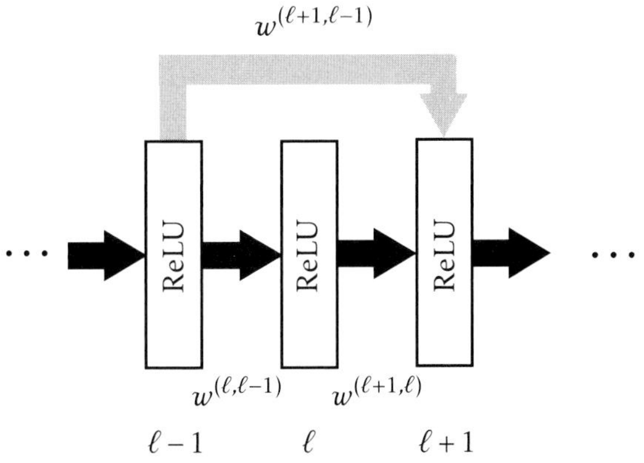

Figure 7.12 Schematic illustration of a network with a short cut that skips one layer (gray arrow). After Figure 1 from Ref. [93]

connections, and the gray arrow indicates a connection that skips one layer. The notation in Figure 7.12 differs somewhat from that of Algorithm 4. The weights from layer $\ell - 1$ to ℓ, for example, are denoted by $w_{jk}^{(\ell,\ell-1)}$, and those from layer $\ell - 1$ to $\ell + 1$ by $w_{ij}^{(\ell+1,\ell-1)}$ (gray arrow in Figure 7.12). Note that the superscripts are ordered in the same way as the subscripts: the *right* index refers to the layer on the *left*. According to Figure 7.12, neuron j in layer $\ell + 1$ computes

$$V_j^{(\ell+1)} = g\Big(\sum_k w_{jk}^{(\ell+1,\ell)} V_k^{(\ell)} + \sum_n w_{jn}^{(\ell+1,\ell-1)} V_n^{(\ell-1)} - \theta_j^{(\ell+1)}\Big). \tag{7.27}$$

As usual, the argument of the activation function is the local field $b_j^{(\ell+1)}$. The weights of all connections are trained in the usual fashion, by stochastic gradient descent.

To illustrate the structure of the resulting formulae, consider a chain of neurons, just one neuron per layer, with short cuts that skip one neuron (Figure 7.13). We calculate the weight increments using Equations (6.15) and (6.16). The recursion (6.17) applies only to standard feed-forward networks without skipping layers. In order to determine how to update the weights for the network shown in Figure 7.13, we need to evaluate the gradients $\partial V^{(L)}/\partial V^{(\ell)}$. To begin with, consider the learning rule for $w^{(L,L-1)}$. Using Equations (6.15) and (6.16), one finds

$$\delta w^{(L,L-1)} = \eta \delta^{(L)} V^{(L-1)} \quad \text{with} \quad \delta^{(L)} = (t - V^{(L)}) g'(b^{(L)}), \tag{7.28}$$

as in Algorithm 4. In the same way, one obtains

$$\delta w^{(L,L-2)} = \eta \delta^{(L)} V^{(L-2)} \quad \text{with} \quad \delta^{(L)} = (t - V^{(L)}) g'(b^{(L)}). \tag{7.29}$$

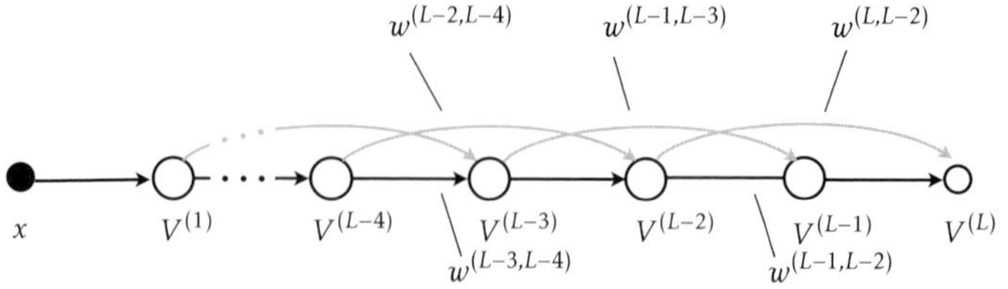

Figure 7.13 Chain of neurons with short cuts (gray arrows) that skip single neurons

Now consider the learning rule for $w^{(L-1,L-2)}$. Using $\partial V^{(L)}/\partial V^{(L-1)} = g'(b^{(L)}) w^{(L,L-1)}$ gives

$$\delta w^{(L-1,L-2)} = \eta \delta^{(L-1)} V^{(L-2)} \quad \text{with} \quad \delta^{(L-1)} = \delta^{(L)} w^{(L,L-1)} g'(b^{(L-1)}), \tag{7.30}$$

as before. But the update for $w^{(L-2,L-3)}$ is different because now the short cuts come into play. The connection from layer $L-2$ to L gives rise to an extra term:

$$\frac{\partial V^{(L)}}{\partial V^{(L-2)}} = \frac{\partial V^{(L)}}{\partial V^{(L-1)}} \frac{\partial V^{(L-1)}}{\partial V^{(L-2)}} + g'(b^{(L)}) w^{(L,L-2)} . \tag{7.31}$$

Evaluating the partial derivatives yields:

$$\begin{aligned} \delta w^{(L-2,L-3)} = \eta \delta^{(L-2)} V^{(L-3)} \quad \text{with} \quad \delta^{(L-2)} &= \delta^{(L-1)} w^{(L-1,L-2)} g'(b^{(L-2})\\ &+ \delta^{(L)} w^{(L,L-2)} g'(b^{(L-2)}) . \end{aligned} \tag{7.32}$$

Iterating further in this way, one finds the following error-backpropagation rule

$$\delta^{(\ell-1)} = \delta^{(\ell)} w^{(\ell,\ell-1)} g'(b^{(\ell-1)}) + \delta^{(\ell+1)} w^{(\ell+1,\ell-1)} g'(b^{(\ell-1)}) \tag{7.33}$$

with $w^{(\ell+1,\ell-1)} = 0$ for $\ell \geq L-1$. The first term is the same as in Algorithm 4. The second term is due to the skipping connections. These connections reduce the vanishing-gradient problem. To see this, note that we can write the error $\delta^{(\ell)}$ as

$$\delta^{(\ell)} = \delta^{(L)} \sum_{\ell_1,\ell_2,\ldots,\ell_n} w^{(L,\ell_n)} g'(b^{(\ell_n)}) \cdots w^{(\ell_2,\ell_1)} g'(b^{(\ell_1)}) w^{(\ell_1,\ell)} g'(b^{(\ell)}), \tag{7.34}$$

where the sum is over all paths $L > \ell_n > \ell_{n-1} > \cdots > \ell_1 > \ell$ back through the network. The structure of the general formula, for networks with more than only one neuron per layer, is analogous to Equation (7.34). According to this equation, the smallest errors, or gradients, in networks with many layers are dominated by the product corresponding to the path with the smallest number of steps (factors). Therefore short cuts tend to increase the small gradients.

Finally, the network described in Ref. [92] used unit weights for the skipping connections. In this case, the local field of $V_j^{(\ell+1)}$ takes the form

$$b_j^{(\ell+1)} = \sum_k w_{jk}^{(\ell+1,\ell)} V_k^{(\ell)} - \theta_j^{(\ell+1)} + V_j^{(\ell-1)} \equiv F + V_j^{(\ell-1)} , \tag{7.35}$$

assuming that the hidden layers have the same number of neurons. Here F is a residual contribution to the local field (when $F = 0$, the inputs $V_j^{(\ell-1)}$ are passed right through to $b_j^{(\ell+1)}$). Therefore such networks are called *residual networks* [92]. But note that the networks described in Ref. [92] use convolution layers (Section 8.1).

7.5 Outputs and Energy Functions

Up until now, we discussed networks that have the same activation functions for all neurons in all layers, either sigmoid or tanh activation functions [Equation (6.19)], or ReLU functions (Sections 1.3 and 7.3). In the output layer, one often uses neurons with a different activation function, so-called *softmax* outputs:

$$O_i = \frac{e^{\alpha b_i^{(L)}}}{\sum_{k=1}^{M} e^{\alpha b_k^{(L)}}} . \tag{7.36}$$

Here $b_i^{(L)} = \sum_j w_{ij}^{(L)} V_j^{(L-1)} - \theta_i^{(L)}$ are the local fields in the output layer. In the limit $\alpha \to \infty$, we see that $O_i = \delta_{ii_0}$ where i_0 is the index of the winning output neuron, the one with the largest value $b_i^{(L)}$ (Chapter 10). For $\alpha = 1$, Equation (7.36) is a *soft* version of this maximum criterion, thus the name *softmax*. We set α to unity from now on.

Two important properties of softmax outputs are, first, that $0 \leq O_i \leq 1$. Second, the values of the outputs sum to unity,

$$\sum_{i=1}^{M} O_i = 1 . \tag{7.37}$$

Therefore the outputs of softmax units can be interpreted as probabilities. Consider classification problems where the inputs must be assigned to one of M classes. In this case, the output $O_i^{(\mu)}$ of softmax unit i is assumed to represent the probability that the input $\boldsymbol{x}^{(\mu)}$ is in class i (in terms of the targets: $t_i^{(\mu)} = 1$ while $t_k^{(\mu)} = 0$ for $k \neq i$). If $O_i^{(\mu)} \approx 1$, we assume that the network is quite certain that input $\boldsymbol{x}^{(\mu)}$ is in class i. On the other hand, if all $O_k^{(\mu)} \approx M^{-1}$, we interpret the network output as uncertain. But note that neural networks may fail like humans sometimes do: their output can be very certain yet wrong (Section 8.6).

Softmax units are used in conjunction with a different energy function,

$$H = -\sum_{i\mu} t_i^{(\mu)} \log O_i^{(\mu)} . \tag{7.38}$$

Here and in the following, *log* stands for the *natural logarithm*. The function (7.38) is minimal when $O_i^{(\mu)} = t_i^{(\mu)}$ (Exercise 7.5). To find the correct backpropagation formula for the energy function (7.38), we need to evaluate

$$\frac{\partial H}{\partial w_{mn}} = -\sum_{i\mu} \frac{t_i^{(\mu)}}{O_i^{(\mu)}} \frac{\partial O_i^{(\mu)}}{\partial w_{mn}} . \tag{7.39}$$

Here the labels denoting the output layer were omitted, and in the following equations the index μ that refers to the input pattern is dropped as well. Using the identities

$$\frac{\partial O_i}{\partial b_l} = O_i(\delta_{il} - O_l) \quad \text{and} \quad \frac{\partial b_l}{\partial w_{mn}} = \delta_{lm} V_n , \tag{7.40}$$

one obtains

$$\frac{\partial O_i}{\partial w_{mn}} = \sum_l \frac{\partial O_i}{\partial b_l} \frac{\partial b_l}{\partial w_{mn}} = O_i(\delta_{im} - O_m) V_n . \tag{7.41}$$

So

$$\delta w_{mn} = -\eta \frac{\partial H}{\partial w_{mn}} = \eta \sum_{i\mu} t_i^{(\mu)} (\delta_{im} - O_m^{(\mu)}) V_n^{(\mu)} = \eta \sum_{\mu} (t_m^{(\mu)} - O_m^{(\mu)}) V_n^{(\mu)} , \tag{7.42}$$

since $\sum_{i=1}^{M} t_i^{(\mu)} = 1$ for the type of classification problem where each input belongs to precisely one class. The corresponding learning rule for the thresholds reads

$$\delta\theta_m = -\eta \frac{\partial H}{\partial \theta_m} = -\eta \sum_{\mu} (t_m^{(\mu)} - O_m^{(\mu)}) . \tag{7.43}$$

Equations (7.42) and (7.43) highlight a further advantage of softmax output neurons (apart from the fact that they allow the output to be interpreted in terms of probabilities). The weight and threshold increments for the output layer derived in Chapter 6 [Equations (6.6a) and (6.11a)] contain factors of derivatives $g'(B_m^{(\mu)})$. As noted earlier, these derivatives tend to zero when the activation function saturates, slowing down the learning. But here the rate at which the neuron learns is simply proportional to the output error, $(t_m^{(\mu)} - O_m^{(\mu)})$, without any possibly small factor $g'(b)$. Softmax units are normally only used in the output layer, because the learning speedup is coupled to the use of the energy function (7.38), and because it is customary to avoid dependencies between the neurons within a hidden layer.

There is an alternative form of the energy function that is very similar to the above, but works with sigmoid activation functions and 0/1 targets. Instead of Equation (7.38), one chooses

$$H = -\sum_{i\mu} \Big[t_i^{(\mu)} \log O_i^{(\mu)} + (1 - t_i^{(\mu)}) \log(1 - O_i^{(\mu)}) \Big] , \tag{7.44}$$

with $O_i = \sigma(b_i)$, $i = 1, \ldots, M$, and where σ denotes the sigmoid function (6.19a). To compute the weight increments, we apply the chain rule

$$\frac{\partial H}{\partial w_{mn}} = -\sum_{i\mu}\Big(\frac{t_i^{(\mu)}}{O_i^{(\mu)}} - \frac{1-t_i^{(\mu)}}{1-O_i^{(\mu)}}\Big)\frac{\partial O_l}{\partial w_{mn}} = -\sum_{i\mu}\frac{t_i^{(\mu)} - O_i^{(\mu)}}{O_i^{(\mu)}(1-O_i^{(\mu)})}\frac{\partial O_l}{\partial w_{mn}}. \tag{7.45}$$

Using Equation (6.20), we obtain

$$\delta w_{mn} = \eta \sum_{\mu}(t_m^{(\mu)} - O_m^{(\mu)})V_n^{(\mu)}, \tag{7.46}$$

identical to Equation (7.42). The thresholds are adjusted in an analogous fashion, Equation (7.43). But now the interpretation of the outputs is slightly different, since the values of the softmax units in the output layers sum to unity, while those with sigmoid activation functions do not. In either case, one can use the definition (6.30) for the classification error.

To conclude this section, we briefly discuss the meaning of the energy functions (7.38) and (7.44). In Section 7.1, we saw how training a neural network corresponds to fitting a target function. For a quadratic energy function, this is reminiscent of regression analysis in mathematical statistics, where the predictive accuracy of a model is improved by minimising the sum over the squared errors. Now consider the energy function (7.44) for a single sigmoid output with targets $t = 0$ and $t = 1$. In this case, the network output is interpreted as the probability $O^{(\mu)} = \text{Prob}(t^{(\mu)} = 1|\boldsymbol{x}^{(\mu)})$ of observing $t^{(\mu)} = 1$. The corresponding likelihood is the joint probability of observing the outcomes $t^{(\mu)}$ for p independent inputs $\boldsymbol{x}^{(\mu)}$:

$$\mathscr{L} = \prod_{\mu=1}^{p}\big(O^{(\mu)}\big)^{t^{(\mu)}}\big(1 - O^{(\mu)}\big)^{1-t^{(\mu)}}, \tag{7.47}$$

under the model determined by the weights and thresholds of the network. Minimising the negative log-likelihood $-\mathscr{L}$ (Section 4.4) corresponds to minimising (7.44). This is just binary logistic regression [94] to predict a binary outcome $t = 0$ or 1. The case $M > 1$ corresponds to a multivariate regression problem [94], with M possibly correlated outcome variables $t_1, \ldots, t_M$.

When the targets describe M mutually exclusive categorical outcomes, $t_i = 0, 1$ with $\sum_{i=1}^{M} t_i = 1$, the softmax output O_i is interpreted as the probability of observing $t_i = 1$. An example is the problem of classifying handwritten digits (Section 8.3). Training the network then corresponds to multinomial regression [94] with the log-likelihood (7.38). Note that Equation (7.44), for $M = 1$, is equivalent to (7.38) for $M = 2$, because $O_2 = 1 - O_1$ and $t_2 = 1 - t_1$. At any rate, these remarks motivate why the energy functions (7.38) and (7.44) are sometimes called log-likelihoods. Equation (7.44) is also referred to as *cross entropy*, because it has the same form as the cross entropy [65] characterising the difference between two Bernoulli distributions: the network output O_i, and the target t_i.

7.6 Regularisation

Deeper networks have more neurons, so the problem of overfitting (Figure 6.6) tends to be more severe for deeper networks. *Regularisation* schemes limit the tendency to overfit. Apart from cross validation (Section 6.4), a number of other regularisation schemes have proved useful for deep networks: *weight decay*, *pruning*, *drop out*, *expansion of the training set*, and *batch normalisation*. This section summarises the most important aspects of these methods.

7.6.1 Weight Decay

Recall Figure 5.17 which shows a solution of the classification problem defined by the Boolean XOR function. In the solution illustrated in this Figure, all weights equal ± 1, and also the thresholds are of order unity. If one uses the backpropagation algorithm to find a solution to this problem, one may find that the weights continue to grow during training. As mentioned above, this can be problematic because it may imply that the local fields become so large that the activation functions saturate. Then training slows down, as explained in Section 7.2.

To prevent the weights from growing, one can reduce them by some factor during training: $w_{ij} \to (1-\varepsilon)w_{ij}$ for $0 < \varepsilon < 1$. This corresponds to adding weight increments of the form

$$\delta w_{mn} = -\varepsilon w_{mn} \quad \text{for} \quad 0 < \varepsilon < 1\,. \tag{7.48}$$

This is achieved by adding a term to the energy function:

$$H = \underbrace{\frac{1}{2}\sum_{i\mu}\Big(t_i^{(\mu)} - O_i^{(\mu)}\Big)^2}_{\equiv H_0} + \frac{\gamma}{2}\sum_{ij} w_{ij}^2\,. \tag{7.49}$$

Gradient descent on H gives:

$$\delta w_{mn} = -\eta \frac{\partial H_0}{\partial w_{mn}} - \varepsilon w_{mn} \tag{7.50}$$

with $\varepsilon = \eta\gamma$. One can include a corresponding term for the thresholds. The scheme summarised here is sometimes called L_2-*regularisation*. An alternative scheme is L_1-*regularisation*. It amounts to

$$H = \frac{1}{2}\sum_{i\mu}\Big(t_i^{(\mu)} - O_i^{(\mu)}\Big)^2 + \frac{\gamma}{2}\sum_{ij} |w_{ij}|\,. \tag{7.51}$$

This gives the learning rule

$$\delta w_{mn} = -\eta \frac{\partial H_0}{\partial w_{mn}} - \varepsilon \operatorname{sgn}(w_{mn}) . \tag{7.52}$$

The discontinuity of the learning rule at $w_{mn} = 0$ is cured by defining $\operatorname{sgn}(0) = 0$. Comparing Equations (7.50) and (7.52), we see that L_1-regularisation puts more weights to zero, compared with the L_2-scheme [5].

An alternative to these two methods is *max-norm regularisation* [95], where the weights are constrained to remain smaller than a given constant: $|w_{ij}| \leq c$. If a $|w_{ij}|$ exceeds the positive constant c, then w_{ij} is rescaled so that $|w_{ij}| = c$.

These weight-decay schemes are referred to as *regularisation* schemes because they tend to help against overfitting. How does this work? Weight decay adds a constraint to the problem of minimising the energy function. The result is a compromise [5] between a small value of H and small weight values. The idea is that a network with smaller weights is more robust to the effect of noise. When the weights are small, then small changes in some of the patterns do not give a substantially different training result. When the network has large weights, by contrast, it may happen that small changes in the input yield significant differences in the training result that are difficult to generalise.

7.6.2 Pruning

The term *pruning* refers to removing unnecessary weights or neurons from the network, to improve its efficiency. The simplest approach is *weight elimination* by *weight decay* [96]. Weights that tend to remain very close to zero during training are removed by setting them to zero and not updating them anymore. Neurons that have zero weights for all incoming connections are effectively removed (*pruned*). Pruning is a regularisation method: by removing unnecessary weights, one reduces the risk of overfitting. As opposed to drop out (Section 7.6.3), where hidden neurons are only temporarily ignored, pruning refers to permanently removing hidden neurons. The idea is to train a large network, and then to prune a large fraction of neurons to obtain a much smaller network. It is usually found that such pruned networks generalise better than small networks that were trained without pruning. Up to 90% of the hidden neurons can be removed in some cases. In general, pruning is an excellent way to create efficient classifiers for real-time applications.

An efficient pruning algorithm is based on the idea to remove weights in such a way that the effect upon the energy function is as small as possible [97]. The idea is to find the optimal weight, to remove it, and to change the other weights in such a way that the energy function increases as little as possible. The algorithm works as

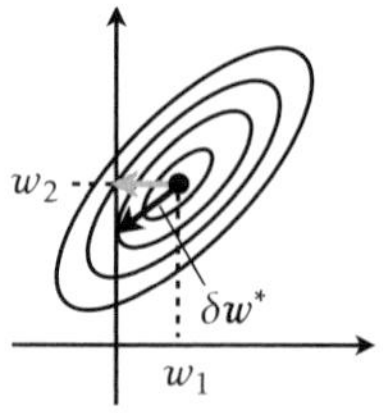

Figure 7.14 Pruning algorithm (schematic). The minimum of H is located at $[w_1, w_2]^{\mathsf{T}}$. The contours of the quadratic approximation to H are represented as solid black lines. The weight change $\delta \boldsymbol{w} = [-w_1, 0]^{\mathsf{T}}$ (gray arrow) leads to a smaller increase in H than $\delta \boldsymbol{w} = [0, -w_2]^{\mathsf{T}}$. The black arrow represents the optimal $\delta \boldsymbol{w}_q^*$, resulting in an even smaller increase in H

follows. Assume that the network was trained, so that it reached a (local) minimum of the energy function H. One expands the energy function around this minimum. To second order, the expansion of H reads:

$$H = H_{\min} + \tfrac{1}{2}\delta \boldsymbol{w} \cdot \mathbb{M}\delta \boldsymbol{w} + \text{higher orders in } \delta \boldsymbol{w}. \tag{7.53}$$

The term linear in $\delta \boldsymbol{w}$ vanishes because we expand around a local minimum. The matrix $\mathbb{M}$ is the *Hessian*, the matrix of second derivatives of the energy function.

For the next step, it is convenient to adopt the following notation [97]. One groups all weights in the network into a long weight vector $\boldsymbol{w}$ (as opposed to grouping them into a weight matrix $\mathbb{W}$, as we did in Chapter 2). A particular component w_q is extracted from the vector $\boldsymbol{w}$ as follows:

$$w_q = \hat{\mathbf{e}}_q \cdot \boldsymbol{w} \quad \text{where} \quad \hat{\mathbf{e}}_q = \begin{bmatrix} \vdots \\ 1 \\ \vdots \end{bmatrix} \leftarrow q\,. \tag{7.54}$$

Here $\hat{\mathbf{e}}_q$ is the Cartesian unit vector in the direction q, with components $[\hat{\mathbf{e}}_q]_j = \delta_{qj}$. In this notation, the elements of $\mathbb{M}$ are $M_{pq} = \partial^2 H / \partial w_p \partial w_q$. Now eliminating the weight w_q amounts to setting

$$\delta w_q = -w_q\,. \tag{7.55}$$

To minimise the damage to the network, we should eliminate the weight that has the least effect upon H, changing the other weights at the same time so that H increases as little as possible (Figure 7.14). This is achieved by minimising

$$\min_q \min_{\delta \boldsymbol{w}} \{\tfrac{1}{2}\delta \boldsymbol{w} \cdot \mathbb{M}\delta \boldsymbol{w}\} \quad \text{subject to the constraint} \quad \hat{\mathbf{e}}_q \cdot \delta \boldsymbol{w} + w_q = 0\,. \tag{7.56}$$

The constant term $H_{\min}$ was dropped because it does not matter. Now we first minimise H w.r.t. $\delta \boldsymbol{w}$, for a given value of q. The linear constraint is incorporated using a *Lagrange multiplier* as in Section 6.3, to form the *Lagrangian*

$$\mathscr{L} = \tfrac{1}{2}\delta\boldsymbol{w} \cdot \mathbb{M}\delta\boldsymbol{w} + \lambda(\hat{\mathbf{e}}_q \cdot \delta\boldsymbol{w} + w_q)\,. \tag{7.57}$$

A necessary condition for a minimum $[\delta\boldsymbol{w}, \lambda]$ satisfying the constraint is

$$\frac{\partial\mathscr{L}}{\partial\delta\boldsymbol{w}} = \mathbb{M}\delta\boldsymbol{w} + \lambda\hat{\mathbf{e}}_q = 0 \quad \text{and} \quad \frac{\partial\mathscr{L}}{\partial\lambda} = \hat{\mathbf{e}}_q \cdot \delta\boldsymbol{w} + w_q = 0\,. \tag{7.58}$$

We denote the solution of these equations by $\delta\boldsymbol{w}^*$ and λ^*. It is obtained by solving the linear system

$$\begin{bmatrix} \mathbb{M} & \hat{\mathbf{e}}_q \\ \hat{\mathbf{e}}_q^{\mathsf{T}} & 0 \end{bmatrix} \begin{bmatrix} \delta\boldsymbol{w}^* \\ \lambda^* \end{bmatrix} = \begin{bmatrix} 0 \\ -w_q \end{bmatrix}. \tag{7.59}$$

If $\mathbb{M}$ is invertible, then the top rows of Equation (7.59) give

$$\delta\boldsymbol{w}^* = -\mathbb{M}^{-1}\hat{\mathbf{e}}_q\lambda^*\,. \tag{7.60}$$

Inserting this result into $\hat{\mathbf{e}}_q^{\mathsf{T}}\delta\boldsymbol{w}^* + w_q = 0$, we find

$$\delta\boldsymbol{w}^* = -\mathbb{M}^{-1}\hat{\mathbf{e}}_q w_q(\hat{\mathbf{e}}_q^{\mathsf{T}}\mathbb{M}^{-1}\hat{\mathbf{e}}_q)^{-1} \text{ and } \lambda^* = w_q(\hat{\mathbf{e}}_q^{\mathsf{T}}\mathbb{M}^{-1}\hat{\mathbf{e}}_q)^{-1}\,. \tag{7.61}$$

We see that $\hat{\mathbf{e}}_q \cdot \delta\boldsymbol{w}^* = -w_q$, so that the weight w_q is eliminated. The other weights are also changed (black arrow in Figure 7.14). The final step is to find the optimal q by minimising

$$\mathscr{L}(\delta\boldsymbol{w}^*, \lambda^*; q) = \frac{1}{2}w_q^2(\hat{\mathbf{e}}_q^{\mathsf{T}}\mathbb{M}^{-1}\hat{\mathbf{e}}_q)^{-1}\,. \tag{7.62}$$

The Hessian of the energy function is expensive to evaluate, and so is the inverse of this matrix. Usually one resorts to an approximate expression for $\mathbb{M}^{-1}$ [97]. One possibility is to set the off-diagonal elements of $\mathbb{M}$ to zero [98]. But in this case, the other weights are not adjusted, because $\hat{\mathbf{e}}_{q'} \cdot \delta\boldsymbol{w}_q^* = 0$ for $q' \neq q$ if $\mathbb{M}$ is diagonal. Here it is necessary to retrain the network after weight elimination.

The algorithm is summarised in Algorithm 5. It succeeds better than elimination by weight decay in removing the unnecessary weights in the network [97]. Weight decay eliminates the smallest weights. One obtains weight elimination of the smallest weights by substituting $\mathbb{M} = \mathbb{I}$ in the algorithm described above [Equation (7.62)]. Since small weights are often needed to achieve a small training error, this is usually not a good approximation.

To illustrate the effect of pruning for neural networks with hidden layers, consider the XOR function. Recall that it can be represented by a hidden layer with two neurons (Figure 5.17). For random initial weights, backpropagation takes a long time to find a valid solution, and networks with many more hidden neurons tend to perform much better [99]. The numerical experiments in Ref. [99] indicate that with two hidden neurons, only about 49% of the networks learned the task in 10,000 training steps of stochastic gradient descent, and networks with more neurons in the hidden layer learn more easily (98.5 % for $n = 10$ hidden neurons).

Algorithm 5 Pruning least important weight

train the network to reach $H_{\min}$;
compute $\mathbb{M}^{-1}$ approximately;
determine q^* as the value of q for which $\mathscr{L}(\delta \boldsymbol{w}^*, \lambda^*; q)$ is minimal;
if $\mathscr{L}(\delta \boldsymbol{w}^*, \lambda^*; q^*) \ll H_{\min}$ **then**
  adjust all weights using $\delta \boldsymbol{w} = -w_{q^*}\mathbb{M}^{-1}\hat{\mathbf{e}}_{q^*}(\hat{\mathbf{e}}_{q^*}^{\mathsf{T}}\mathbb{M}^{-1}\hat{\mathbf{e}}_{q^*})^{-1}$;
  goto 2;
else
  end;
end if

The data in Ref. [98] also show that pruned networks, initially trained with $n = 10$ hidden neurons, still show excellent training success (83.3 % if only $n = 2$ hidden neurons remain). The networks were pruned iteratively during training, removing the neurons with the largest average magnitude. After training, the weights and threshold were reset to their initial values, the values before training began.

One can draw three conclusions from the numerical experiments described in Ref. [99]. First, iterative pruning during training singles out neurons in the hidden layer that had initial weights and thresholds resulting in the correct decision boundaries (*lottery-ticket* effect [99]). Second, the pruned network with two hidden neurons has much better training success than the network that was trained with only two hidden neurons. Third, despite pruning more than 50% of the hidden neurons, the network with $n = 4$ hidden neurons performs almost as well as the one with $n = 10$ hidden neurons (97.9 % training success). When training deep networks, it is common to start with many neurons in the hidden layers and prune up to 90% of them. This results in small trained networks that can classify efficiently and reliably.

7.6.3 Drop Out

In this regularisation scheme, some hidden neurons are ignored during training [95]. In each step of the training algorithm (for each mini batch or for each individual pattern), one disregards at random a fraction q of neurons from each hidden layer by setting their outputs to zero, and by updating only the weights and thresholds of the remaining neurons. For the next step in the training algorithm, the ignored neurons are put back, and another set of hidden neurons is disregarded. Once the training is completed, all hidden neurons are activated. Their outputs are multiplied by $1 - q$, to ensure that the local fields are independent of q, on average.

This method is motivated by noting that the performance of machine-learning algorithms is usually improved by combining the results of several learning attempts [5, 95], for instance by separately training several networks with different layouts and averaging over their outputs. For deep networks, this is computationally very expensive. Drop out is an attempt to achieve the same goal more efficiently. The idea is that dropout corresponds to effectively training a large number of different networks. If there are k hidden neurons, then there are 2^k different combinations of neurons that are turned on or off. The hope is that the network learns more robust features of the input data in this way and that this reduces overfitting. In practice, the method is applied together with max-norm regularisation (Section 7.6.1).

7.6.4 Expanding the Training Set

If one trains a network with a fixed number of hidden neurons on larger training sets, one observes that the network generalises with higher accuracy (better classification success). The reason is that overfitting is reduced when the training set is larger. Thus, a way of avoiding overfitting is to *expand* or *augment* the training set. It is sometimes argued that the recent success of deep neural networks in image recognition and object recognition is in large part due to larger training sets. One example is ImageNet, a database of more than 10^7 hand-classified images, into more than 20,000 categories [100]. It is expensive to improve training sets in this way because it requires manual annotation. An alternative is to expand a training set *artificially*. For digit recognition (Figure 2.1), for example, one could create more input patterns by shifting, rotating, and shearing the digits, or by adding noise.

7.6.5 Batch Normalisation

Batch normalisation [101] can significantly speed up the training of deep networks with backpropagation. The idea is to shift and normalise the input data for each hidden layer, not only for the input patterns (Section 6.3). This is done separately for each mini batch, and for each component of the inputs into the given layer (Algorithm 6). Denoting the states of the neurons feeding into the layer in question by $V_j^{(\mu)}$, $j = 1, \ldots, j_{\max}$, one calculates the average and variance over each mini batch

$$\overline{V}_j = \frac{1}{m_B} \sum_{\mu=1}^{m_B} V_j^{(\mu)} \quad \text{and} \quad \sigma_B^2 = \frac{1}{m_B} \sum_{\mu=1}^{m_B} (V_j^{(\mu)} - \overline{V}_j)^2 , \tag{7.63}$$

subtracts the mean from the $V_j^{(\mu)}$, and divides by $\sqrt{\sigma_B^2 + \epsilon}$. The parameter $\epsilon > 0$ is added to the denominator to avoid division by zero when σ_B^2 evaluates to zero.

Algorithm 6 Batch normalisation

for $j = 1, \ldots, j_{\max}$ **do**
  calculate mean $\overline{V}_j \leftarrow \frac{1}{m_B}\sum_{\mu=1}^{m_B} V_j^{(\mu)}$
  calculate variance $\sigma_B^2 \leftarrow \frac{1}{m_B}\sum_{\mu=1}^{m_B}(V_j^{(\mu)} - \overline{V}_j)^2$
  normalise $\hat{V}_j^{(\mu)} \leftarrow (V_j^{(\mu)} - \overline{V}_j)/\sqrt{\sigma_B^2 + \epsilon}$
  calculate outputs as: $g(\gamma_j \hat{V}_j^{(\mu)} + \beta_j)$
end for

There are two additional parameters in Algorithm 6, γ_j and β_j. They are learnt by backpropagation, just like the weights and thresholds. In general, the new parameters are allowed to differ from layer to layer, $\gamma_j^{(\ell)}$ and $\beta_j^{(\ell)}$.

Batch normalisation was originally motivated by arguing that it reduces possible covariate shifts faced by hidden neurons in layer ℓ: as the parameters of the neurons in the preceding layer $\ell - 1$ change, their outputs shift thus forcing the neurons in layer ℓ to adapt. However, Ref. [102] argues that batch normalisation does not reduce the internal covariate shift but speeds up the training by effectively smoothing the energy landscape.

Batch normalisation helps combat the *vanishing-gradient problem* because it prevents local fields of hidden neurons to grow. This makes it possible to use sigmoid functions in deep networks, because the distribution of inputs remains normalised.

It is sometimes argued that batch normalisation has a regularising effect, and it has been suggested [101] that batch normalisation can replace drop out (Section 7.6.3). It is also argued that batch normalisation may help the network to generalise better, in particular if each mini batch contains randomly picked inputs. Then batch normalisation corresponds to randomly transforming the inputs to each hidden neuron (by the randomly changing means and variances). This may help to make the learning more robust. There is no theory that proves either of these claims, but it is an empirical fact that batch normalisation often speeds up the training.

7.7 Summary

Neural networks with many layers of hidden neurons are called deep networks. Error backpropagation in deep networks suffers from the vanishing-gradient problem. It can be reduced by using ReLU units, by initialising the weights in certain ways, and with networks containing connections that skip layers. Yet vanishing or exploding gradients remain a fundamental difficulty, slowing learning down in the initial phase of training. Nevertheless, deep neural networks have become immensely successful in object recognition, outperforming other algorithms significantly.

Since deep networks contain many free parameters, deep networks tend to overfit the training data, so that they must be regularised. Apart from cross validation, there are other ways of regularising the problem: weight decay, drop out, pruning, and data-set augmentation.

7.8 Further Reading

Deep networks suffer from *catastrophic forgetting*: when we train a network on a new input distribution that is quite different from the one the network was originally trained on, then the network tends to forget what it learned initially. A good starting point for further reading is Ref. [103].

The stochastic gradient-descent algorithm (with or without minibatches) samples the input-data distribution uniformly randomly. As mentioned in Section 6.3, it may be advantageous to sample those inputs more frequently that initially cause larger output errors. More generally, the algorithm may use other criteria to choose certain input data more often, with the goal to speed up learning. It may even suggest how to augment a given training set most efficiently, by asking to specifically label certain types of input data (*active learning*) [104].

Another question concerns the structure of the energy landscape for multilayer perceptrons. It seems that local minima are perhaps less important for deep networks than for Hopfield networks, because the energy functions of deep networks tend to have more saddle points than minima [105], just like Gaussian random functions [106]. A recent study explores the relation between the multilayer layout of the perceptron network and the properties of the energy landscape [107].

Finally, training tends to work best when all input patterns appear with roughly the same frequency in the training set. Unlike humans, neural networks tend to struggle with rare input patterns. Special techniques, however, allow networks to recognise rare patterns, by comparison with features of the input distribution that are well represented (*few-shot* learning) [108]. Standard algorithms for few-shot learning use elements of unsupervised learning (Chapter 10).

7.9 Exercises

7.1 Pruning. Show that the expression (7.61) for the weight increment $\delta \boldsymbol{w}^*$ minimises the Lagrangian (7.57) subject to the constraint (7.55).

7.2 Decision boundaries for XOR problem. Figure 7.5 shows the layout of a network that solves the Boolean XOR problem. Draw the decision boundaries for the four hidden neurons in the input plane, and label the boundaries and the regions as in Figure 5.15.

7.3 Vanishing-gradient problem. Train the network shown in Figure 7.8 on the iris data set, available from the machine-learning repository of the University of California Irvine [70]. Measure the effects upon the neurons in the different layers by numerically evaluating the derivative of the energy function H w.r.t. their thresholds.

7.4 Residual network. Derive Equation (7.34) for the error $\delta^{(\ell)}$ in layer ℓ of the chain of neurons shown in Figure 7.13.

7.5 Log-likelihood. The log-likelihood function (7.38) is an energy function for softmax output neurons. Show that the function has a global minimum at $\boldsymbol{O}^{(\mu)} = \boldsymbol{t}^{(\mu)}$.

7.6 Cross-entropy function. The cross-entropy function (7.44) is an energy function for sigmoid output neurons. Write down a cross-entropy function for tanh-output neurons, and show that it has a global minimum at $\boldsymbol{O}^{(\mu)} = \boldsymbol{t}^{(\mu)}$, where the function takes the value zero.

7.7 Softmax outputs. Consider a network with L layers with softmax outputs $O_i^{(\mu)}$. Compute the derivative of $O_i^{(\mu)}$ with respect to the local field $b_m^{(L,\mu)}$ of output neuron m. The network is trained by gradient descent on the negative log-likelihood function $H = -\sum_{i\mu} t_i^{(\mu)} \log\big(O_i^{(\mu)}\big)$. The targets $t_i^{(\mu)}$ satisfy the constraint $\sum_i t_i^{(\mu)} = 1$, for all patterns μ. Derive the stochastic gradient-descent learning rule for the weights $w_{mn}^{(L)}$ in layer L.

7.8 Generalised XOR function. The parity function can be viewed as a generalisation of the XOR function to $N > 2$ input dimensions, because it becomes the XOR function for $N = 2$. Another way to generalise the XOR function to $N > 2$-dimensional inputs is to define a Boolean function that gives unity if exactly one of its inputs equals unity. Otherwise, the function evaluates to zero. Construct networks that represent this function, for $N = 3$ and $N = 4$.

8
Convolutional Networks

Convolutional networks have been around since the 1980s. They became widely used after Krizhevsky et al. [109] won the ImageNet challenge (Section 8.5) with a convolutional net. One reason for the recent success of convolutional networks is that they have fewer connections than fully connected networks with the same number of neurons. This has two advantages. Firstly, such networks are obviously cheaper to train. Secondly, as pointed out above, reducing the number of connections regularises the network, it reduces the risk of overfitting.

Convolutional neural networks are designed for object recognition and image classification. They take images as inputs (Figure 8.1), not just a list of attributes (Figure 5.1). Convolutional networks have important properties in common with networks of neurons in the visual cortex of the human brain [4]. First, there is a spatial array of input terminals. For image analysis, this is the two-dimensional array of bits shown in Figure 8.2(**a**). Second, neurons are designed to detect local features of the image (such as edges or corners for instance). The maps learned by such neurons, from inputs to output, are referred to as *feature maps*. Since these features occur in different parts of the image, one uses the same *kernel* (or *filter*)

Figure 8.1 Images of iris flowers. From left to right: Iris setosa (copyright T. Monto), iris versicolor (copyright R. A. Nonenmacher), and iris virginica. Copyright A. Westermoreland. All images are copyrighted under the creative commons licence. Originals in colour

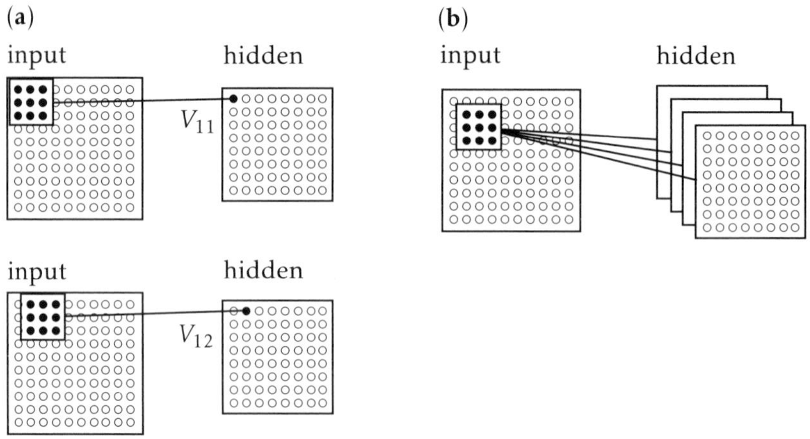

Figure 8.2 (**a**) Feature map, kernel, and receptive field (schematic). A feature map (the 8×8 array of hidden neurons) is obtained by translating a kernel (filter) with a 3×3 receptive field over the input image, a 10×10 array of pixels. (**b**) A convolution layer consists of a number of feature maps, each corresponding to a kernel that detects a certain feature in the input image. After a figure in Ref. [5]

for different parts of the image, always with the same weights and thresholds for different parts of the image. Since these kernels are local, and since they act in a translational-invariant way, the number of neurons from the two-dimensional input array is greatly reduced, compared with fully connected networks. Feature maps are obtained by convolution of the kernel with the input image. Therefore, layers consisting of a number of feature maps corresponding to different kernels are also referred to as *convolution layers*, Figure 8.2(**b**).

Convolutional networks can have many convolution layers. The idea is that the additional layers can learn more abstract features (Section 8.7). Apart from feature maps, convolutional networks contain other types of layers. *Pooling layers* perform local averages over the output of the convolution layers, to speed up learning by reducing the number of variables. Convolutional networks may also contain fully connected layers.

8.1 Convolution Layers

Figure 8.2(**a**) illustrates how a feature map is obtained by convolution of the input image with a kernel which reads a 3×3 part of the input image [5]. In analogy with the terminology used in neuroscience, this 3×3 array is called the *local receptive field* of the kernel. The outputs of the kernel from different parts of the input image make up the feature map, here an 8×8 array of hidden neurons: neuron V_{11} connects to the 3×3 area in the upper left-hand corner of the input image. Neuron V_{12} connects to a shifted area, as illustrated in Figure 8.2(**a**), and so forth. Since the input has 10×10 pixels, the dimension of the feature map is 8×8 in this example. The important point is that the neurons V_{11} and V_{12}, and all other neurons in this convolution layer, share their weights and the threshold. In the example shown in Figure 8.2(**a**), there are thus only nine independent weights, and one threshold. Since the different hidden neurons share weights and thresholds, their computation rule is a discrete *convolution* [4]:

$$V_{ij} = g\Big(\sum_{p=1}^{3}\sum_{q=1}^{3} w_{pq} x_{p+i-1,q+j-1} - \theta\Big). \tag{8.1}$$

In Figure 8.2(**a**), the local receptive field is shifted by one pixel at a time. Sometimes it is useful to use a larger *stride* $[s_1, s_2]$, to shift the receptive field by s_1 pixels horizontally and by s_2 pixels vertically. Also, the local receptive regions need not have size 3×3. If we assume that their size is $Q \times P$, and that $s_1 = s_2 = s$, the rule (8.1) reads instead

$$V_{ij} = g\Big(\sum_{p=1}^{P}\sum_{q=1}^{Q} w_{pq} x_{p+s(i-1),q+s(j-1)} - \theta\Big). \tag{8.2}$$

Figure 8.2(**a**) depicts a two-dimensional input array. For colour images, there are three colour *channels*; in this case, the input array is three-dimensional and the input bits are labelled by three indices: two for position and the last one for colour, x_{pqr}. Usually one connects several feature maps with different kernels to the input layer, as shown in Figure 8.2(**b**). The different kernels detect different features of the input image, one detects edges for example, and another one detects corners, and so forth. To account for these extra dimensions, one groups weights (and thresholds) into higher-dimensional arrays (*tensors*). The convolution takes the form

$$V_{ijk} = g\Big(\sum_{p=1}^{P}\sum_{q=1}^{Q}\sum_{r=1}^{R} w_{pqrk} x_{p+s(i-1),q+s(j-1),r} - \theta_k\Big) \tag{8.3}$$

(see Figure 8.3). All neurons in a given convolution layer have the same threshold. The software package *TensorFlow* [110] is designed to efficiently perform tensor operations as in Equation (8.3).

If one couples several convolution layers together, the number of neurons in these layers decreases as one moves to the right. To avoid this, one can *pad* the image (and the convolution layers) by adding rows and columns of bits equal to zero [4]. In Figure 8.2(**a**), for example, one obtains a convolution layer of the same dimension as the original image by adding one column each on the left-hand and

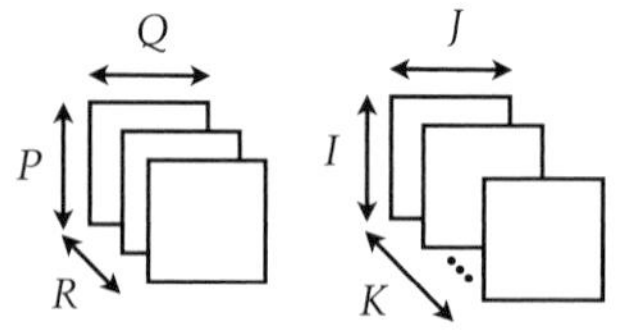

Figure 8.3 Illustration of summation in Equation (8.3). Each feature map has a receptive field of dimension $P \times Q \times R$. There are K feature maps, each of dimension $I \times J$

right-hand sides of the image, as well as two rows, one at the bottom and one at the top. In general, the numbers of rows and columns need not be equal, so the amount of padding is specified by four numbers $[p_1, p_2, p_3, p_4]$.

Convolution layers are trained with backpropagation. Consider the simplest case, Equation (8.1). As usual, we use the chain rule to evaluate the gradients

$$\frac{\partial V_{ij}}{\partial w_{mn}} = g'(b_{ij}) \frac{\partial b_{ij}}{\partial w_{mn}} \tag{8.4}$$

with local field $b_{ij} = \sum_{pq} w_{pq} x_{p+i-1,q+j-1} - \theta$. The derivative of b_{ij} is evaluated by applying rule (5.25):

$$\frac{\partial b_{ij}}{\partial w_{mn}} = \sum_{pq} \delta_{mp} \delta_{nq} x_{p+i-1,q+j-1} . \tag{8.5}$$

In this way, one can train networks with several stacked convolution layers too. It is important to keep track of the summation boundaries. To that end, it helps to pad out the image and the convolution layers, so that the upper bounds remain the same in different layers.

Details aside, the fundamental principle of feature maps is that the map is applied in the same form to different parts of the image (*translational invariance*). In this way, each weight in a given feature map is trained on different parts of the image. This effectively increases the training set for the feature map and combats overfitting.

8.2 Pooling Layers

Pooling layers process the output of convolution layers. A neuron in a pooling layer takes the outputs of several neighbouring feature maps and compresses their outputs into a single number [5]. There are no weights or thresholds associated with

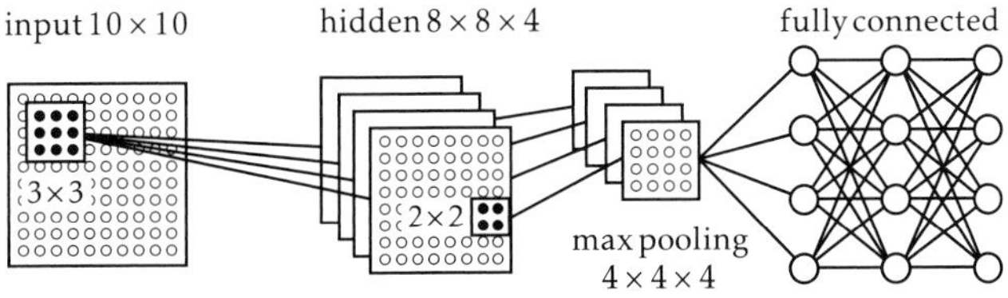

Figure 8.4 Layout of a convolutional neural network for object recognition and image classification (schematic). The inputs are stored in a 10×10 array. They connect to a convolution layer with four different feature maps with 3×3 kernels, stride [1, 1], and zero padding. Each convolution layer connects to a 2×2 max-pooling layer with stride [2, 2] and zero padding. Between these and the output layer are two fully connected hidden layers. After a figure in Ref. [5].

pooling layers. *Max-pooling units*, for example, take the maximum over several nearby feature-map outputs. Instead, one may compute the root-mean square of the map values (*L_2-pooling*). Just as for convolution layers, we need to specify strid and padding. Other ways of pooling are discussed in Ref. [4].

Usually, several feature maps are connected to the input. Pooling is performed independently for each feature map [5]. The network layout looks like the one shown schematically in Figure 8.4. In this figure, the pooling layers connect to a number of fully connected hidden layers that connect to the output neurons. There are as many output neurons as there are classes to be recognised. This layout is similar to the layout used by Krizhevsky et al. [109] in the ImageNet challenge (see Section 8.5).

8.3 Learning to Read Handwritten Digits

Figure 8.5 shows patterns from the MNIST data set of handwritten digits [111]. The data set derives from a data set compiled by the National Institute of Standards and Technology (NIST), of digits handwritten by high-school students and employees of the United States Census Bureau. The data contains 60,000 images of digits, each with 28×28 pixels, and a *test set* of 10,000 digits. The images are grayscale with 8-bit resolution, so each pixel contains a value ranging from 0 to 255. The images in the database were preprocessed. The procedure is described on the MNIST home page (http://yann.lecun.com/exdb/mnist/). Each original binary image from the National Institute of Standards and Technology was represented as a 20×20 gray-scale image, preserving the aspect ratio of the digit. The resulting image was placed in a 28×28 image so that the centre-of-mass of the image coincided with its geometrical centre. These preprocessing steps improve the performance of the learning algorithm.

Figure 8.5 Examples of digits from the MNIST data set of handwritten digits [111] (http://yann.lecun.com/exdb/mnist/). The images were produced using MATLAB (https://se.mathworks.com/products/matlab.html). Copyright for the data set: Y. LeCun and C. Cortes

The goal of this section is to show how the principles introduced so far allow neural networks to learn the MNIST data with low classification error, following Ref. [5]. As described in Chapter 6, one divides the data set into a *training set* and a *validation set*, here with 50,000 digits and 10,000 digits, respectively [5]. The validation set is used for cross validation. The test data set allows us to measure the classification error after training. For this purpose, we must use a data set that was not involved in the training. As described in Section 6.3, the inputs are preprocessed further by subtracting the mean image averaged over the whole training set from each input image [Equation (6.21)].

To find good parameter values and network layouts is one of the main difficulties when training a neural network, and it usually requires experimenting. There are recipes for finding certain parameters [112], but the general approach is still trial and error [5]. Consider first a network with one hidden layer with ReLU activation functions (Section 7.3), and a softmax output layer (Section 7.5) with 10 outputs O_i and energy function (7.38). Output O_i is interpreted as the probability that the pattern fed to the network falls into category i. The network is trained with stochastic gradient descent with momentum, Equation (6.31). The learning rate is set to $\eta = 0.001$, and the momentum constant to $\alpha = 0.9$. The mini-batch size [Equation (6.18)] equals 8192. Cross validation and early stopping are implemented as follows: during training, the algorithm keeps track of the smallest validation error observed so far. Training stops when the validation error becomes larger than the minimum for a specified number of times, equal to 5 in this case.

Figure 8.6 shows how the training and the validation energies decrease during training, for networks with 30 and 100 hidden neurons. One epoch corresponds to applying p patterns or $p/m_{\mathrm{B}} = 50000/8192$ iterations [5] (Section 6.1). The energies are a little lower for the network with 100 hidden neurons. But one

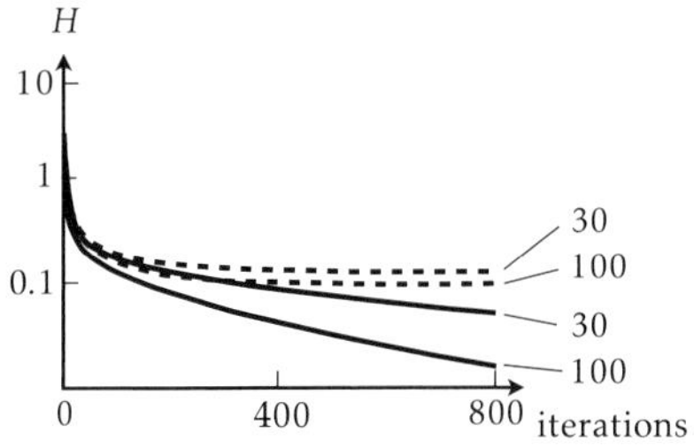

Figure 8.6 Energy functions for the MNIST training set (solid lines) and for the validation set (dashed lines) for a fully connected hidden layer with 30 neurons, and for a similar algorithm, but with 100 neurons in the hidden layer. The data were smoothed and the plot is schematic. The x-axis shows iterations. One iteration corresponds to feeding one minibatch of patterns. One epoch consists of $50000/8192 \approx 6$ iterations. Based on simulations performed by Oleksandr Balabanov

observes overfitting in both cases: after many training steps the validation energy is much higher than the training energy. Early stopping caused the training of the larger network to abort after 135 epochs, this corresponds to 824 iterations. The resulting classification accuracy is about 97.2% for the network with 100 hidden neurons. It is difficult to increase the classification accuracy by adding more hidden layers, most likely because the network overfits the data (Section 6.4). This problem becomes more acute as one adds more hidden neurons. The tendency of the network to overfit is reduced by regularisation (Section 7.6). For the network with one hidden layer with 100 ReLU neurons, L_2-regularisation improves the classification accuracy to almost 98%.

Convolutional networks can be optimised to yield higher classification accuracies than those quoted above. A convolutional network with one convolution layer with 20 feature maps, a max-pooling layer, and a fully connected hidden layer with 100 ReLU neurons, similar to the network shown schematically in Figure 8.7, gives classification accuracy only slightly above 98% after training for 60 epochs. Adding a second convolution layer and batch normalisation (Section 7.6.5) gives a classification accuracy is 98.99% after 30 epochs (this layout is similar to a layout described in MathWorks [113]). The accuracy can be improved further by tuning parameters and network layout, and by using ensembles of convolutional neural networks [111]. The best classification accuracy found in this way is 99.77% [114]. Several of the MNIST digits are difficult to recognise for humans too (Figure 8.8). It is not surprising that the network fails on these digits.

The above examples show also that it takes some experimenting to find the right parameters and network layout, as well as long training times to reach the best classification accuracies. It could be argued that one reaches a stage of *diminishing*

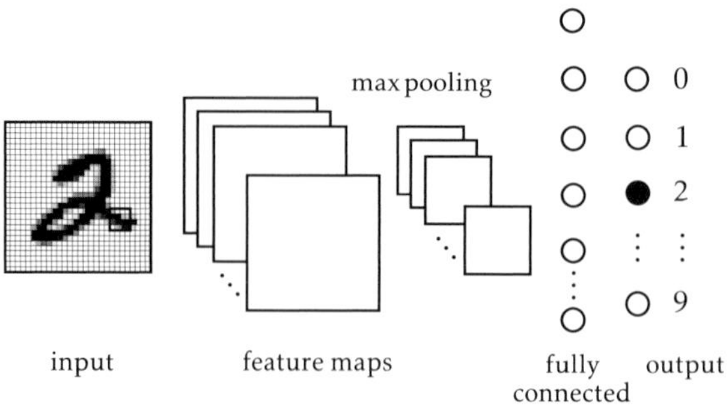

Figure 8.7 Convolutional network that classifies handwritten digits (schematic)

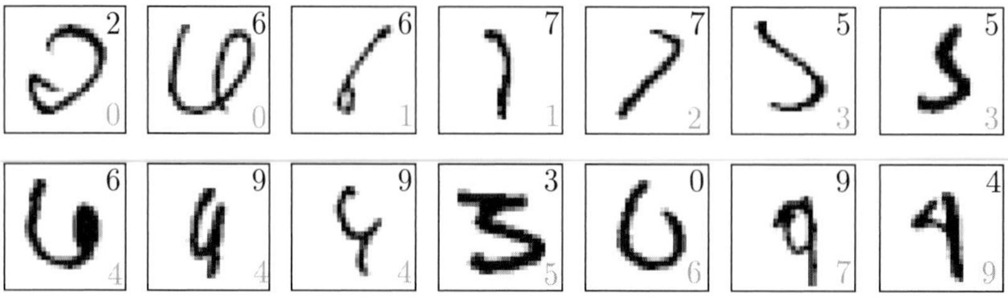

Figure 8.8 Some handwritten digits from the MNIST test set misclassified by a convolutional network that achieved an overall classification accuracy of 98%. Target (top right), network output (bottom right). Data from Oleksandr Balabanov. After a figure in Ref. [6], see also Fig. 2(c) in Ref. [112]

returns as the classification error falls below a fraction of a percent.

8.4 Coping with Deformations of the Input Distribution

How well does a MNIST-trained convolutional network classify your own handwritten digits? Figure 8.9(**a**) shows examples of digits drawn by colleagues at the University of Gothenburg, preprocessed in the same way as the MNIST data. Using a MNIST-trained convolutional network on these digits yields a classification accuracy of about 90%, substantially lower than the classification errors quoted in the previous section.

A possible cause is that the digits in Figure 8.9(**a**) have a more slender stroke than those in Figure 8.5. It was suggested in Ref. [115] that differences in line thickness can confuse algorithms designed to read handwritten text [116]. There are different methods for normalising the line thickness of handwritten text. Applying the method proposed in Ref. [116] to our digits results in Figure 8.9(**b**). The algorithm has a free parameter, T, that specifies the line thickness. In Figure 8.9(**b**), it was taken to be $T = 10$, close to the average line thickness of the MNIST digits,

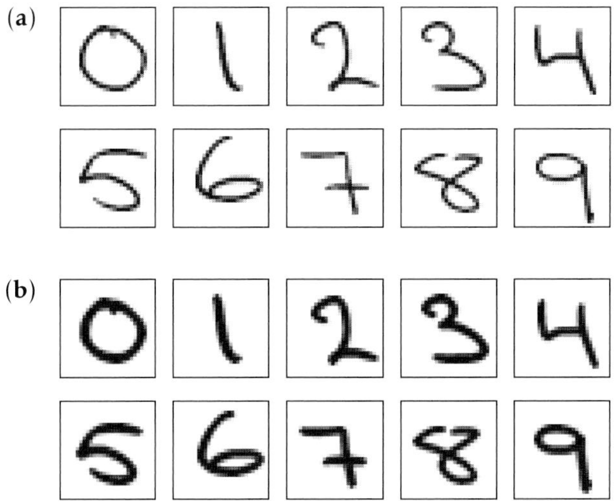

Figure 8.9 (**a**) Non-MNIST handwritten digits, preprocessed like the MNIST digits. (**b**) Same digits, except that the thickness of the stroke was normalised (see the text). Data from Oleksandr Balabanov

which is approximately $T \approx 9.7$. If we run a MNIST-trained convolutional network on a data set of 60 digits with normalised line thickness, it fails on only two digits. This corresponds to a classification accuracy of roughly 97%, not so bad – yet not as good as the best results in Section 8.3. But note that this estimate of the classification accuracy is not very precise, because the test set had only 60 digits. To obtain a better estimate, more test digits are needed.

A question is of course whether there are other significant differences between our non-MNIST handwritten digits and those in the MNIST data. At any rate, the results of this section raise a point of fundamental importance. We have seen that convolutional networks can be trained to represent a distribution of input patterns with very high accuracy. But the network may not work as well on a data set with a slightly different input distribution, perhaps because the patterns were preprocessed differently or because they were slightly deformed in other ways.

8.5 Deep Learning for Object Recognition

Deep learning has become so popular in the last few years because deep convolutional networks are good at recognising objects in images. Figure 8.10 shows a frame from a movie taken by a data-collection vehicle. A convolutional network was trained to recognise objects, and to localise them in the image by means of bounding boxes around the objects.

Figure 8.10 Object recognition using a deep convolutional network. Shown is a frame from a movie recorded by a data-collection vehicle of the company Zenseact (https://zenseact.com/). The neural network recognises pedestrians, cars, and lorries, and localises them in the image by bounding boxes. Copyright © Zenseact AB 2020. Reproduced with permission. Original in colour

Convolutional networks excel at this task, as demonstrated by the *ImageNet large-scale visual recognition challenge* (ILSVRC) [117], a competition for object recognition and localisation in images, based upon the ImageNet database [100]. The challenge is based on a subset of ImageNet. The training set contains more than 10^6 images manually classified into one of 1,000 classes. There are approximately 1,000 images for each class. The validation set contains 50,000 images.

The ILSVRC challenge consists of several tasks. One task is *image classification*, to list the object classes found in the image. A common measure for accuracy is the so-called *top-5 error* for this classification task. The algorithm lists the five object classes with the highest softmax outputs. The result is defined to be correct if the annotated class is among these five. The error equals the fraction of incorrectly classified images. Why does one not simply judge whether the most probable class is the correct one? The reason is that the images in the ImageNet database are annotated by a single-class identifier. Often this is not unique. The image in Figure 8.5, for example, shows not only a car but also trees, yet the image is annotated with the class label *car*. The resulting classification ambiguity is reduced by considering the top five softmax outputs, and checking whether the annotated class is among them.

The tasks in the ILSVRC challenge are significantly more difficult than the digit recognition described in Section 8.3. One reason is that the ImageNet classes are organised into a deep hierarchy of subclasses. This results in highly specific

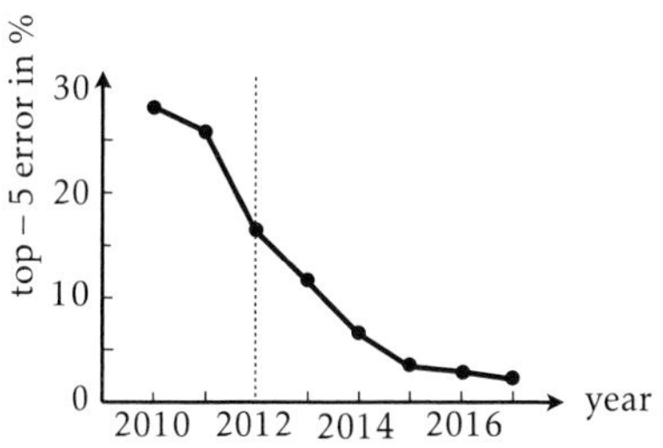

Figure 8.11 Smallest classification error for the ImageNet challenge [117]. The data up to 2014 comes from Ref. [117]. The data for 2015 comes from Ref. [92], for 2016 from Ref. [118], and for 2017 from Ref. [119]. From 2012 onwards, the smallest error was achieved by convolutional neural networks. After Figure 1.12 in Goodfellow et al. [4]

subclasses that can be difficult to tell apart. The algorithm must be very sensitive to small differences between similar subclasses. We say that the algorithm must have high *inter-class variability* [120]. Different images in the same subclass, on the other hand, may look quite different. The algorithm should nevertheless recognise them as similar, belonging to the same class. In other words, the algorithm should have small *intra-class variability* [120].

Since 2012, algorithms based on deep convolutional networks won the ILSVRC challenge. Figure 8.11 shows that the error has significantly decreased until 2017, the last year of the challenge in the form described above. We saw in previous sections that deep networks are difficult to train. So how can these algorithms work so well? It is generally argued that the recent success of deep convolutional networks is mainly due to three factors.

First, there are now much larger and better-annotated training sets available. ImageNet is an example. Excellent training data is now recognised as one of the most important factors. Companies developing software for self-driving cars and systems that help avoid accidents understand that good training sets are indispensable. At the same time, it is a challenge to create high-quality training data, because one must *manually* collect and annotate the data (Figure 8.12). This is costly, also because it is important to have as large data sets as possible, in order to reduce overfitting. In addition one must aim for a large variability in the collected data. Second, the hardware is much better today. Deep networks are nowadays implemented on single or multiple GPUs. There are even dedicated chips for this purpose. Third, improved regularisation techniques (Section 7.6) help to fight overfitting, and skipping connections (Section 7.4) render the networks less susceptible to the vanishing-gradient problem (Section 7.2).

The winning algorithm for 2012 was based on a network with five convolution layers and three fully connected layers, using drop out, ReLU activation functions, and data-set augmentation [109]. The algorithm was implemented on GPU

Figure 8.12 Reproduced from http://xkcd.com/1897 under the creative commons attribution-noncommercial 2.5 license

processors. The 2013 ILSVRC challenge was also won by a convolutional network [121], with 22 layers. Nevertheless, the network has substantially fewer free parameters (weights and thresholds) than the 2012 network: 4×10^6 instead of 60×10^6. In 2015, the winning algorithm [92] had 152 layers. One significant new element in the layout was the idea to allow connections that skip layers (Section 7.4). The best algorithms in 2016 [122] and 2017 [119] used ensembles of convolutional networks, where the classification is based on the ensemble average of the outputs.

8.6 Summary

Convolutional networks can be trained to recognise objects in images with high accuracy. An advantage of convolutional networks is that they have fewer weights than fully connected networks with the same number of neurons, and that the weights of a given feature map are trained on different parts of the input images, effectively increasing the size of the training set. This helps against overfitting. Another view is that the hidden neurons are forced to agree on a particular choice of weights, they must compromise. This yields a more robust training result.

It is sometimes stated that convolutional networks are now better than humans, in that they recognise objects with lower classification errors than humans [123]. This and similar statements refer to an experiment showing that the human classification error in recognising objects in the ImageNet database is about 5.1% [124], worse than the most recent convolutional neural-network algorithms (Figure 8.11).

This notion is not unproblematic, for several reasons. To begin with, the article [123] refers to the 2015 ILSVRC competition, where the top scores were quite similar, and it has been debated whether interpreting the rules of the competition in different ways allowed competitors to gain an advantage. Second, and more importantly, it is clear that these algorithms learn in quite a different way from humans. The algorithms can detect local features, but since these convolutional networks rely on translational invariance, they do not easily understand global features, and can mistake a leopard-patterned sofa for a leopard [125]. It may help to include more leopard-patterned sofas in the training set, but the essential difficulty remains: translational invariance imposes constraints on what convolutional networks can learn [125]. More fundamentally one may argue that humans learn differently, by abstraction instead of going through very large training sets.

We have also seen that convolutional networks are sensitive to small changes in the input data. Convolutional networks excel at learning the properties of a given input distribution, but they may have difficulties in recognising patterns sampled from a slightly different distribution, even if the two distributions appear to be very similar to the human eye. Note also that this problem cannot be solved by cross validation, because training and validation sets are drawn from the same input distribution, but here we are concerned with what happens when the network is applied to an input distribution different from the one it was trained on.

Here is another example illustrating this point: the authors of Ref. [126] trained a convolutional network on perturbed grayscale images from the ImageNet database, adding a little bit of noise independently to each pixel (*white noise*) before training. This network failed to recognise images that were weakly perturbed in a different way, by setting a small number of pixels to white or black. But when we look at the images, we have no difficulties seeing through the noise.

Refs. [127, 128] illustrate intriguing failures of convolutional networks [5]. Szegedy et al. [127] demonstrate that the way convolutional networks partition input space can lead to unexpected results. The authors took an image that the network classifies correctly with high confidence, and perturbed it slightly. The perturbation was not random but specifically designed to push the input pattern over a decision boundary. The difference between the original and perturbed images (*adversarial images*) is undetectable to the human eye, yet the network misclassifies the perturbed image with high confidence [127]. This reflects the fact that decision boundaries are always close in high-dimensional input space.

Figure 1 in Ref. [128] shows images that are completely unrecognisable to the human eye. Yet a convolutional network classifies these images with high confidence. This illustrates that there is no telling what a network may do if the input is far away from the training distribution. Unfortunately, the network can sometimes be highly confident yet wrong. Nevertheless, despite these problems,

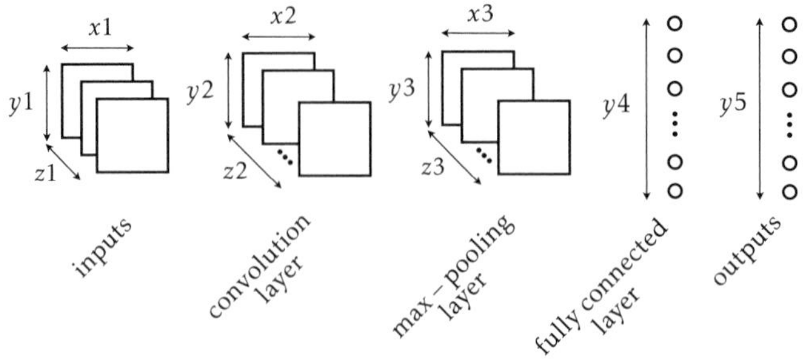

Figure 8.13 Layout of convolutional network for Exercise 8.1

deep convolutional networks have enjoyed tremendous success in image classification during the past years, and they have found widespread use in industry and science.

Finally, the fundamental mechanisms of deep learning are quite well understood, but many open questions remain. It is fair to say that the theory of deep learning has somewhat lagged behind the associated practical successes, although progress has been made in recent years.

8.7 Further Reading

The online book of Nielsen [5] is an excellent introduction to convolutional neural networks and guides the reader through all the steps required to program a convolutional network to recognise handwritten digits. Nielsen's chapter *Deep Learning* [5] is the main source for Section 8.3.

What do the hidden layers in a convolutional layer actually compute? Feature maps that are directly coupled to the inputs detect local features, such as edges or corners. Yet it is unclear precisely how hidden convolutional layers help the network to learn. To which input features do the neurons of a certain hidden layer react most strongly? Input patterns chosen to maximise the outputs of neurons in a given layer [129, 130] reveal intricate geometric structures that defy straightforward interpretation. An example is shown on the cover of this book, see also Exercise 8.7.

It has been suggested that more general models, normally used for natural-language processing, may outperform convolutional nets in image-process-ing tasks when there is enough data [131]. An advantage is that these models do not rely on translational invariance, unlike convolutional networks.

8.8 Exercises

8.1 Number of parameters of a convolutional network. A convolutional network has the following layout (Figure 8.13): an input layer of size $21 \times 21 \times 3$,

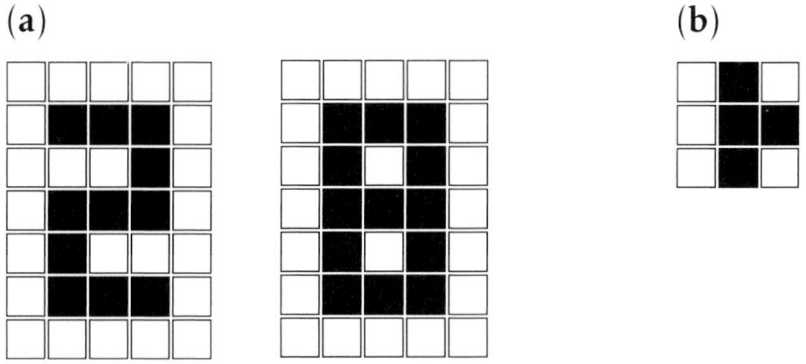

Figure 8.14 (**a**) Input patterns with 0/1 bits (□ corresponds to $x_i = 0$ and ■ to $x_i = 1$). (**b**) 3×3 kernel of a feature map. ReLU neurons, zero threshold, weights either 0 or 1 (□ corresponds to $w = 0$ and ■ to $w = 1$). See Exercise 8.3

a convolutional layer with ReLU activations with 16 kernels with local receptive fields of size 2×2, stride $[1, 1]$, and padding $[0, 0, 0, 0]$, a max-pooling layer with local receptive field of size 2×2, stride $= [2, 2]$, padding $= [0, 0, 0, 0]$, a fully connected layer with 20 neurons with sigmoid activations, and a fully connected output layer with 10 neurons. In one or two sentences, explain the function of each of the layers. Enter the values of the parameters $x1, y1, z1, x2, \ldots, y5$ into Figure 8.13 and determine the number of trainable parameters (weights and thresholds) for the connections into each layer of the network.

8.2 Convolutional network. Figure 8.4 shows the schematic layout of a convolutional net. Explain how a *convolution layer* works. In your discussion, refer to the terms *convolution*, *colour channel*, *receptive field*, *feature map*, *stride*, and explain the meaning of the parameters in the computation rule

$$V_{ij} = g\Big(\sum_{p=1}^{P}\sum_{q=1}^{Q} w_{pq} x_{p+s(i-1),q+s(j-1)} - \theta\Big). \tag{8.6}$$

Explain how a *pooling layer* works, and why it is useful.

8.3 Feature map. The two patterns shown in Figure 8.14(a) are processed by a very simple convolutional network that has one convolution layer with one single 3×3 kernel with ReLU neurons, zero threshold, weights as given in Figure 8.14(b), and stride [1,1]. The resulting feature map is fed into a 3×3 max-pooling layer with stride [1,1]. Finally, there is a fully connected classification layer with two output neurons with Heaviside activation functions (binary threshold unit). For both patterns, determine the resulting feature map and the output of the max-pooling layer. Find weights and thresholds of the classification layer that allow one to classify the two patterns into different classes.

8.4 Distorted MNIST digits. Train a convolutional network on the MNIST data set. Distort the patterns in the test set by adding noise, in two different ways. First, choose q pixels randomly and make them black (or leave them black). Second, choose q pixels randomly and make them white (or leave them white). Vary q and investigate the performance of the convolutional network for both noisy test sets.

8.5 CIFAR-10 data set. Train a fully connected network with two hidden layers on the CIFAR-10 data set [132], and minimise the classification error by optimising the network parameters. Compare with the performance of an optimised convolutional network.

8.6 Bars-and-stripes data set. Construct a convolutional network to classify the patterns of the bars-and-stripes data set (Figure 4.4) into patterns with bars (black columns) and stripes (black rows). Use a convolution layer with at most four 2×2 kernels, one pooling layer, and one fully connected layer for classification. Give all parameters of the network (weights, thresholds, padding, and stride, where relevant).

8.7 Visualise activations of hidden neurons. Train a convolutional network on the CIFAR-10 data set [132], and for each feature map, construct the input patterns that achieve maximal activation. Use gradient ascent to modify the values of the input pixels in order to find these patterns. See the cover image of this book, along with Refs. [129, 130].

9

Supervised Recurrent Networks

The layout of the perceptrons analysed in the previous chapters is special. All connections are one way, and only to the layer to the right, so that the update rule for the i-th neuron in layer ℓ becomes, for example,

$$V_i^{(\ell)} = g\Big(\sum_j w_{ij}^{(\ell)} V_j^{(\ell-1)} - \theta_i^{(\ell)}\Big). \tag{9.1}$$

The backpropagation algorithm relies on this *feed-forward* layout. It means that the derivatives $\partial V_j^{(\ell-1)}/\partial w_{mn}^{(\ell)}$ vanish. This ensures that the outputs are nested functions of the inputs, which in turn implies the simple iterative structure of the backpropagation algorithm (Chapter 6).

In some cases, it is necessary or convenient to use networks that do not have this simple layout. The Hopfield networks discussed in Part I are examples where all connections are symmetric. More general networks may have a feed-forward layout with *feedbacks*, as shown in Figure 9.1. Such networks are called *recurrent networks*. There are many different ways in which the feedbacks can act: from the output layer to hidden neurons for example, or there could be connections between the neurons in a given layer. Neurons 3 and 4 in Figure 9.1 are output neurons; they are associated with targets just as in Chapters 5 to 7. The layout of recurrent networks is very general, but because of the feedbacks, we must consider how such networks can be trained.

Unlike multilayer perceptrons that represent an input-to-output mapping in terms of nested activation functions, recurrent networks are used as *dynamical networks*, where the iteration index t replaces the layer index ℓ:

$$V_i(t) = g\Big(\sum_j w_{ij}^{(vv)} V_j(t-1) + \sum_k w_{ik}^{(vx)} x_k - \theta_i^{(v)}\Big) \quad \text{for} \quad t = 1, 2, \ldots \tag{9.2}$$

See Figure 9.1 for the definition of the different weights. The parameters $\theta_i^{(v)}$ are thresholds. Equation (9.2) is analogous to the deterministic McCulloch-Pitts

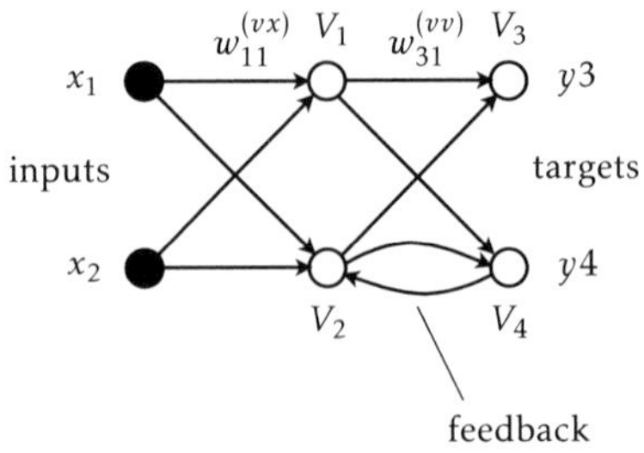

Figure 9.1 Network with a feedback connection. Neurons 1 and 2 are hidden neurons. The weights from the input x_k to the neurons V_i are denoted by $w^{(vx)}_{ik}$, the weight from neuron V_j to neuron V_i is $w^{(vv)}_{ij}$. Neurons 3 and 4 are output neurons, with prescribed target values y_i. To avoid confusion with the iteration index t, the targets are denoted by y in this chapter

dynamics of Hopfield networks and Boltzmann machines [c.f. Equation (1.5)]. As in the case of Hopfield networks (Exercise 2.10), one may also consider a continuous network dynamics:

$$\tau \frac{\mathrm{d}V_i}{\mathrm{d}t} = -V_i + g\Big(\sum_j w^{(vv)}_{ij} V_j(t) + \sum_k w^{(vx)}_{ik} x_k - \theta^{(v)}_i\Big)\,, \tag{9.3}$$

with time constant τ. We shall see in a moment why it is convenient to assume that the dynamics is continuous in t, as in Equation (9.3).

Recurrent networks can learn in different ways. One possibility is to use a training set of pairs $[\boldsymbol{x}^{(\mu)}, \boldsymbol{y}^{(\mu)}]$ with $\mu = 1, \ldots, p$. To avoid confusion with the iteration index t, the targets are denoted by y in this chapter. One feeds a pattern from this set and runs the dynamics (9.2) or (9.3) for the given $\boldsymbol{x}^{(\mu)}$ until it reaches a steady state $\boldsymbol{V}^*$ (if this does not happen, the training fails). Then one adjusts the weights by one gradient-descent step using the energy function

$$H = \frac{1}{2}\sum_k (E^*_k)^2 \quad \text{where } E^*_k = \begin{cases} y^{(\mu)}_k - V^*_k & \text{if } V_k \text{ is an output neuron,} \\ 0 & \text{otherwise.} \end{cases} \tag{9.4}$$

The asterisk in this equation indicates that all variables are evaluated in the steady state, at $\boldsymbol{V} = \boldsymbol{V}^*$. Iterating these steps, one feeds another pattern $\boldsymbol{x}^{(\mu)}$, finds the steady state $\boldsymbol{V}^*$, adjusts the weights, and so forth. One continues to iterate these steps until the steady-state outputs yield the correct targets for all input patterns. This is reminiscent of the algorithms discussed in Chapters 5 to 7. Instead of defining the energy function in terms of the mean squared output error, one could also use the negative log-likelihood function (7.44).

Another possibility is that inputs and targets change as functions of time t while the network dynamics runs. This allows one to solve *temporal classification tasks*. The network is trained on a set of input sequences $\boldsymbol{x}(t)$ and corresponding target sequences $\boldsymbol{y}(t)$. In this way, recurrent networks can translate written text or recognise speech. The network can be trained by unfolding its dynamics in time as explained in Section 9.2, although this algorithm suffers from the vanishing-gradient problem discussed in Chapter 7.

9.1 Recurrent Backpropagation

This section summarises how to generalise Algorithm 4 to recurrent networks with feedback connections. Recall the recurrent network shown in Figure 9.1. The neurons V_i have smooth activation functions, and they are connected by weights $w_{ij}^{(vv)}$. Several neurons may be linked to inputs $x_k^{(\mu)}$, with weights $w_{ik}^{(vx)}$. Other neurons are output units with associated target values $y_i^{(\mu)}$.

One takes the dynamics to be continuous in time, Equation (9.3), and assumes that $\boldsymbol{V}(t)$ runs into a steady state

$$\boldsymbol{V}(t) \to \boldsymbol{V}^* \quad \text{so that} \quad \frac{\mathrm{d}V_i^*}{\mathrm{d}t} = 0\,. \tag{9.5}$$

Equation (9.3) implies

$$V_i^* = g\Big(\sum_j w_{ij}^{(vv)} V_j^* + \sum_k w_{ik}^{(vx)} x_k - \theta_i^{(v)}\Big)\,, \tag{9.6}$$

and it is assumed that $\boldsymbol{V}^*$ is a *linearly stable* steady state of the dynamics (9.3), so that small perturbations $\delta\boldsymbol{V}$ away from $\boldsymbol{V}^*$ decay with time.

The synchronous discrete dynamics (9.2) can exhibit undesirable stable periodic solutions [133], as mentioned in Section 1.3. This is a reason for using the continuous dynamics (9.3), yet convergence to the steady state is not guaranteed in this case either.

Equation (9.6) is a non-linear self-consistent condition for the components of $\boldsymbol{V}^*$, in general difficult to solve. However, if the steady state $\boldsymbol{V}^*$ is stable, we can use the dynamics (9.3) to automatically pick out the steady-state solution $\boldsymbol{V}^*$. This solution depends on the pattern $\boldsymbol{x}^{(\mu)}$. Note that the superscript (μ) is left out in Equations (9.5) and (9.6), and also in the remainder of this section.

The goal is to find weights so that the outputs give the correct target values in the steady state. To this end, one uses gradient descent on the energy function (9.4). Following Ref. [1], consider first how to adjust the weights $w_{ij}^{(vv)}$:

$$\delta w_{mn}^{(vv)} = -\eta \frac{\partial H}{\partial w_{mn}^{(vv)}} = \eta \sum_k E_k^* \frac{\partial V_k^*}{\partial w_{mn}^{(vv)}}\,. \tag{9.7}$$

One calculates the gradients of $\boldsymbol{V}^*$ by differentiating Equation (9.6):

$$\frac{\partial V_i^*}{\partial w_{mn}^{(vv)}} = g'(b_i^*)\frac{\partial b_i^*}{\partial w_{mn}^{(vv)}} = g'(b_i^*)\Big(\delta_{im}V_n^* + \sum_j w_{ij}^{(vv)}\frac{\partial V_j^*}{\partial w_{mn}^{(vv)}}\Big). \tag{9.8}$$

Here $b_i^* = \sum_j w_{ij}^{(vv)}V_j^* + \sum_k w_{ik}^{(vx)}x_k - \theta_i^{(v)}$ is the local field in the steady state. Equation (9.8) is a self-consistent equation for the gradient, as opposed to the explicit expressions we found in Chapter 6. The reason for the difference is that the recurrent network has feedbacks.

Since Equation (9.8) is linear in the gradients, it can be solved by matrix inversion, at least formally. In terms of the matrix $\mathbb{L}$ with elements

$$L_{ij} = \delta_{ij} - g'(b_i^*)w_{ij}^{(vv)}, \tag{9.9}$$

Equation (9.8) can be written as

$$\sum_j L_{ij}\frac{\partial V_j^*}{\partial w_{mn}^{(vv)}} = \delta_{im}g'(b_i^*)V_n^*. \tag{9.10}$$

If $\mathbb{L}$ is invertible, one applies $\sum_i \left(\mathbb{L}^{-1}\right)_{ki}$ to both sides. Using the fact that $\sum_i \left(\mathbb{L}^{-1}\right)_{ki} L_{ij} = \delta_{kj}$ one finds

$$\frac{\partial V_k^*}{\partial w_{mn}^{(vv)}} = \left(\mathbb{L}^{-1}\right)_{km} g'(b_m^*)V_n^*. \tag{9.11}$$

Inserting this result into (9.7), one obtains

$$\delta w_{mn}^{(vv)} = \eta\sum_k E_k^* \left(\mathbb{L}^{-1}\right)_{km} g'(b_m^*)V_n^*. \tag{9.12}$$

This learning rule can be written in the form of the backpropagation rule (6.10) by introducing the error

$$\Delta_m^* = g'(b_m^*)\sum_k E_k^* \left(\mathbb{L}^{-1}\right)_{km}. \tag{9.13}$$

Then the learning rule (9.12) takes the form

$$\delta w_{mn}^{(vv)} = \eta\Delta_m^* V_n^*. \tag{9.14}$$

If there are no recurrent connections, then $L_{ij} = \delta_{ij}$. In this case, Equation (9.13) reduces to the standard expression (6.6b), Exercise 9.1.

The learning rule for the weights $w_{mn}^{(vx)}$ is derived in an analogous fashion. The result is

$$\delta w_{mn}^{(vx)} = \eta\Delta_m^* x_n. \tag{9.15}$$

The learning rules (9.14) and (9.15) are well-defined only if the matrix $\mathbb{L}$ is invertible. Otherwise, the solution (9.11) does not exist. Also, matrix inversion is an expensive operation. As described in Chapter 5, one can try to avoid the problem by finding the inverse iteratively. The trick [1] is to write down a dynamical equation for Δ_i that has a steady state at the solution of Equation (9.13):

$$\tau \tfrac{\mathrm{d}}{\mathrm{d}t}\Delta_j = -\Delta_j + g'(b_j^*)E_j^* + \sum_i \Delta_i w_{ij}^{(vv)} g'(b_j^*)\,. \tag{9.16}$$

It is left as an exercise (Exercise 9,2) to verify that the dynamics (9.16) admits a steady state satisfying Equation (9.13). Equation (9.16) is written in a form to stress that (9.16) and (9.3) exhibit the same *duality* as Algorithm 4, between forward propagation of states of neurons and backpropagation of errors. The sum in Equation (9.16) has the same form as the recursion for the errors in Algorithm 4, except that there are no layer indices ℓ here.

Equation (9.16) admits the steady state (9.13). But does $\Delta_i(t)$ converge to Δ_i^*? For convergence, it is necessary that the steady state is linearly stable. Whether or not this is the case is determined by *linear stability analysis* [85]. One asks: does a small deviation from the steady state increase or decrease under Equation (9.16)? In other words, if one writes

$$\boldsymbol{V}(t) = \boldsymbol{V}^* + \delta\boldsymbol{V}(t) \quad \text{and} \quad \boldsymbol{\Delta}(t) = \boldsymbol{\Delta}^* + \delta\boldsymbol{\Delta}(t)\,, \tag{9.17}$$

do $\delta\boldsymbol{V}(t)$ and $\delta\boldsymbol{\Delta}(t)$ grow in magnitude? To answer this question, one inserts this *ansatz* into (9.3) and (9.16), and linearises:

$$\tau \tfrac{\mathrm{d}}{\mathrm{d}t}\delta V_i = -\,\delta V_i + g'(b_i^*)\sum_j w_{ij}^{(vv)}\delta V_j \approx -\sum_j L_{ij}\delta V_j\,, \tag{9.18a}$$

$$\tau \tfrac{\mathrm{d}}{\mathrm{d}t}\delta\Delta_j = -\delta\Delta_j + \sum_i \delta\Delta_i w_{ij}^{(vv)} g'(b_j^*) \approx -\sum_i \delta\Delta_i g'(b_i^*)L_{ij}/g'(b_j^*)\,. \tag{9.18b}$$

Equation (9.18a) shows: whether or not the norm of $\delta\boldsymbol{V}(t)$ grows is determined by the eigenvalues of the matrix $\mathbb{L}$. We say that $\boldsymbol{V}^*$ is a linearly stable steady state of Equation (9.3) if all eigenvalues of $\mathbb{L}$ have negative real parts. In this case, $|\delta\boldsymbol{V}(t)| \to 0$. If at least one eigenvalue has a positive real part then $|\delta\boldsymbol{V}|$ grows. In this case, we say that $\boldsymbol{V}^*$ is linearly unstable. Since the matrix with elements $g'(b_i^*)L_{ij}/g'(b_j^*)$ has the same eigenvalues as $\mathbb{L}$, $\boldsymbol{\Delta}^*$ is a stable steady state of (9.16) if $\boldsymbol{V}^*$ is a stable steady state of (9.3). If the steady states are unstable, the algorithm does not converge.

In summary, recurrent backpropagation is analogous to backpropagation (Algorithm 4) for layered feed-forward networks, save for two differences. First, the non-linear network dynamics is no longer a simple input-to-output mapping with nested activation functions, but a non-linear dynamics that may (or may not) converge to a steady state. Second, the feedbacks give rise to linear self-consistent equations for the steady-state gradients $\partial V_j^*/\partial w_{mn}$, which can be viewed as steady-state conditions for a dual dynamics of the errors.

The main conclusion of this section is that convergence of the training is not guaranteed if the network has feedback connections (for a layered feed-forward network without feedbacks, recurrent backpropagation simplifies to stochastic gradient descent, Algorithm 4, see Exercise 9.1). This explains why stochastic gradient descent is used mostly for multilayer networks with feed-forward layouts: the algorithm tends to fail for networks with feedbacks. However, it is possible to get rid of the feedbacks in recurrent networks by unfolding the dynamics in time. This is described in the next section.

9.2 Backpropagation through Time

Recurrent networks can be used to learn sequential inputs, as in speech recognition and machine translation. The training set consists of time sequences $[\boldsymbol{x}(t), \boldsymbol{y}(t)]$ of inputs and targets. The network is trained on the sequences and learns to predict the targets. In this context, the layout differs from the one described in the previous section. There are two main differences. Firstly, the inputs and targets depend on t and one uses a discrete-time update rule. Secondly, separate output neurons $O_i(t)$ are added to the layout. The update rule takes the form

$$V_i(t) = g\Big(\sum_j w_{ij}^{(vv)} V_j(t-1) + \sum_k w_{ik}^{(vx)} x_k(t) - \theta_i^{(v)}\Big), \tag{9.19a}$$

$$O_i(t) = g\Big(\sum_j w_{ij}^{(ov)} V_j(t) - \theta_i^{(o)}\Big). \tag{9.19b}$$

The activation function of the output neurons O_i can be different from that of the hidden neurons V_j. One possibility is to use the softmax function for the outputs [134, 135]. For the hidden neurons, one often uses tanh activations.

To train recurrent networks with time-dependent inputs and with the dynamics (9.19), one uses *backpropagation through time*. The idea is to unfold the network in time to get rid of the feedbacks. The price paid is that one obtains large networks in this way, with as many copies of the original neurons as there are time steps.

The procedure is illustrated in Figure 9.2 for a recurrent network with one hidden neuron, one input terminal, and one output neuron. The unfolded network has T inputs and outputs. It can be trained in the usual way with *stochastic gradient descent*. The errors are calculated using backpropagation as in Algorithm 4, but here the error is propagated back in time, not from layer to layer. The energy function is the squared error summed over all time steps

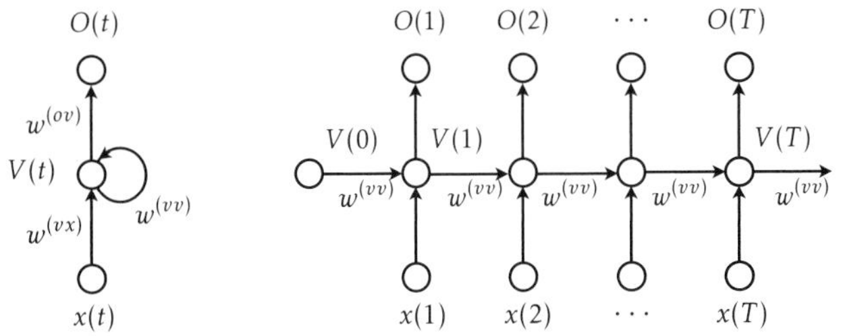

Figure 9.2 Left: recurrent network with one input terminal, one hidden neuron, and one output neuron. Right: same network but unfolded in time. The weights $w^{(vv)}$ remain unchanged as drawn, also the weights $w^{(vx)}$ and $w^{(ov)}$ remain unchanged (not drawn). After Figures 7 and 8 in Ref. [135].

$$H = \frac{1}{2}\sum_{t=1}^{T} E_t^2 \quad \text{with} \quad E_t = y_t - O_t . \tag{9.20}$$

One could use the negative log-likelihood function (7.38), but here we use the squared output-error function (9.20). There is only one hidden neuron in our example, and the inputs and outputs are also one-dimensional. Here and in the following, we write the time argument as a subscript, O_t instead of $O(t)$ and so forth, because there is no risk of confusing the time index with other subscripts.

Consider first how to adjust the weight $w^{(vv)}$. Gradient descent (5.24) yields

$$\delta w^{(vv)} = \eta \sum_{t=1}^{T} E_t \frac{\partial O_t}{\partial w^{(vv)}} = \eta \sum_{t=1}^{T} \Delta_t w^{(ov)} \frac{\partial V_t}{\partial w^{(vv)}} . \tag{9.21a}$$

Here

$$\Delta_t = E_t g'(B_t) \tag{9.22}$$

is an output error, and $B_t = w^{(ov)} V_{t-1} - \theta^{(o)}$ is the local field of the output neuron at time t [Equation (9.19)]. Equation (9.21a) is similar to the learning rule for recurrent backpropagation, Equations (9.7) and (9.8), but the derivative $\partial V_t / \partial w^{(vv)}$ is evaluated differently. Equation (9.19a) yields the recursion

$$\frac{\partial V_t}{\partial w^{(vv)}} = g'(b_t)\left(V_{t-1} + w^{(vv)} \frac{\partial V_{t-1}}{\partial w^{(vv)}}\right) \tag{9.23}$$

for $t \geq 1$. Since $\partial V_0 / \partial w^{(vv)} = 0$, Equation (9.23) implies

$$\begin{aligned}
\frac{\partial V_1}{\partial w^{(vv)}} &= g'(b_1) V_0 , \\
\frac{\partial V_2}{\partial w^{(vv)}} &= g'(b_2) V_1 + g'(b_2) w^{(vv)} g'(b_1) V_0 ,
\end{aligned}$$

$$\frac{\partial V_3}{\partial w^{(vv)}} = g'(b_3)V_2 + g'(b_3)w^{(vv)}g'(b_2)V_1 + g'(b_3)w^{(vv)}g'(b_2)w^{(vv)}g'(b_1)V_0$$

$$\vdots$$

$$\frac{\partial V_{T-1}}{\partial w^{(vv)}} = g'(b_{T-1})V_{T-2} + g'(b_{T-1})w^{(vv)}g'(b_{T-2})V_{T-3} + \ldots$$

$$\frac{\partial V_T}{\partial w^{(vv)}} = g'(b_T)V_{T-1} + g'(b_T)w^{(vv)}g'(b_{T-1})V_{T-2} + \ldots$$

Equation (9.21a) says that we must sum over t. Regrouping the terms in this sum yields

$$\begin{aligned}
&\Delta_1\frac{\partial V_1}{\partial w^{(vv)}} + \Delta_2\frac{\partial V_2}{\partial w^{(vv)}} + \Delta_3\frac{\partial V_3}{\partial w^{(vv)}} + \ldots \\
&= [\Delta_1 g'(b_1) + \Delta_2 g'(b_2)w^{(vv)}g'(b_1) + \Delta_3 g'(b_3)w^{(vv)}g'(b_2)w^{(vv)}g'(b_1) + \ldots]V_0 \\
&\quad + [\Delta_2 g'(b_2) + \Delta_3 g'(b_3)w^{(vv)}g'(b_2) + \Delta_4 g'(b_4)w^{(vv)}g'(b_3)w^{(vv)}g'(b_2) + \ldots]V_1 \\
&\quad + [\Delta_3 g'(b_3) + \Delta_4 g'(b_4)w^{(vv)}g'(b_3) + \Delta_5 g'(b_5)w^{(vv)}g'(b_4)w^{(vv)}g'(b_3) + \ldots]V_2 \\
&\quad\ \vdots \\
&\quad + [\Delta_{T-1} g'(b_{T-1}) + \Delta_T g'(b_T)w^{(vv)}g'(b_{T-1})]V_{T-2} \\
&\quad + [\Delta_T g'(b_T)]V_{T-1}\,.
\end{aligned}$$

To write the learning rule in the usual form, we define *errors* δ_t recursively:

$$\delta_t = \begin{cases} \Delta_T w^{(ov)} g'(b_T) & \text{for } t = T, \\ \Delta_t w^{(ov)} g'(b_t) + \delta_{t+1} w^{(vv)} g'(b_t) & \text{for } 0 < t < T. \end{cases} \tag{9.24}$$

Then the learning rule (9.21a) takes the form

$$\delta w^{(vv)} = \eta \sum_{t=1}^{T} \delta_t V_{t-1}\,, \tag{9.25}$$

just like Equation (6.9), or like the error recursion in Algorithm 4. The factor $w^{(vv)}g'(b_{t-1})$ in the recursion (9.24) gives rise to a product of many such factors in δ_t when T is large, exactly as described in Section 7.2 for multilayer perceptrons. This means that the training of recurrent networks suffers from *unstable gradients*, as backpropagation of multilayer perceptrons does: if the factors $|w^{(vv)}g'(b_p)|$ are smaller than unity, then the errors δ_t become very small when t becomes small (*vanishing-gradient problem*). This means that the early states of the hidden neuron no longer contribute to the learning, causing the network to forget what it has learned about early inputs. When $|w^{(vv)}g'(b_p)| > 1$, on the other hand, exploding gradients make learning impossible. In summary, *unstable gradients* in recurrent neural networks occur much in the same way as in multilayer perceptrons. The resulting difficulties for training recurrent neural networks are discussed in more detail in Section 9.3; see also Ref. [136].

A slight variation of the above algorithm (*truncated backpropagation through time*) suffers less from the exploding-gradient problem. The idea is that the exploding gradients are tamed by truncating the memory. This is achieved by limiting the error propagation backwards in time, errors are computed back to $T-\tau$ and not further, where τ is the truncation time [2]. Naturally, this implies that long-time correlations cannot be learnt.

The learning rules for the weights $w^{(vx)}$ are obtained in a similar fashion. Equation (9.19a) yields the recursion

$$\frac{\partial V_t}{\partial w^{(vx)}} = g'(b_t)\left(x_t + w^{(vv)}\frac{\partial V_{t-1}}{\partial w^{(vx)}}\right). \tag{9.26}$$

This looks just like Equation (9.23), except that V_{t-1} is replaced by x_t. As a consequence we have

$$\delta w^{(vx)} = \eta \sum_{t=1}^{T} \delta_t x_t\,. \tag{9.27}$$

The learning rule for $w^{(ov)}$ is simpler to derive. From Equation (9.19b) we find by differentiation w.r.t. $w^{(ov)}$:

$$\delta w^{(ov)} = \eta \sum_{t=1}^{T} E_t g'(B_t) V_t = \eta \sum_{t=1}^{T} \Delta_t V_t\,. \tag{9.28}$$

How are the thresholds $\theta^{(v)}$ adjusted? Going through the above derivation we see that we must replace V_{t-1} in Equation (9.25) by -1. It works in the same way for the output threshold.

In order to keep the formulae simple, we derived the algorithm for a single hidden neuron, a single output neuron, and one-component inputs, so that we could leave out the indices referring to different hidden neurons, and different input and output components. If we consider several hidden and output neurons and multi-dimensional inputs, the structure of the equations remains exactly the same, except for a number of extra sums over those indices:

$$\delta w_{mn}^{(vv)} = \eta \sum_{t=1}^{T} \delta_m^{(t)} V_n^{(t-1)} \tag{9.29}$$

$$\delta_j^{(t)} = \begin{cases} \sum_i \Delta_i^{(T)} w_{ij}^{(ov)} g'(b_j^{(T)}) & \text{for } t = T\,, \\ \sum_i \Delta_i^{(t)} w_{ij}^{(ov)} g'(b_j^{(t)}) + \sum_i \delta_i^{(t+1)} w_{ij}^{(vv)} g'(b_j^{(t)}) & \text{for } 0 < t < T\,. \end{cases}$$

The second term in the recursion for $\delta_j^{(t)}$ is analogous to the error recursion in Algorithm 4. The time index t here plays the role of the layer index ℓ in Algorithm 4.

Algorithm 7 Backpropagation through time

initialise weights $w_{mn}^{(vv)}$, $w_{mn}^{(vx)}$, $w_{mn}^{(ov)}$ and thresholds $\theta_m^{(v)}$, $\theta_m^{(o)}$;
for $\tau = 1, \ldots, \tau_{\max}$ **do**
  choose input sequence $\boldsymbol{x}(1), \ldots, \boldsymbol{x}(T)$;
  initialise $V_j(0) = 0$;
  for $t = 1, \ldots, T$ **do**
    propagate forward:
    $b_i(t) \leftarrow \sum_j w_{ij}^{(vv)} V_j(t-1) + \sum_k w_{ik}^{(vx)} x_k(t) - \theta_i^{(v)}$ and $V_i(t) \leftarrow g[b_i(t)]$;
    compute outputs:
    $B_i(t) \leftarrow \sum_j w_{ij}^{(ov)} V_j(t) - \theta_i^{(o)}$ and $O_i(t) \leftarrow g[B_i(t)]$;
  end for
  compute errors for $t = T$ (targets y_i):
  $\Delta_i(T) \leftarrow [y_i - O_i(T)] g'[B_i(T)]$ and $\delta_j(T) \leftarrow \sum_i \Delta_i(T) w_{ij}^{(ov)} g'[b_j(T)]$;
  for $t = T, \ldots, 2$ **do**
    propagate backwards: $\Delta_i(t) = [y_i - O_i(t)] g'[B_i(t)]$ and
    $\delta_j(t-1) \leftarrow \sum_i \Delta_i^{(t)} w_{ij}^{(ov)} g'(b_j^{(t)}) + \sum_i \delta_i^{(t+1)} w_{ij}^{(vv)} g'(b_j^{(t)})$;
  end for
  $\delta w_{mn}^{(vv)} = 0$, $\delta w_{mn}^{(vx)} = 0$, $\delta w_{mn}^{(ov)} = 0$, $\delta\theta^{(v)} = 0$, $\delta\theta^{(o)} = 0$;
  for $t = 1, \ldots, T$ **do**
    $\delta w_{mn}^{(vv)} = \delta w_{mn}^{(vv)} + \eta \delta_m(t) V_n(t-1)$;
    $\delta w_{mn}^{(vx)} = \delta w_{mn}^{(vx)} + \eta \delta_m(t) x_n(t)$;
    $\delta w_{mn}^{(ov)} = \delta w_{mn}^{(ov)} + \eta \Delta_m(t) V_n(t)$;
    $\delta\theta_m^{(v)} = \delta\theta_m^{(v)} - \eta \delta_m(t)$;
    $\delta\theta_m^{(o)} = \delta\theta_m^{(o)} - \eta \Delta_m(t)$;
  end for
  adjust weights and thresholds: $w_{mn}^{(vv)} = w_{mn}^{(vv)} + \delta w_{mn}^{(vv)}, \ldots$;
end for

A difference is that the weights in Equation (9.29) are the same for all time steps. The algorithm is summarised in Algorithm 7.

In conclusion, we see that backpropagation through time for recurrent networks is similar to backpropagation for multilayer perceptrons. After the recurrent network is unfolded to get rid of the feedback connections, it can be trained by backpropagation. The time index t takes the role of the layer index ℓ. Backpropagation through time is the standard approach for training recurrent networks, despite the fact that it suffers from the vanishing-gradient problem. The next section describes how improvements to the layout make it possible to more efficiently train recurrent networks.

9.3 Vanishing Gradients

Hochreiter and Schmidhuber [137] suggested replacing the hidden neurons of the recurrent network with computation units that are specially designed to reduce the vanishing-gradient problem. The method is referred to as *long-short-term memory* (LSTM). The basic ingredient is the same as in *residual networks* (Section 7.4): short cuts reduce the vanishing-gradient problem. For our purposes, we can think of LSTMs as units that replace the hidden neurons. For a detailed description of LSTMs, see Ref. [138].

Gated recurrent units [139] serve the same purpose as LSTMs, and they function in a similar way. It has been argued that LSTMs outperform gated recurrent units for certain tasks, but since they are simpler than LSTMs, the remainder of this section focuses on gated recurrent units. As illustrated in Figure 9.3, these units replace the hidden neurons of a recurrent neural network [with update rule (9.19a)] with a new rule:

$$z_m(t) = \sigma\Big(\sum_k w_{mk}^{(zx)} x_k(t) + \sum_j w_{mj}^{(zv)} V_j(t-1)\Big), \tag{9.30a}$$

$$r_n(t) = \sigma\Big(\sum_k w_{nk}^{(rx)} x_k(t) + \sum_j w_{nj}^{(rv)} V_j(t-1)\Big), \tag{9.30b}$$

$$h_i(t) = g\Big(\sum_k w_{ik}^{(hx)} x_k(t) + \sum_j w_{ij}^{(hv)} r_j(t) V_j(t-1)\Big), \tag{9.30c}$$

$$V_i(t) = [1 - z_i(t)]h_i(t) + z_i(t)V_i(t-1). \tag{9.30d}$$

The first two equations are referred to as *gates* because they regulate how the values of the hidden state variables V_i are passed through the unit. Here $\sigma(b)$ is the sigmoid function (6.19a). If $z_m(t) = 0$ for all m, and $r_n(t) = 1$ for all n, Equation (9.30) coincides with the standard update rule (9.19a), save for the thresholds which were left out in Equation (9.30). As explained previously, the resulting

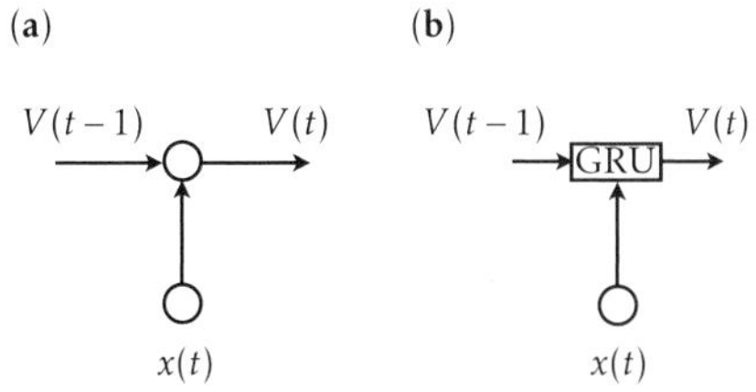

Figure 9.3 Gated recurrent unit. (**a**) The symbol refers to the standard recursion (9.19a) for the hidden variable, as in the right panel of Figure 9.2. (**b**) Gated recurrent unit (9.30).

recurrent network suffers from the vanishing-gradient problem. This means that states in the past history, $\boldsymbol{V}(0)$, $\boldsymbol{V}(1), \ldots$, have little effect upon the present state $\boldsymbol{V}(t)$ for $t \gg 1$. The recurrent network forgets early inputs, so that it cannot learn from them.

If by contrast $z_m(t) = 1$ for all m, then the input is passed right through the unit. Since $\partial V_i(t)/\partial V_j(t-1) = \delta_{ij}$ in this case, the gradients do not decrease as the dynamics explores the history. However, since $\boldsymbol{V}(t-1) = \boldsymbol{V}(t)$, the recurrent network reproduces previous states. This is analogous to skipping layers in a residual network, although comparison with Equation (7.35) reveals some differences in detail.

For the recurrent network to learn in a meaningful way from past inputs, the weights in Equation (9.30) (and the thresholds) are adjusted so that the gated recurrent unit operates between these two extreme limits. This is achieved by including the weights and thresholds of the gated recurrent unit in the gradient-descent minimisation of the energy function (9.20). The learning rules for the weights (and thresholds) are calculated in the same as before. Using Equation (9.19b), one has

$$\delta w_{mn}^{(ab)} = \eta \sum_{t=1}^{T} \sum_{i} \Delta_i(t) w_{ij}^{(ov)} \frac{\partial V_j(t)}{\partial w_{mn}^{(ab)}}, \tag{9.31}$$

where $w^{(ab)}$ stands for $w^{(zx)}$, $w^{(zv)}$, $w^{(rx)}, \ldots$. The derivatives $\partial V_j(t)/\partial w_{mn}^{(ab)}$ are evaluated using the chain rule and Equations (9.30).

It is instructive to inspect the values of $z_i(t)$ and $r_i(t)$ when the recurrent network operates after training. Suppose that a unit assumes small values of $z_i(t)$ and $r_i(t)$. This means that the update of the state variable $V_i(t)$ is determined entirely by the instantaneous inputs $x_k(t)$. Since the unit does not refer to the past history of the hidden-state variables, it truncates the dynamical memory. In the opposite limit, when $z_i(t) \approx r_i(t) \approx 1$, the unit can contribute to building up long-term dynamical memory. These arguments suggest that a unit with just one gate may achieve the same goal [140]:

$$z_m(t) = \sigma\Big(\sum_k w_{mk}^{(zx)} x_k(t) + \sum_j w_{mj}^{(zv)} V_j(t-1)\Big), \tag{9.32a}$$

$$h_i(t) = g\Big(\sum_k w_{ik}^{(hx)} x_k(t) + \sum_j w_{ij}^{(hv)} z_j(t) V_j(t-1)\Big), \tag{9.32b}$$

$$V_i(t) = (1 - z_i) h_i(t) + z_i(t) V_i(t-1). \tag{9.32c}$$

This unit is easier to train because it has fewer parameters than the standard gated recurrent unit (9.30). Yet the additional parameters in Equation (9.30) may help to represent and exploit correlations on different time scales. LSTMs have even more parameters. How this tradeoff between ease of training and accurate representation of time correlations works out may well depend on the problem at hand. In the following section, we describe recurrent networks with LSTM units, following Refs. [134, 135].

9.4 Recurrent Networks for Machine Translation

Recurrent networks are used for machine translation [134, 135]. How does this work? The networks are trained using backpropagation through time. The vanishing-gradient problem is dealt with by using LSTMs (Section 9.3).

How are the network inputs and outputs coded? For machine translation, one represents all words in a given dictionary in terms of a code. The conceptually simplest code is one where 100... represents the first word in the dictionary, 010... the second word, and so forth. The drawback of this scheme is that it does not account for the fact that two given words might be more or less closely related to each other. Other encoding schemes are described in Ref. [135].

Each input to the recurrent network is a vector with as many components as there are words in the dictionary. A sentence corresponds to a sequence $\boldsymbol{x}_1, \boldsymbol{x}_2, \ldots, \boldsymbol{x}_T$. Each sentence ends with an end-of-sentence tag, <EOS>. This tag tells the network when the input sentence ends. This is necessary because the number of words per sentence is not fixed. Now suppose that a possible translation reads $\boldsymbol{x}'_1, \boldsymbol{x}'_2, \ldots, \boldsymbol{x}'_{T'}$. The task of the network is to determine the probability $p(\boldsymbol{x}'_1, \ldots, \boldsymbol{x}'_{T'}|\boldsymbol{x}_1, \ldots \boldsymbol{x}_T)$ that the translation is correct. The idea is to estimate this probability recursively as

$$p(\boldsymbol{x}'_1, \ldots, \boldsymbol{x}'_{T'}|\boldsymbol{x}_1, \ldots \boldsymbol{x}_T) = \prod_{t=1}^{T'} p(\boldsymbol{x}'_t|\boldsymbol{x}'_1, \ldots, \boldsymbol{x}'_{t-1}; \boldsymbol{x}_1, \ldots, \boldsymbol{x}_T)\,. \tag{9.33}$$

Sutskever *et al.* [134] describe how to achieve this with a recurrent network with two hidden LSTMs. The network uses softmax outputs $\boldsymbol{O}_t$, where j-th component of $\boldsymbol{O}_t$ is interpreted as the probability that the j-th component of $\boldsymbol{x}'_t$ is the correct word at position t in the translated sentence. As shown in Figure 9.4, the first LSTM processes the input sentence $\boldsymbol{x}_1, \ldots, \boldsymbol{x}_T$, encoding its contents in the hidden states. When the <EOS> tag appears, the second LSTM takes over, using the information encoded by the first LSTM as an input. The second LSTM recursively outputs the translated sentence word by word, using Equation (9.33).

There is a large number of recent papers on machine translation with recurrent neural networks. Most studies are based on the training algorithm described in Section 9.2, backpropagation through time. The different approaches mainly differ in their network layouts. Google's machine translation system uses a deep network

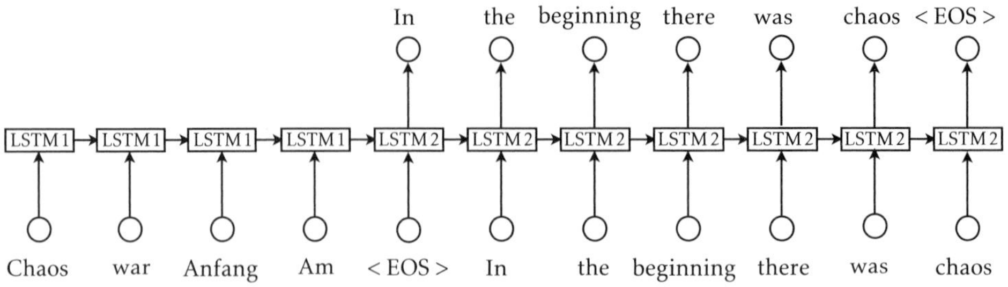

Figure 9.4 Schematic illustration of an unfolded recurrent network for machine translation, after Refs. [134, 135]. The rectangular boxes represent the hidden states in the form of two long-short-term memory (LSTM) units, see Section 9.3. Sutskever et al. [134] found that the network translates much better if the sentence is read in reverse order, from the end. <EOS> is an end-of-sequence tag. After Figure 1 in Ref. [134].

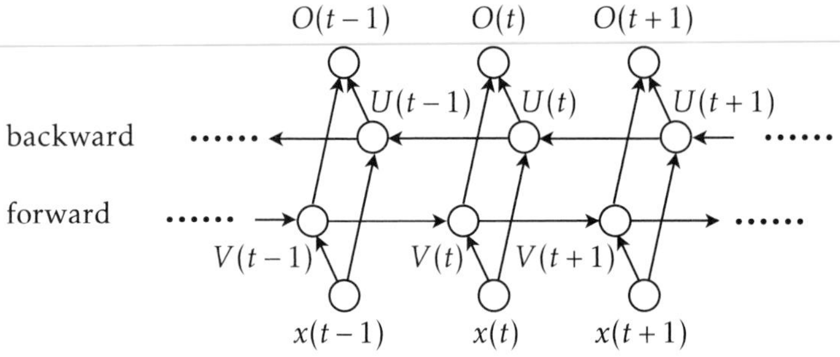

Figure 9.5 Schematic illustration of a bidirectional recurrent network. The network consists of two hidden neurons, $U(t)$ and $V(t)$, that are unfolded in different ways. After Figure 12 in Ref. [135].

with several layers of hidden units arranged in a bidirectional layout [141]. In such bidirectional networks, different hidden units are unfolded forward as well as backwards in time, as shown schematically in Figure 9.5. For several hidden and output neurons and multi-dimensional inputs, the bidirectional network has the dynamics

$$
\begin{aligned}
V_i(t) &= g\Big(\sum_j w_{ij}^{(vv)} V_j(t-1) + \sum_k w_{ik}^{(vx)} x_k(t) - \theta_i^{(v)}\Big),\\
U_i(t) &= g\Big(\sum_j w_{ij}^{(uu)} U_j(t+1) + \sum_k w_{ik}^{(ux)} x_k(t) - \theta_i^{(u)}\Big),\\
O_i(t) &= g\Big(\sum_j w_{ij}^{(ov)} V_j(t) + \sum_j w_{ij}^{(ou)} U_j(t) - \theta_i^{(o)}\Big),
\end{aligned}
\tag{9.34}
$$

where the hidden neurons $\boldsymbol{V}$ are updated forward in time, while $\boldsymbol{U}$ are updated backwards in time. It is natural to use bidirectional networks for machine translation because correlations go either way in a sentence, forward and backwards. In German, for example, the finite verb form is usually at the end of the sentence. In practice, the hidden states are represented by LSTMs [134, 135, 141], instead of hidden neurons as in Equation (9.34).

Different schemes for scoring the accuracy of a translation are described by Lipton et al. [135]. One difficulty is that there are often several different valid translations of a given sentence, and the score must compare the machine translation with all of them. More recent papers on machine translation usually use the so-called BLEU score to evaluate the translation accuracy. The acronym stands for *bilingual evaluation understudy*. The scheme was proposed by Papieni et al. [142]. It is argued to evaluate the accuracy of a translation not too differently from humans.

9.5 Reservoir Computing

An alternative to backpropagation through time for recurrent networks is *reservoir computing* [143]. This method has been used with success to predict chaotic dynamics [144, 145] and rare transitions in stochastic bi-stable systems [146].

Consider input data in the form of a time series $\boldsymbol{x}(0), \ldots \boldsymbol{x}(T-1)$ of N-dimensional vectors $\boldsymbol{x}(t)$ and a corresponding series of M-dimensional targets $\boldsymbol{y}(t)$. The goal is to train the recurrent network so that its outputs $\boldsymbol{O}(t)$ approximate the targets as precisely as possible, by minimising the energy function $H = \frac{1}{2}\sum_{t=\tau}^{T-1}\sum_{i=1}^{N}[E_i(t)]^2$, where $E_i(t) = y_i(t) - O_i(t)$ is the output error, and τ represents an initial transient that is disregarded.

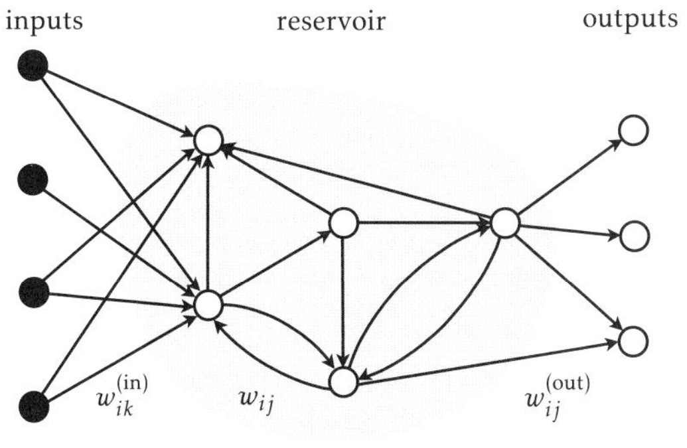

Figure 9.6 Reservoir computing (schematic). Not all connections are drawn. There can be connections from all inputs to all neurons in the reservoir, and from all reservoir neurons to all output neurons.

Figure 9.6 shows the layout for this task. There are N input terminals. They are connected with weights $w_{jk}^{(\text{in})}$ to a reservoir of hidden neurons with state variables $r_j(t)$. The reservoir is linked to M linear output units $O_i(t)$ with weights $w_{ij}^{(\text{out})}$. The reservoir itself is a large recurrent network with weights w_{ij}. The update rule is similar to Equation (9.19). There are many different versions that differ in detail [147]. One possibility is [146]

$$r_i(t+1) = g\Big(\sum_j w_{ij} r_j(t) + \sum_{k=1}^{N} w_{ik}^{(\text{in})} x_k(t)\Big), \tag{9.35a}$$

$$O_i(t+1) = \sum_j w_{ij}^{(\text{out})} r_j(t+1), \tag{9.35b}$$

for $t = 0, \ldots, T-1$ with initial conditions $r_j(0)$.

The main difference compared with the training algorithms described in the previous sections of this chapter is that the input weights $w_{jk}^{(\text{in})}$ and the reservoir weights w_{jk} are randomly initialised and then kept constant throughout the computation. Only the output weights $w_{jk}^{(\text{out})}$ are trained. The idea is that the dynamics of a sufficiently large reservoir finds non-linear, high-dimensional representations of the input data [143], not unlike sparse representations of binary classification problems embedded in a high-dimensional space that become linearly separable in this way (Section 5.4).

In addition, and this is a difference to the problem described in Section 5.4, the reservoir is a dynamical memory. This requires that the reservoir states faithfully represent the input sequence: similar input sequences should yield similar reservoir activations, provided one iterates it long enough.

However, for random weights, the recurrent reservoir dynamics can be unstable, or chaotic. In this case, the state of the reservoir after many iterations bears no relation to the input sequence. To avoid this, one requires that the reservoir dynamics is linearly stable. Linearising the reservoir dynamics (9.35a) gives

$$\delta \boldsymbol{r}(t+1) = \mathbb{D}(t+1)\mathbb{W}\delta \boldsymbol{r}(t), \tag{9.36}$$

where $\mathbb{D}(t+1)$ is a diagonal matrix with entries $D_{ii}(t+1) = g'[b_i(t+1)]$, and where $b_i(t+1) = \sum_j w_{ij} r_j(t) + \sum_k w_{ik}^{(\text{in})} x_k(t)$. Whether or not $\delta \boldsymbol{r}$ grows is then determined by the singular values of $\mathbb{J}_t = \mathbb{D}(t)\mathbb{W}\mathbb{D}(t-1)\mathbb{W}\cdots\mathbb{D}(1)\mathbb{W}$, as in Section 7.2. The singular values of $\mathbb{J}_t$ are denoted by $\Lambda_1(t) \geq \Lambda_2(t) \geq \cdots$. At large times, when driven with a stationary input series, the maximal Lyapunov exponent $\lambda_1 = \lim_{t\to\infty} t^{-1} \log \Lambda_1(t)$ must be negative to ensure that the reservoir dynamics (9.35) is stable:

$$\lambda_1 < 0. \tag{9.37}$$

Sometimes the stability criterion is quoted in terms of the maximal eigenvalue of $\mathbb{W}$. If one uses tanh activation-functions and if the local fields $b_i(t)$ remain small, then the diagonal elements of $\mathbb{D}(t)$ remain close to unity. In this case, the stability condition for the reservoir dynamics is given by the weight matrix $\mathbb{W}$ alone. In general the singular values of $\mathbb{W}$ are different from its eigenvalues, but what matters here is that the maximal singular value of $\mathbb{W}^t$ approaches $e^{t \log|\nu_1|}$, where ν_1 is the eigenvalue of $\mathbb{W}$ with largest modulus (Exercise 9.8).

For inputs with long time correlations, the reservoir dynamics must not decay too quickly, so that it can represent the dynamical correlations in the input sequence. There is no precise mathematical theory that says how to optimise the reservoir. In practice one adjusts the maximal Lyapunov exponent by trial and error. Its optimal value depends on the properties of the input series, for instance upon whether it is chaotic or not.

There are many different recipes for how to set up a reservoir. Usually the reservoir is sparse, with only a small fraction of weights non-zero. The elements of the resulting weight matrix $\mathbb{W}$ are rescaled to adjust λ_1 [145]. The weight matrix $\mathbb{W}^{(\text{in})}$ is commonly taken to be a full matrix, and its elements are drawn from the same distribution as those of the reservoir. Lukosevicius [147] gives a practical overview over different schemes for setting up reservoir computers.

For time-series prediction, one trains the network on many input series $\boldsymbol{x}(0), \ldots, \boldsymbol{x}(T-1)$ with targets $\boldsymbol{y}(t) = \boldsymbol{x}(t)$. After training, one continues to iterate the network dynamics with inputs $\boldsymbol{x}(T+k) = \boldsymbol{O}(T+k)$ to predict $\boldsymbol{x}(T+k+1)$, for $k = 0, 1, 2, \ldots$. In order to represent complex spatio-temporal patterns, Pathak et al. [144] found it necessary to use several parallel reservoirs. Lim et al. [146] used a chain of reservoirs, replacing Equation (9.35a) by a set of nested update rules.

Tanaka et al. [148] describe different physical implementations of reservoir computers, based on electronic RC-circuits, optical cavities or resonators, spin-torque oscillators, or mechanical devices.

9.6 Summary

It is sometimes said that recurrent networks learn *dynamical systems*, while multilayer perceptrons learn *input-output maps*. This emphasises a difference in how these networks are usually used, but we should bear in mind that they are trained in similar ways, by backpropagation. Neither is it given that the tasks must differ: recurrent networks are also used to learn time-independent data. It is true, however, that tools from *dynamical-systems theory* help analyse the dynamics of recurrent networks [136, 149].

Recurrent neural networks can be trained by stochastic gradient descent after unfolding the network in time, to get rid of feedback connections. This algorithm suffers from the vanishing-gradient problem. To overcome this difficulty, the hidden neurons in the recurrent network are replaced by composite units that are trained to sometimes act as residual connections, passing the signal right through, and sometimes as non-linear units that can learn correlations in a meaningful way. There are different versions, long-short-term memory units and gated recurrent units. They all work in similar ways. Successful layouts for machine translation use deep bidirectional networks with layers of LSTMs.

An alternative scheme is reservoir computing, where a large reservoir of hidden neurons is used to represent correlations in the input data, and a set of linear output units is trained to learn the original sequence from such representations. The idea is that it is easier to learn intricate features of an input sequence from a high-dimensional, sparse representation of the data.

9.7 Further Reading

The training of recurrent networks is discussed in Chapter 15 of Ref. [2]; see also Refs. [150, 151]. Recurrent backpropagation is described by Hertz, Krogh, and Palmer [1], for a very similar network layout. How LSTMs combat the vanishing-gradient problem is explained in Ref. [138]. For a recent review of recurrent neural networks, see Ref. [135]. This webpage [152] gives a very enthusiastic overview of what recurrent networks can do. A more pessimistic view is expressed in this blog. For a review of reservoir computing, see Ref. [143].

9.8 Exercises

9.1 Recurrent backpropagation. Derive Equation (9.15) for the weight increments $\delta w_{mn}^{(vx)}$ in recurrent backpropagation. Show how the recurrent-backpropagation algorithm simplifies to Algorithm 4 for layered feed-forward networks when there are no feedbacks.

9.2 Steady state in recurrent backpropagation. Verify that the dynamics (9.3) has a steady state satisfying Equation (9.6). Determine its linear stability.

9.3 Learning rules for backpropagation through time. Derive the learning rules (9.27) and (9.28) from Equation (9.19).

9.4 Recurrent network. Figure 9.7 shows a simple recurrent network with one hidden neuron $V(t)$, one input component $x(t)$, and one output $O(t)$. The network

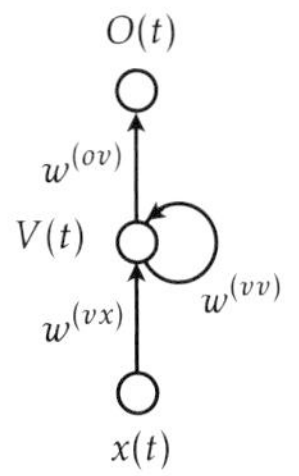

Figure 9.7 Recurrent network with one input terminal $x(t)$, one hidden neuron $V(t)$, and one output neuron $O(t)$. See Exercise 9.4.

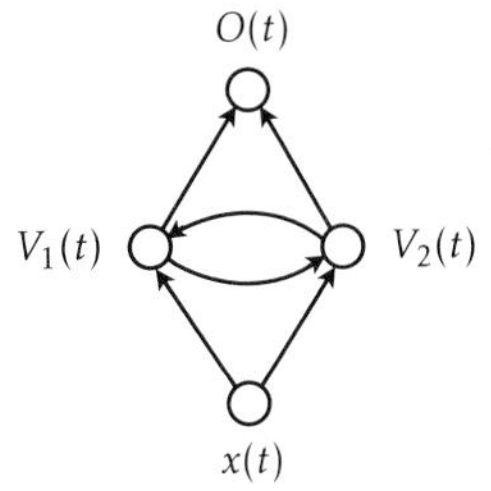

Figure 9.8 Recurrent network used in Exercise 9.5. After Figure 3 in Ref. [135].

learns a time series of input-output pairs $[x(t), y(t)]$ for $t = 1, 2, 3, \ldots, T$. Here t is a discrete time index and $y(t)$ is the target value at time t. This network can be trained by backpropagation by *unfolding it in time*. Write down the dynamical rules for this network, the rules that determine $V(t)$ in terms of $V(t-1)$ and $x(t)$, and $O(t)$ in terms of $V(t)$. Assume that both $V(t)$ and $O(t)$ have the same activation function $g(b)$. Derive the learning rule for $w^{(ov)}$ using gradient descent on the energy function $H = \frac{1}{2}\sum_{t=1}^{T} E(t)^2$ with $E(t) = y(t) - O(t)$. Denote the learning rate by η.

9.5 Backpropagation through time. A recurrent network with two hidden neurons is shown in Figure 9.8. Write down the dynamical rules for this network. Assume that all neurons have the same activation function $g(b)$. Draw the unfolded network. Derive the learning rules for the weights.

9.6 Backpropagation through time for thresholds. Derive learning rules for the thresholds $\theta_j^{(v)}$ and $\theta_i^{(0)}$ for the backpropagation-through-time algorithm.

9.7 Dual dynamics for recurrent backpropagation. Show that the dynamics (9.16) admits a steady state satisfying (9.13).

9.8 Eigenvalues and singular values. Compute the eigenvalues and the singular values of the matrices

$$\mathbb{A}_1 = \begin{bmatrix} 1 & 2 \\ 2 & 2 \end{bmatrix}, \quad \mathbb{A}_2 = \begin{bmatrix} 1 & 1 \\ -1 & 1 \end{bmatrix}, \quad \mathbb{A}_3 = \begin{bmatrix} 1 & 2 \\ 0 & 2 \end{bmatrix}. \tag{9.38}$$

These examples illustrate that the singular values Λ_α of a symmetric matrix equal its eigenvalues ν_α. For a normal matrix, $\Lambda_\alpha = |\nu_\alpha|$. In general, singular values and eigenvalues differ. Then show that for all three matrices, the maximal singular value $\Lambda_1(t)$ of $\mathbb{A}^t$ converges to $\mathrm{e}^{t \log|\nu_1|}$, where ν_1 is the eigenvalue of $\mathbb{A}$ with largest modulus.

9.9 Time series prediction. Ikeda modelled the ray dynamics of light in an optical resonator by the map [151]

$$x_1(t+1) = 1 + u\big[x_1(t)\cos(\tau) - x_2(t)\sin(\tau)\big], \tag{9.39a}$$

$$x_2(t+1) = u\big[x_1(t)\cos(\tau) + x_2(t)\sin(\tau)\big], \tag{9.39b}$$

with $\tau = 0.4 - 6/(1 + |\boldsymbol{x}(t)|^2)$, and $u = 0.8$. In this case the time series generated by Equation (9.39) are chaotic and therefore difficult to predict. Train a reservoir computer on a data set of time series generated by numerical solution of Equation (9.39). Evaluate how well the reservoir computer manages to predict the time series. For background on time-series prediction, refer to *Nonlinear Time Series Analysis* by Kantz and Schreiber [154].

Part III

Learning without Labels

Chapters 5 to 9 describe supervised learning of labelled data with neural networks. The network is trained to reproduce the correct labels (targets) for each input pattern. The analysis of unlabelled data requires different methods. Machine learning can be applied with success to large data sets of high-dimensional unlabelled data. The machine can for instance mark patterns that are typical for the given distribution, or detect outliers. Other tasks are to detect similarity, to find clusters in the data (Figure 10.1), and to determine non-linear, low-dimensional representations of high-dimensional data. More recently, such *unsupervised* learning algorithms have been used to generate synthetic data, patterns that resemble those in a certain data set. One possible application is data-set augmentation for supervised learning.

Learning without labels is called unsupervised learning, because there are no targets that tell the network whether it has learnt correctly or not. There is no obvious function to fit, or dynamics to learn. Instead, the network organises the input data in relevant ways. This requires *redundancy* in the input data [1]. It is sometimes said that unsupervised learning corresponds to learning without a teacher, implying that the network itself discovers suitable ways of organising the input data. This is inaccurate, because unsupervised networks usually operate with a pre-determined learning rule, like Hopfield networks.

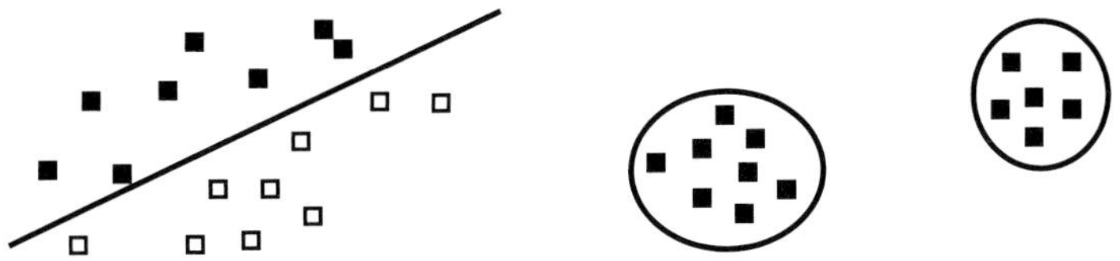

Figure 10.1 Supervised learning finds decision boundaries for labelled data, like in the binary classification problem shown on the left. Unsupervised learning can find clusters in the input data (right)

Part III of this book is organised as follows: Chapter 10 describes unsupervised-learning algorithms, starting with unsupervised Hebbian learning to detect familiarity and similarity of input patterns (Sections 10.1 and 10.2). Related algorithms can be used to find low-dimensional non-linear projections of high-dimensional input data (self-organising maps, Section 10.3). In Section 10.4, these algorithms are compared and contrasted with a standard unsupervised clustering algorithm, K-means clustering. Section 10.5 introduces radial basis-function networks, they learn using a hybrid algorithm with supervised and unsupervised elements. Section 10.6 explains how to use layered feed-forward networks for unsupervised learning.

Chapter 11 deals with learning tasks that lie in between supervised and unsupervised learning, problems where the machine receives partial feedback on its performance in the shape of a penalty or a reward. Such tasks can be solved by *reinforcement-learning* algorithms that allow a neural network or more generally an agent to learn to reproduce outputs that tend to give positive rewards. Several algorithms for reinforcement learning are described, the associative reward-penalty algorithm (Section 11.1), temporal difference learning (Section 11.2), and Q-learning (Section 11.3). The Q-learning algorithm is illustrated by demonstrating how it allows two players to learn to compete in the board game tic-tac-toe.

10

Unsupervised Learning

10.1 Oja's Rule

A simple example for an unsupervised-learning algorithm uses a single McCulloch-Pitts neuron with linear activation function (Figure 10.2). The neuron computes[1] $y = \boldsymbol{w} \cdot \boldsymbol{x}$ with weight vector $\boldsymbol{w} = [w_1, \ldots, w_N]^{\mathsf{T}}$. Now consider a distribution $P_{\text{data}}(\boldsymbol{x})$ of input patterns $\boldsymbol{x} = [x_1, \ldots, x_N]^{\mathsf{T}}$ with continuous-valued components x_i. Patterns are drawn from this distribution at random and fed one after another to the net. For each pattern $\boldsymbol{x}$, the weights $\boldsymbol{w}$ are adjusted as follows:

$$\boldsymbol{w}' = \boldsymbol{w} + \delta\boldsymbol{w} \quad \text{with} \quad \delta\boldsymbol{w} = \eta y \boldsymbol{x} \,. \tag{10.1}$$

This rule is also called *Hebbian unsupervised learning* rule [1] because it is reminiscent of Hebb's rule (Chapter 2). As usual, η is the learning rate.

What can this rule learn about the input distribution $P_{\text{data}}(\boldsymbol{x})$? Since we keep adding multiples of the pattern vectors $\boldsymbol{x}$ to the weights (just as described in Section 5.2), the magnitude of the output $|y|$ becomes the larger the more often the input pattern occurs in the distribution $P_{\text{data}}(\boldsymbol{x})$. So the most familiar pattern produces the largest output. In this way, the network can detect how *familiar* certain input patterns are.

A problem is that the components of the weight vector continue to grow as we keep adding. This means that the simple Hebbian learning rule (10.1) does not converge to a steady state. To analyse learning outcomes we want the learning to converge. This is achieved by adding a weight-decay term with coefficient proportional to y^2 (Section 7.6.1):

$$\delta\boldsymbol{w} = \eta y (\boldsymbol{x} - y\boldsymbol{w}) \,. \tag{10.2}$$

[1] In this chapter, we follow a common convention [1] and denote the output of unsupervised-learning algorithms by y.

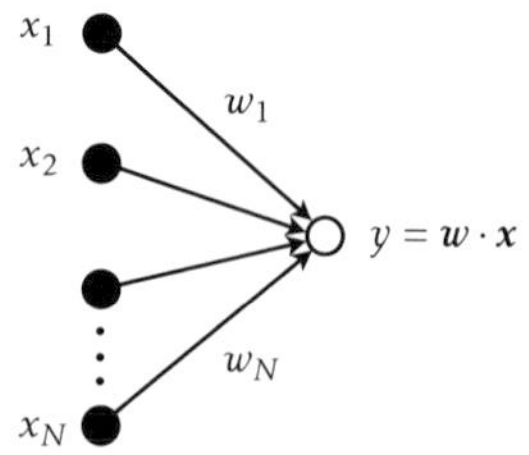

Figure 10.2 Unsupervised Hebbian learning with a single linear output unit that has weight vector $\boldsymbol{w}$. The network output is denoted by y in this chapter

Making use of $y = \boldsymbol{w} \cdot \boldsymbol{x} = \boldsymbol{w}^{\mathsf{T}}\boldsymbol{x} = \boldsymbol{x}^{\mathsf{T}}\boldsymbol{w}$, Equation (10.2) can be rewritten in the following form:

$$\delta \boldsymbol{w} = \eta\big\{\boldsymbol{x}\boldsymbol{x}^{\mathsf{T}}\boldsymbol{w} - [\boldsymbol{w} \cdot (\boldsymbol{x}\boldsymbol{x}^{\mathsf{T}})\boldsymbol{w}]\boldsymbol{w}\big\}\,. \tag{10.3}$$

This learning rule is called *Oja's rule* [155]. Equation (10.3) ensures that $\boldsymbol{w}$ remains normalised. To see why, consider an analogy: a vector $\boldsymbol{q}$ that obeys the differential equation

$$\tfrac{\mathrm{d}}{\mathrm{d}t}\boldsymbol{q} = \mathbb{A}(t)\boldsymbol{q}\,. \tag{10.4}$$

For a general matrix $\mathbb{A}(t)$, the norm $|\boldsymbol{q}|$ may increase or decrease, depending on the singular values of $\mathbb{A}$. We can ensure that $\boldsymbol{q}$ remains normalised by adding a term to Equation (10.4):

$$\tfrac{\mathrm{d}}{\mathrm{d}t}\boldsymbol{w} = \mathbb{A}(t)\boldsymbol{w} - [\boldsymbol{w} \cdot \mathbb{A}(t)\boldsymbol{w}]\boldsymbol{w}\,. \tag{10.5}$$

The vector $\boldsymbol{w}$ turns in the same way as $\boldsymbol{q}$, and if we set $|\boldsymbol{w}| = 1$ initially, then $\boldsymbol{w}$ remains normalised, $\boldsymbol{w} = \boldsymbol{q}/|\boldsymbol{q}|$ (Exercise 10.1). Equation (10.5) describes the dynamics of the normalised orientation vector of a small rod in turbulence [156], where $\mathbb{A}(t)$ is the matrix of fluid-velocity gradients.

Returning to Equation (10.2), we note that the dynamics of (10.2) and (10.5) is the same in the limit of small learning rates η. Therefore we conclude that $\boldsymbol{w}$ remains normalised under (10.3) when the learning rate is small enough. Oja's algorithm is summarised in Algorithm 8. One draws a pattern $\boldsymbol{x}$ from the distribution $P_{\mathrm{data}}(\boldsymbol{x})$ of input patterns, applies it to the network, and updates the weights as prescribed in Equation (10.2). This is repeated many times. In the following, we denote the average over T input patterns as $\langle \cdots \rangle = \frac{1}{T}\sum_{t=1}^{T} \cdots$.

While the rule (10.1) does not have a steady state, Oja's rule (10.3) does. For zero-mean input data, its steady state $\boldsymbol{w}^*$ corresponds to the principal component of the input data, as illustrated in Figure 10.3. This can be seen by analysing the steady-state condition

$$0 = \langle \delta \boldsymbol{w} \rangle_{\boldsymbol{w}^*}\,. \tag{10.6}$$

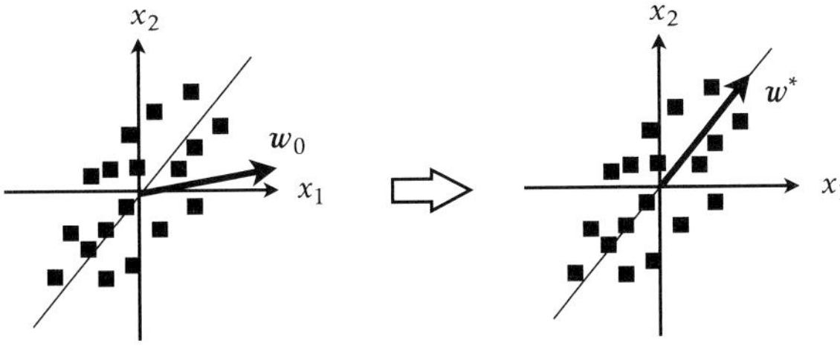

Figure 10.3 Oja's rule finds the principal component of zero-mean data (schematic). The initial weight vector is $\boldsymbol{w}_0$; the steady-state weight vector is $\boldsymbol{w}^*$

Here $\langle\cdots\rangle_{\boldsymbol{w}^*}$ is an average over iterations of the learning rule (10.3) at fixed $\boldsymbol{w}^*$, the steady state. Equation (10.6) says that the weight increments $\delta\boldsymbol{w}$ must average to zero in the steady state, to ensure that the weights neither grow nor decrease in the long run. Equation (10.6) is a condition upon $\boldsymbol{w}^*$. Using the learning rule (10.3), it can be written as

$$0 = \mathbb{C}'\boldsymbol{w}^* - (\boldsymbol{w}^* \cdot \mathbb{C}'\boldsymbol{w}^*)\boldsymbol{w}^* \quad \text{with} \quad \mathbb{C}' = \langle \boldsymbol{x}\boldsymbol{x}^{\mathsf{T}}\rangle\,. \tag{10.7}$$

Equation (10.7) shows that $\boldsymbol{w}^*$ must be an eigenvector of the matrix[2] $\mathbb{C}'$, normalised to unity, $|\boldsymbol{w}^*| = 1$. But which one?

We denote the eigenvectors and eigenvalues of $\mathbb{C}'$ by $\boldsymbol{u}_\alpha$ and λ_α, and investigate the stability of $\boldsymbol{w}^* = \boldsymbol{u}_\alpha$ for different values of α by linear stability analysis, just as in Section 9.1. To this end, consider a small perturbation $\boldsymbol{\varepsilon}_t$ away from $\boldsymbol{w}^* = \boldsymbol{u}_\alpha$:

$$\boldsymbol{w}_t = \boldsymbol{u}_\alpha + \boldsymbol{\varepsilon}_t\,. \tag{10.8}$$

A difference to the analysis in Section 9.1 is that the dynamics is now discrete in time. The perturbation at the next time step, $\boldsymbol{\varepsilon}_{t+1}$, is defined by $\boldsymbol{w}_{t+1} = \boldsymbol{u}_\alpha + \boldsymbol{\varepsilon}_{t+1}$. A second difference is that the sequence of weight increments depends on the randomly chosen input patterns. In order to determine the linear stability one should iterate and then linearise the dynamics (10.3), to see whether $\boldsymbol{\varepsilon}_t$ grows or

Algorithm 8 Oja's rule

initialise weights randomly;
for $t = 1,\ldots,T$ **do**
  draw an input pattern $\boldsymbol{x}$ from $P_{\text{data}}(\boldsymbol{x})$;
  adjust all weights using $\delta\boldsymbol{w} = \eta y(\boldsymbol{x} - y\boldsymbol{w})$;
end for

[2] For zero-mean input data, $\mathbb{C}'$ equals the data-covariance matrix, Equation (6.24).

not. However, in the limit of small learning rate, it is sufficient to average over $\boldsymbol{x}$ before iterating (Exercise 10.4). To linear order in $\boldsymbol{\varepsilon}_t$, one finds

$$\boldsymbol{\varepsilon}_{t+1} \approx \boldsymbol{\varepsilon}_t + \eta\Big[\mathbb{C}'\boldsymbol{\varepsilon}_t - 2\boldsymbol{u}_\alpha(\boldsymbol{u}_\alpha \cdot \mathbb{C}'\boldsymbol{\varepsilon}_t) - (\boldsymbol{u}_\alpha \cdot \mathbb{C}\boldsymbol{u}_\alpha)\boldsymbol{\varepsilon}_t\Big] = \mathbb{M}^{(\alpha)}\boldsymbol{\varepsilon}_t\,, \tag{10.9}$$

where the last equality sign defines the matrix $\mathbb{M}^{(\alpha)}$. The steady state $\boldsymbol{w}^* = \boldsymbol{u}_\alpha$ is linearly stable if all eigenvalues of $\mathbb{M}^{(\alpha)}$ have real parts with magnitudes smaller than unity.[3] To determine the eigenvalues of $\mathbb{M}^{(\alpha)}$, we use the fact that $\mathbb{M}^{(\alpha)}$ has the same eigenvectors as $\mathbb{C}'$. Since $\mathbb{C}'$ is symmetric, these eigenvectors form an orthonormal basis, $\boldsymbol{u}_\alpha \cdot \boldsymbol{u}_\beta = \delta_{\alpha\beta}$. As a consequence, the eigenvalues of $\mathbb{M}^{(\alpha)}$ are simply given by

$$\Lambda_\beta^{(\alpha)} = \boldsymbol{u}_\beta \cdot \mathbb{M}^{(\alpha)}\boldsymbol{u}_\beta = 1 + \eta[(\lambda_\beta - \lambda_\alpha) - 2\lambda_\alpha\delta_{\alpha\beta}]\,. \tag{10.10}$$

Since $\mathbb{C}'$ is a positive-semidefinite matrix (its eigenvalues λ_α cannot be negative), Equation (10.10) shows that there are eigenvalues with $|\Lambda_\beta^{(\alpha)}| > 1$ unless $\boldsymbol{w}^*$ is the leading eigenvector of $\mathbb{C}'$, the one corresponding to its largest eigenvalue. This means that Algorithm 8 finds the principal component of zero-mean data, and it also implies that the algorithm maximises $\langle y^2\rangle$ over all $\boldsymbol{w}$ with $|\boldsymbol{w}| = 1$, see Section 6.3. Note that $\langle y\rangle = 0$ for zero-mean input data.

Now consider inputs with non-zero mean. In this case, Algorithm 8 still finds the maximal-eigenvalue direction of $\mathbb{C}'$. But for inputs with non-zero mean, this direction is different from the principal direction. Figure 10.4 illustrates this difference. The figure shows three data points in a two-dimensional input plane. The elements of $\mathbb{C}' = \langle \boldsymbol{x}\boldsymbol{x}^{\mathsf{T}}\rangle$ are

$$\mathbb{C}' = \frac{1}{3}\begin{bmatrix} 2 & 1 \\ 1 & 2 \end{bmatrix}, \tag{10.11}$$

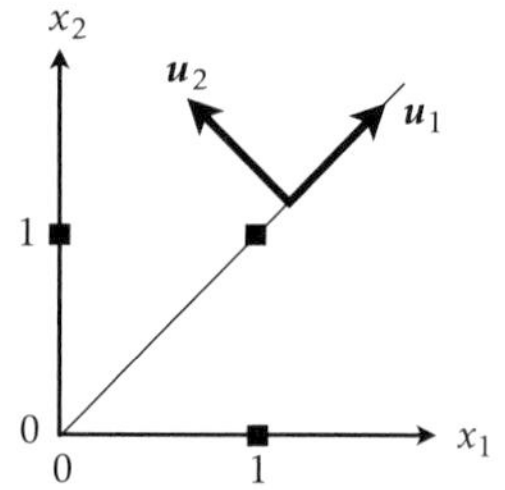

Figure 10.4 Input data with non-zero mean. Algorithm 8 converges to $\boldsymbol{u}_1$, but the principal direction is $\boldsymbol{u}_2$

[3] For time-continuous dynamics (Section 9.1), linear stability is ensured when all eigenvalues have negative real parts, for discrete dynamics their magnitudes must all be smaller than unity [85].

with eigenvalues and eigenvectors

$$\lambda_1 = 1\,, \quad \boldsymbol{u}_1 = \frac{1}{\sqrt{2}}\begin{bmatrix}1\\1\end{bmatrix} \quad \text{and} \quad \lambda_2 = \tfrac{1}{3}\,, \quad \boldsymbol{u}_2 = \frac{1}{\sqrt{2}}\begin{bmatrix}-1\\1\end{bmatrix}. \tag{10.12}$$

So the maximal-eigenvalue direction of $\mathbb{C}'$ is $\boldsymbol{u}_1$. To compute the principal direction of the data, we must determine the data-covariance matrix $\mathbb{C}$, Equation (6.24). Its maximal-eigenvalue direction is $\boldsymbol{u}_2$, and this is the principal component of the data shown in Figure 10.4.

Oja's rule can be generalised to determine principal components of zero-mean input data using M output neurons that compute $y_i = \boldsymbol{w}_i \cdot \boldsymbol{x}$ for $i = 1, \ldots, M$:

$$\delta w_{ij} = \eta y_i \Big(x_j - \sum_{k=1}^{M} y_k w_{kj}\Big). \tag{10.13}$$

This is called Oja's M-rule [1]. For $M = 1$, Equation (10.13) simplifies to Oja's rule.

10.2 Competitive Learning

Oja's M-rule (10.13) results in neurons that are activated simultaneously. Any input usually causes several outputs to assume non-zero values $y_i \neq 0$ at the same time. In Sections 4.5 and 7.1, we encountered the notion of a winning neuron where the weights are trained in such a way that each pattern activates only a single neuron, and different patterns activate different winning neurons. This allows one to represent a distribution of input patterns with a neural network.

Unsupervised learning algorithms can categorise or cluster input data in this way: similar inputs are classified to belong to the same category and activate the same winning neuron. This is called *competitive learning* [1]. Figure 10.5(**a**) shows an example: input patterns on the unit circle that cluster into two distinct clusters. The idea is to find weight vectors $\boldsymbol{w}_i$ that point into the direction of the clusters. To this end, we take M linear output units i with weight vectors $\boldsymbol{w}_i$, $i = 1, \ldots, M$. We feed a pattern $\boldsymbol{x}$ from the distribution $P_{\rm data}(\boldsymbol{x})$ and *define* the *winning neuron* i_0 as the one that has a minimal angle between its weight and the pattern vector $\boldsymbol{x}$. This is illustrated in Figure 10.5(**b**), where $i_0 = 2$. Then only this weight vector is updated by adding a little bit of the difference $\boldsymbol{x} - \boldsymbol{w}_{i_0}$ between the pattern vector and the weight of the winning neuron. The other weights remain unchanged:

$$\delta \boldsymbol{w}_i = \begin{cases} \eta(\boldsymbol{x} - \boldsymbol{w}_i) & \text{for} \quad i = i_0(\boldsymbol{x}, \boldsymbol{w}_1 \ldots \boldsymbol{w}_M)\,, \\ 0 & \text{otherwise.} \end{cases} \tag{10.14}$$

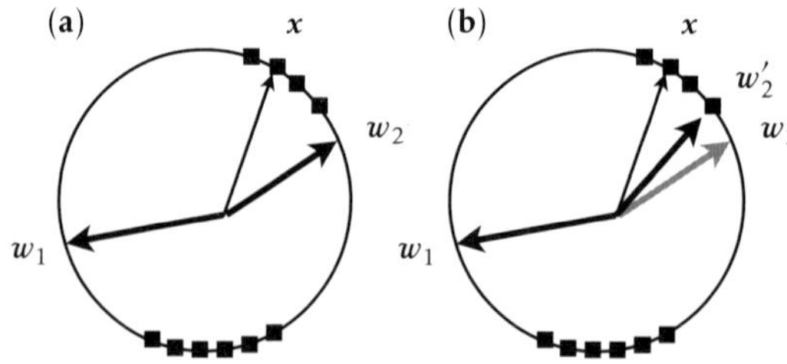

Figure 10.5 Detection of clusters by unsupervised learning. (**a**) Distribution of input patterns on the unit circle and two weight vectors initialised to random angles. The winning neuron for pattern $\boldsymbol{x}$ is the one with weight vector $\boldsymbol{w}_2$. (**b**) Updating $\boldsymbol{w}_2' = \boldsymbol{w}_2 + \delta\boldsymbol{w}$ moves this weight vector closer to $\boldsymbol{x}$

In other words, only the winning neuron is updated, $\boldsymbol{w}'_{i_0} = \boldsymbol{w}_{i_0} + \delta\boldsymbol{w}_{i_0}$. Equation (10.14) is called *competitive-learning* rule.

The learning rule (10.14) has the following geometrical interpretation: the weight of the winning neuron is drawn towards the pattern $\boldsymbol{x}$. Upon iterating (10.14), the weight vectors are drawn to *clusters* of inputs. If the input patterns are normalised as in Figure 10.5, the weights end up normalised on average, even though $|\boldsymbol{w}_{i_0}| = 1$ does not imply that $|\boldsymbol{w}_{i_0} + \delta\boldsymbol{w}_{i_0}| = 1$, in general. The algorithm for competitive learning is summarised in Algorithm 9. When weight and input vectors are normalised, then the winning neuron i_0 is the one with the largest scalar product $\boldsymbol{w}_i \cdot \boldsymbol{x}$. For linear output units $y_i = \boldsymbol{w}_i \cdot \boldsymbol{x}$ (Figure 10.2), this is simply the unit with the largest output. Equivalently, the winning neuron is the one with the smallest distance $|\boldsymbol{w}_i - \boldsymbol{x}|$. Output units with $\boldsymbol{w}_i$ that are very far away from any pattern may never be updated (*dead units*). There are several strategies to avoid this [1]. One possibility is to initialise the weights to directions found in the inputs. Also, how to choose the number of weight vectors is a matter of trial and error. Clearly, it is better to start with too many rather than too few.

Algorithm 9 Competitive learning (Figure 10.5)

initialise weights to vectors with random angles and norm $|\boldsymbol{w}_i| = 1$;
for $t = 1, \ldots, T$ **do**
  draw a pattern $\boldsymbol{x}$ from $P_{\text{data}}(\boldsymbol{x})$;
  find the winning neuron i_0 (smallest angle between $\boldsymbol{w}_{i_0}$ and $\boldsymbol{x}$);
  adjust only the weight of the winning neuron $\delta\boldsymbol{w}_{i_0} = \eta(\boldsymbol{x} - \boldsymbol{w}_{i_0})$;
end for

Finally, consider the relation between the competitive learning rule (10.14) and Oja's rule (10.13). If we define

$$y_i = \delta_{ii_0} = \begin{cases} 1 & \text{for } i = i_0\,, \\ 0 & \text{otherwise,} \end{cases} \tag{10.15}$$

then the rule (10.14) can be written in the form of Oja's M-rule:

$$\delta w_{ij} = \eta y_i \Big(x_j - \sum_{k=1}^{M} y_k w_{kj} \Big) \,. \tag{10.16}$$

Equation (10.16) is reminiscent of Hebb's rule (Chapter 2) with weight decay.

10.3 Self-Organising Maps

In order to analyse high-dimensional data, it is often useful to map the high-dimensional input patterns to a low-dimensional output space, to obtain a low-dimensional representation of the input distribution. Principal-component analysis (Section 6.3) does just that. However, it does not necessarily preserve distance. To visualise clusters or other arrangements of the input patterns, similar patterns or patterns that are close in input space should be mapped to nearby points in output space, and patterns that are far apart should be mapped to outputs that are far from each other. Maps with this property are called *semantic* or *topographic* maps.

Moreover, principal-component analysis is a linear method. As explained in Section 6.3, it projects the data to the space spanned by the leading eigenvectors of the correlation matrix. In many cases, however, the data may not lie in a linear subspace, as illustrated in Figure 10.6. In order to project the data onto the non-linear *principal manifold* (solid line), a non-linear map is needed.

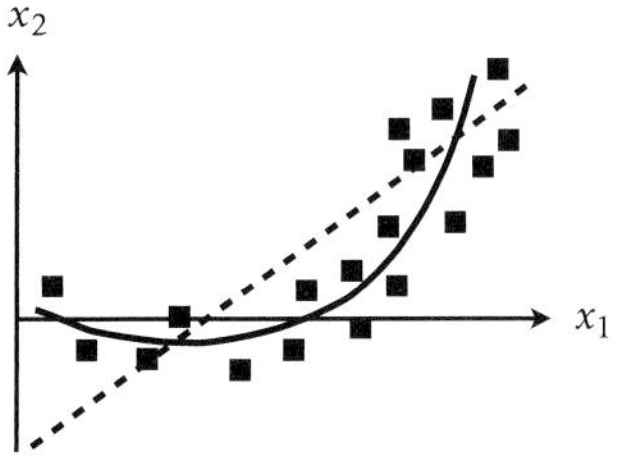

Figure 10.6 Principal-component analysis (Section 6.3) finds the linear principal direction (dashed line) of the data (■). A self-organising map can instead find the *principal manifold* (solid line), a non-linear approximation to the data

In neuroscience, the term topographic map refers to the relation between the spatial arrangement of stimuli and the activation patterns in certain parts of the mammalian brain. Similar patterns of visual stimuli on the retina, for instance, activate close-by regions in the visual cortex [157]. Other cognitive stimuli, auditory and sensory, are mapped in analogous ways. The complex neural networks in the mammalian cortex contain large numbers of such maps, arranged in a hierarchical fashion. They represent local stimuli in terms of spatially localised neural activation. How did this complex structure arise? One possibility is that the mappings are coded in the genetic sequence, that the connections are hard wired, so to speak. However, it is observed that such maps can change over time [158], leading to the hypothesis that they are learned and that our DNA merely encodes a set of fairly simple *learning rules*.

This motivated Kohonen [158, 159] and others to propose and analyse learning rules for topographic maps. The term *self-organising map* [18, 160] emphasises that the mapping develops in response to the stimuli it maps, that it learns in an unsupervised fashion. Kohonen's model for a non-linear self-organising map relies on an ordered array of output neurons, as illustrated in Figure 10.7. The map learns to activate nearby output neurons for similar inputs. This is achieved using a competitive learning rule, similar to the learning rule (10.14) described in the previous section. In order to represent the proximity or similarity of inputs, the rule is endowed with the notion of distance in the output array, by updating not only the

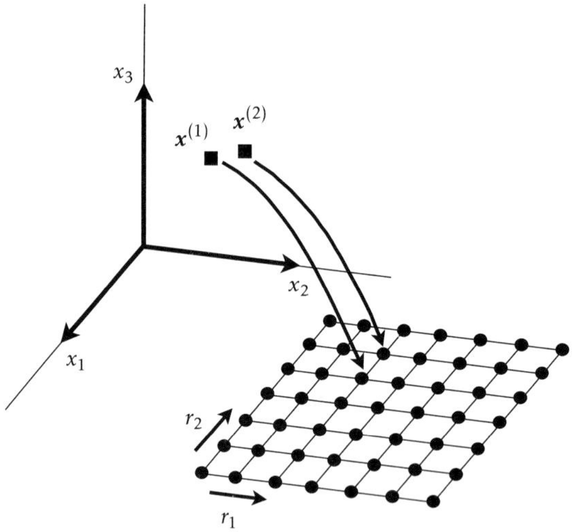

Figure 10.7 Kohonen's self-organising map. If patterns $\boldsymbol{x}^{(1)}$ and $\boldsymbol{x}^{(2)}$ are close in input space, then the two patterns activate neighbouring winning neurons in the output array with coordinates $\boldsymbol{r} = [r_1, r_2]^{\mathsf{T}}$. Often the dimension of the output array is much lower than that of input space

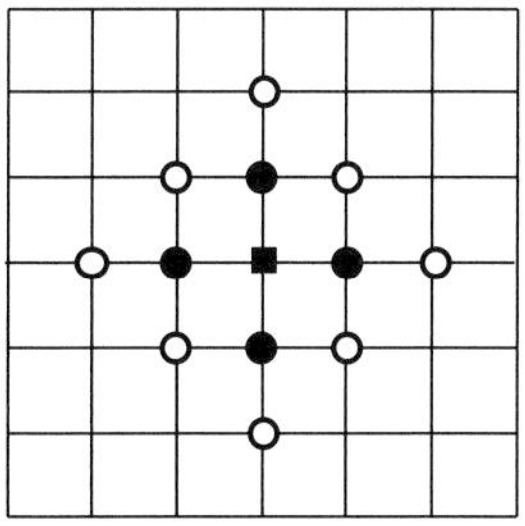

Figure 10.8 Nearest neighbours (•) and next-nearest neighbours (○) to the neuron at the centre (■) of the output array

winning neuron but also those that are neighbours in the output array. To this end, one replaces the competitive-learning rule (10.14) by

$$\delta \boldsymbol{w}_i = \eta h(i, i_0)(\boldsymbol{x} - \boldsymbol{w}_i)\,, \tag{10.17}$$

where i_0 is the index of the winning neuron, the one with weight vector closest to the input $\boldsymbol{x}$. The *neighbourhood function* $h(i, i_0)$ depends on the distance of the neurons i and i_0 in the output array. The neighbourhood function has a maximum at $i = i_0$ and decreases as the distance between i and i_0 increases. One possibility is to assign decreasing values to $h(i, i_0)$ for nearest neighbours, next-nearest neighbours, and so forth (Figure 10.8). Another possibility is to use a Gaussian function of the Euclidean distance $|\boldsymbol{r}_i - \boldsymbol{r}_{i_0}|$ in the output array [1]:

$$h(i, i_0) = \exp\Big(-\tfrac{1}{2\sigma^2}|\boldsymbol{r}_i - \boldsymbol{r}_{i_0}|^2\Big)\,. \tag{10.18}$$

Here $\boldsymbol{r}_i$ is the position of neuron i in the output array (Figure 10.7). Different normalisations of the Gaussian [2] can be subsumed in different learning rates.

Kohonen's rule has two parameters: the learning rate η, and the width σ of the neighbourhood function. Usually one adjusts these parameters as the learning proceeds. Typically one begins with large values for η and σ (*ordering phase*), and then reduces these parameters as the elastic network evolves (*convergence phase*): quickly at first and then in smaller steps, until the algorithm converges [1, 2, 158].

According to Equations (10.17) and (10.18), similar patterns activate nearby neurons in output space, and their weight vectors change in similar ways. Kohonen's rule drags the winning weight vector $\boldsymbol{w}_{i_0}$ towards $\boldsymbol{x}$, just as the competitive learning rule (10.14), but it also drags the neighbouring weight vectors along. Figure 10.9 illustrates a geometrical interpretation of Kohonen's rule [18]. We can think of the weight vectors as pointing to the nodes of an *elastic net* that has the same layout

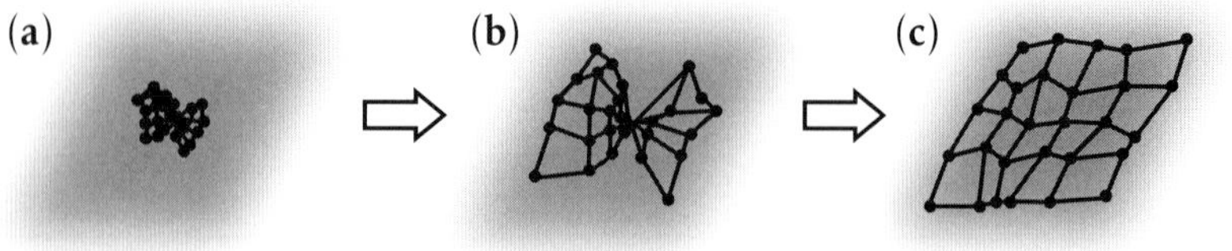

Figure 10.9 Learning a distribution $P_{\text{data}}(\boldsymbol{x})$ (gray) of two-dimensional real-valued inputs $\boldsymbol{x}$ with Kohonen's algorithm. Illustration of the dynamics of the self-organising map in terms of an *elastic net*. (**a**) Initial condition. (**b**) Intermediate stage. (**c**) In the steady state, the elastic network resembles the shape defined by the input distribution $P_{\text{data}}(\boldsymbol{x})$. After Figure 3.4 in Ref. [157]

as the output array. As one feeds patterns from the input distribution, the weights are updated, causing the nodes of the network to move. This changes the shape of the elastic network until it resembles the shape defined by the distribution of input patterns.

Figure 10.6 shows an example where the dimensionality of the output array (one-dimensional) is lower than that of the input space (two-dimensional). The algorithm finds a non-linear approximation to the data, the *principal manifold*. As opposed to the principal direction in principal-component analysis, the principal manifold need not be linear. Therefore it can approximate the data more precisely, leading to a smaller residual variance (Exercise 10.7).

In summary, Kohonen's algorithm learns by distributing the weight vectors of the output neurons to reflect the distribution of input patterns. This works well in general, but problems occur at the boundaries. Why this happens is quite clear (Figure 10.9): since the density of patterns outside the parallelogram is low, the elastic network cannot be drawn very close to the boundary. To analyse how the boundaries affect learning for Kohonen's rule, consider the steady-state condition

$$\langle \delta \boldsymbol{w}_i \rangle = \frac{\eta}{T} \sum_{t=1}^{T} h(i, i_0) \left(\boldsymbol{x}^{(t)} - \boldsymbol{w}_i^* \right) = 0 \,. \tag{10.19}$$

This condition is more complicated than it looks at first sight, because i_0 depends on the weights and on the patterns, as mentioned above. The steady-state condition (10.19) is very difficult to analyse in general. One of the reasons is that global geometric information is difficult to learn. It is usually much easier to learn local structures. This is particularly true in the *continuum limit* where we can analyse local learning progress using Taylor expansions.

The analysis of condition (10.19) in the continuum limit is due to Ritter and Schulten [161], and it is described in detail by Hertz, Krogh, and Palmer [1]. One assumes that there is a very dense network of output neurons, so that one can approximate $i \to \boldsymbol{r}$, $i_0 \to \boldsymbol{r}_0$, $\boldsymbol{w}_i \to \boldsymbol{w}(\boldsymbol{r})$, $h(i, i_0) \to h\big(\boldsymbol{r} - \boldsymbol{r}_0(\boldsymbol{x})\big)$, and $\frac{1}{T}\sum_t \to \int \mathrm{d}\boldsymbol{x}\, P_{\text{data}}(\boldsymbol{x})$. In this continuum limit, Equation (10.19) reads

$$\int \mathrm{d}\boldsymbol{x}\; P_{\text{data}}(\boldsymbol{x})\, h\big(\boldsymbol{r} - \boldsymbol{r}_0(\boldsymbol{x})\big)\big[\boldsymbol{x} - \boldsymbol{w}^*(\boldsymbol{r})\big] = 0\,. \tag{10.20}$$

This is a condition for the steady-state learning outcome, the function $\boldsymbol{w}^*(\boldsymbol{r})$.

In the continuum limit, the position $\boldsymbol{r}_0(\boldsymbol{x})$ of the winning neuron in the output array for pattern $\boldsymbol{x}$ is given by

$$\boldsymbol{w}^*(\boldsymbol{r}_0) = \boldsymbol{x}\,. \tag{10.21}$$

We use this relation to write Equation (10.20) as

$$\int \mathrm{d}\boldsymbol{x}\; P_{\text{data}}(\boldsymbol{x})\, h\big(\boldsymbol{r} - \boldsymbol{r}_0(\boldsymbol{x})\big)\big[\boldsymbol{w}^*(\boldsymbol{r}_0(\boldsymbol{x})) - \boldsymbol{w}^*(\boldsymbol{r})\big] = 0\,. \tag{10.22}$$

Equation (10.21) defines a mapping $\boldsymbol{r}_0(\boldsymbol{x})$ from input space to output space, the self-organising map (Figure 10.7). Assuming that this mapping is one-to-one, we change integration variable from $\boldsymbol{x}$ to $\boldsymbol{r}_0$

$$\int \mathrm{d}\boldsymbol{r}_0\, |\det\mathbb{J}|\; Q(\boldsymbol{r}_0)\, h(\boldsymbol{r} - \boldsymbol{r}_0)\big[\boldsymbol{w}^*(\boldsymbol{r}_0) - \boldsymbol{w}^*(\boldsymbol{r})\big] = 0\,, \tag{10.23}$$

where $Q(\boldsymbol{r}_0) \equiv P_{\text{data}}\big(\boldsymbol{x}(\boldsymbol{r}_0)\big)$, and where the determinant represents the volume element of the variable transformation. Using Equation (10.21), the Jacobian $\mathbb{J}$ of the transformation has elements

$$J_{ij} = \frac{\partial w_i(\boldsymbol{r}_0)}{\partial r_j}\,. \tag{10.24}$$

The neighbourhood function is sharply peaked at $\boldsymbol{r} = \boldsymbol{r}_0$, and this makes it possible to evaluate the steady-state condition (10.23) approximately, expanding the integrand in $\delta\boldsymbol{r} = \boldsymbol{r}_0 - \boldsymbol{r}$, assuming that $\boldsymbol{w}^*(\boldsymbol{r})$ is a smooth function. This is illustrated in Figure 10.10, for one-dimensional inputs and outputs. We consider this special case not only to simplify the notation, but also because it is one of the few cases that admit mathematical analysis (Exercise 10.9). Expanding $w^*(r + \delta r)$ as shown in Figure 10.10 yields

$$w^*(r + \delta r) - w^*(r) = \tfrac{\mathrm{d}}{\mathrm{d}r} w^*(r)\delta r + \tfrac{1}{2}\tfrac{\mathrm{d}^2}{\mathrm{d}r^2} w^*(r)\delta r^2 + \ldots\,. \tag{10.25}$$

The other factors in Equation (10.23) are expanded in a similar way:

$$J(r + \delta r) = \tfrac{\mathrm{d}w^*}{\mathrm{d}r} + \tfrac{\mathrm{d}^2 w^*}{\mathrm{d}r^2}\delta r + \ldots\,, \tag{10.26a}$$

$$Q(r + \delta r) = P_{\text{data}}(w^*) + \delta r \tfrac{\mathrm{d}w^*}{\mathrm{d}r}\tfrac{\mathrm{d}}{\mathrm{d}w} P_{\text{data}}(w^*)\,. \tag{10.26b}$$

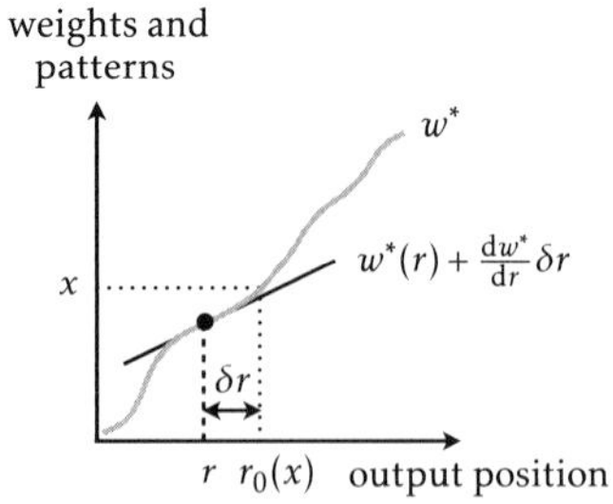

Figure 10.10 In order to find out how the steady-state map $w^*(r)$ varies near r (gray line), one expands w^* in δr around r, $w^*(r+\delta r) = w^*(r) + \frac{dw^*}{dr}\delta r + \frac{1}{2}\frac{d^2w^*}{dr^2}\delta r^2 + \ldots$

Inserting these expressions into Equation (10.20), discarding terms of order higher than δr^2, and changing the integration variable to δr, one finds

$$0 = w'[\tfrac{3}{2}w'' P_{\text{data}}(w^*) + (w')^2 \tfrac{\mathrm{d}}{\mathrm{d}w} P_{\text{data}}(w^*)] \int_{-\infty}^{\infty} \mathrm{d}\delta r\, \delta r^2 h(\delta r)\,, \tag{10.27}$$

where we introduced the short-hand notation $w' = \frac{\mathrm{d}}{\mathrm{d}r} w^*(r)$, dropped the asterisk, and used that the neighbourhood function (10.18) is symmetric, $h(-\delta r) = h(\delta r)$. Since the integral in Equation (10.27) is non-zero, we must either have

$$w' = 0 \quad \text{or} \quad \frac{3}{2} w'' P_{\text{data}}(w) + (w')^2 \tfrac{\mathrm{d}}{\mathrm{d}w} P_{\text{data}}(w) = 0\,. \tag{10.28}$$

The first solution can be excluded because it corresponds to a singular weight distribution that does not contain any geometrical information about the input distribution P_{data}. The second solution gives

$$\frac{w''}{w'} = -\frac{2}{3} \frac{w' \frac{\mathrm{d}}{\mathrm{d}w} P_{\text{data}}(w)}{P_{\text{data}}(w)}\,. \tag{10.29}$$

In other words, $\frac{\mathrm{d}}{\mathrm{d}r} \log |w'| = -\frac{2}{3}\frac{\mathrm{d}}{\mathrm{d}r} \log P_{\text{data}}(w)$, and this means that $|w'| \propto [P_{\text{data}}(w)]^{-\frac{2}{3}}$. The density of output weights can be computed as

$$\varrho(w) = \int \mathrm{d}r \delta[w - w^*(r)]\,, \tag{10.30}$$

where $\delta(w)$ is the Dirac δ-function [162]. Changing variables in the δ-function

$$\delta[w - w^*(r)] = \sum_{j | w = w^*(r_j)} \frac{1}{|w'|} \delta(r - r_j)\,, \tag{10.31}$$

and assuming that the function $w^*(r)$ is one-to-one, one finds

$$\varrho(w) = \frac{1}{|w'|} = [P_{\text{data}}(w)]^{\frac{2}{3}} . \tag{10.32}$$

This tells us that the self-organising map learns the input distribution in the following way: the distribution of output weights in the steady state reflects the distribution of input patterns. Equation (10.32) shows that the two distributions are not equal (equality would have been a perfect outcome). The distribution of weights is instead proportional to $\left[P_{\text{data}}(w)\right]^{\frac{2}{3}}$. Little is known in higher dimensions, but the general idea is that the elastic network has difficulties reaching the corners and edges of the domain where the input distribution is non-zero.

The output of a self-organising map can be interpreted in different ways. For low-dimensional inputs and outputs, one can simply plot the map $w^*(\boldsymbol{r})$, as in Figure 10.7. Dense regions of weights point to regions in input space with a high density of inputs. Often the output dimension is taken to be much lower than the dimension of input space. In this case, the self-organising map performs non-linear *dimensionality reduction*, and it can be used to find clusters in high-dimensional input data [163]. The analysis proceeds in two steps. First, one runs Kohonen's algorithm until the map has converged to a steady state. Second, one feeds all inputs into the net, and for each input one determines the location of the winning neuron in the output array. The spatial activation patterns in the output array represent clusters of similar inputs. This is illustrated in Figure 10.11, which shows how a self-organising map represents handwritten digits from the MNIST data set (Section 8.3). To reveal the semantic map, the figure labels clusters of outputs that correspond to the same digits (as determined by the labels in the training set). We see that the self-organising map groups the same digits together, but it has some difficulty distinguishing the digits 3 and 8, and also 4 and 9.

10.4 *K*-Means Clustering

Sections 10.2 and 10.3 described different ways of finding clusters in input data. In particular, it was shown how self-organising maps can find clusters in high-dimensional input data, and represent them in a low-dimensional, non-linear projection. K-means clustering [2] is an alternative unsupervised-learning algorithm for finding clusters in the input data. Let us compare and contrast this algorithm with Kohonen's self-organising map. The goal is to cluster input patterns $\boldsymbol{x}^{(\mu)}$, $\mu = 1, \ldots, p$ into K clusters. Usually K is much smaller than the number of inputs, p, and than the input dimension N.

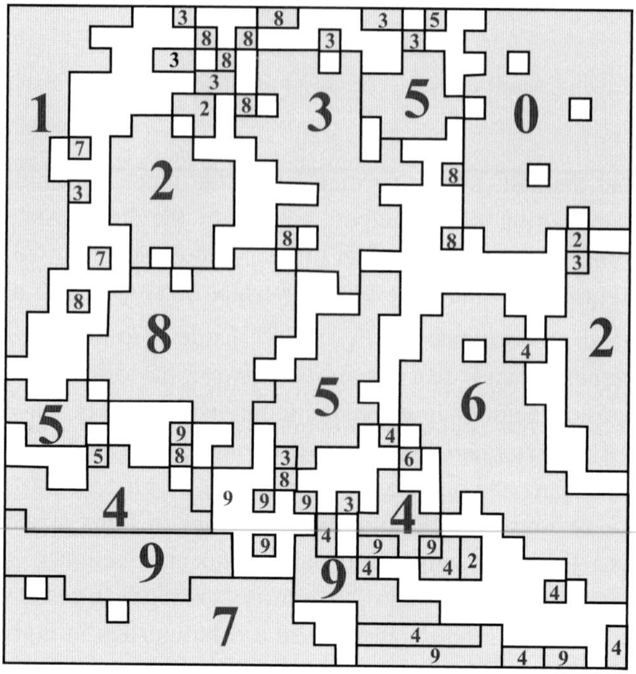

Figure 10.11 Clustering of handwritten digits (MNIST data set) with a self-organising map with a 30×30 output array. In the shaded regions the outputs are quite certain: here the winning neurons are activated by the indicated digit in 80% of the cases. The white regions correspond to outputs where the majority digit appears in less than 80% of the cases, or to outputs that are never activated, or only once. Schematic, based on simulations performed by Juan Diego Arango

A solution of the clustering task is a mapping $k(\mu)$ that associates each input $\boldsymbol{x}^{(\mu)}$ with one of the clusters $k = 1, \ldots, K$. The function $k(\mu)$ is determined by minimising the energy function

$$H(\boldsymbol{w}_1, \ldots, \boldsymbol{w}_K) = \frac{1}{2}\sum_{k=1}^{K}\Big(\sum_{\mu\,|\,k(\mu)=k} |\boldsymbol{x}^{(\mu)} - \boldsymbol{w}_k|^2\Big)\,. \tag{10.33}$$

The second sum is over all values of μ that satisfy $k(\mu) = k$. The vector $\boldsymbol{w}_k$ becomes the average of all pattern vectors in cluster k, and the expression in the parentheses is the variance associated with this cluster:

$$\sigma_k^2 = \sum_{\mu\,|\,k(\mu)=k} |\boldsymbol{x}^{(\mu)} - \boldsymbol{w}_k|^2\,. \tag{10.34}$$

In other words, H measures the sum of the cluster variances σ_k^2. A solution to the clustering problem corresponds to a local minimum of H. To determine the cluster

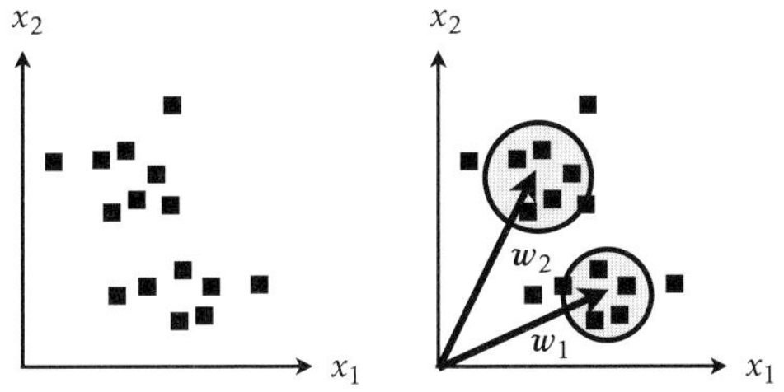

Figure 10.12 Schematic illustration of the K-means clustering algorithm with two weight vectors $\boldsymbol{w}_1$ and $\boldsymbol{w}_2$. The radii of the disks equal s_1 and s_2. See Equation (10.34)

vectors $\boldsymbol{w}_k$ and the corresponding variances σ_k^2, one starts from an initial guess for $k(\mu)$. For each cluster, one begins by adjusting $\boldsymbol{w}_k$ to minimise the cluster variance:

$$\underset{\boldsymbol{w}_k}{\arg\min} = \sum_{\mu|k(\mu)=k} |\boldsymbol{x}^{(\mu)} - \boldsymbol{w}_k|^2 . \tag{10.35}$$

In a second step, one optimises the encoding function

$$k(\mu) = \underset{1\leq k\leq K}{\arg\min} |\boldsymbol{x}^{(\mu)} - \boldsymbol{w}_k|^2 , \tag{10.36}$$

given the vectors $\boldsymbol{w}_k$. These steps are repeated until a satisfactory solution is found (Figure 10.12). The solution is not unique; usually the algorithm converges to a local minimum of H. In practice, one should try different random initialisations to find the best local minimum.

All three algorithms, competitive learning, the self-organising map, and K-means clustering move weight vectors towards clusters in input space. A difference between the self-organising map and the other two algorithms is that the self-organising map uses a neighbourhood function (so that similar inputs activate close-by neurons in the output array) and updates their weight vectors in a similar fashion. In this way, a self-organising map with a large output array can find a smooth parameterisation of the principal manifold. If we shrink the neighbourhood function to the centre point in Figure 10.8, all geometric information is lost and the self-organising map becomes equivalent to competitive learning (Algorithm 9). Essentially, competitive learning and K-means clustering are sequential and batch versions of the same algorithm [1]. So the self-organising map becomes equivalent to K-means clustering when the neighbourhood range tends to zero.

10.5 Radial Basis Functions

Problems that are not linearly separable can be solved by perceptrons with hidden layers, as we saw in Chapter 5. Figure 5.13(**b**), for example, shows a piecewise linear decision boundary parameterised by hidden neurons.

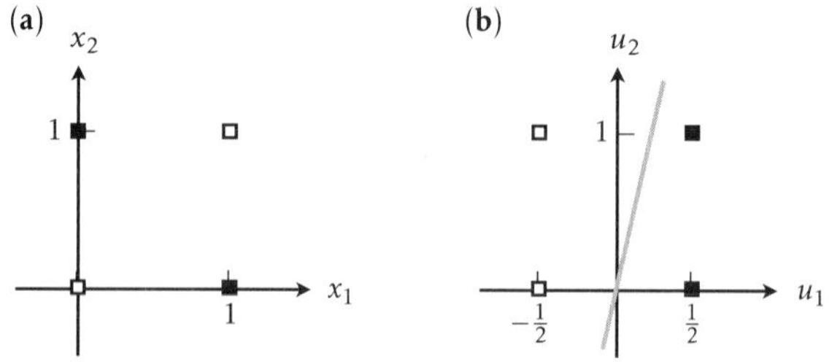

Figure 10.13 Linear separation of the XOR function by the non-linear mapping (10.37). (**a**) In the input plane, the problem is not linearly separable. (**b**) In the u_1-u_2 plane, the problem is homogeneously linearly separable

Separability can also be achieved by a non-linear transformation of input space. Figure 10.13 shows how the XOR problem can be transformed into a linearly separable problem by the transformation

$$u_1(\boldsymbol{x}) = (x_2 - x_1)^2 - \tfrac{1}{2} \quad \text{and} \quad u_2(\boldsymbol{x}) = x_2\,. \tag{10.37}$$

The figure shows the non-separable problem in the x_1-x_2 plane and in the new coordinates u_1 and u_2. Since the problem is homogeneously linearly separable in the u_1-u_2 plane, we can solve it by a single McCulloch-Pitts neuron with weights $\boldsymbol{W}$ and zero threshold, parameterising the decision boundary as $\boldsymbol{W} \cdot \boldsymbol{u}(\boldsymbol{x}) = 0$.

It is even better to map the patterns (non-linearly) to a space of higher dimension, because Cover's theorem (Section 5.4) says that it is easier to separate the patterns there: consider a set $\boldsymbol{u}(\boldsymbol{x}) = [u_1(\boldsymbol{x}), \ldots, u_m(\boldsymbol{x})]^{\mathsf{T}}$ of m polynomial functions of finite order that embed N-dimensional input space in an m-dimensional space. Then the probability that a problem with p points $\boldsymbol{x}^{(\mu)}$ in N-dimensional input space is separable by a polynomial decision boundary is given by $P(p, m)$ [Equation (5.29)] [2, 72]. Note that this probability is independent of the dimension N of input space.

The question is of course how to find the non-linear mapping $\boldsymbol{u}(\boldsymbol{x})$. One possibility is to use *radial basis functions*. The idea is to parameterise the functions $u_j(\boldsymbol{x})$ in terms of weight vectors $\boldsymbol{w}_j$, and to use an unsupervised-learning algorithm to find weights that separate the input data. A common choice [2] is to use radial basis functions of the form:

$$u_j(\boldsymbol{x}) = \exp\Big(-\frac{1}{2s_j^2}\big|\boldsymbol{x} - \boldsymbol{w}_j\big|^2\Big). \tag{10.38}$$

Note that these functions are not of the finite-order polynomial form that was assumed above. Thus, strictly speaking, we cannot invoke Cover's theorem. In practice, the mapping $u_j(\boldsymbol{x})$ works nevertheless quite well. The parameters s_j parameterise the *widths* of the radial basis functions. In the simplest version of the algorithm, they are set to unity. Hertz, Krogh, and Palmer [1] discuss radial basis-function networks with *normalised* radial basis functions

$$u_j(\boldsymbol{x}) = \frac{\exp\left(-\frac{1}{2s_j^2}|\boldsymbol{x}-\boldsymbol{w}_j|^2\right)}{\sum_{k=1}^{m}\exp\left(-\frac{1}{2s_k^2}|\boldsymbol{x}-\boldsymbol{w}_k|^2\right)}. \tag{10.39}$$

Other choices for radial basis functions are given by Haykin [2].

Figure 10.14 shows a radial basis-function network for $N = 2$ and $m = 4$. The four neurons in the hidden layer stand for the four radial basis functions (10.38) that map the inputs to four-dimensional $\boldsymbol{u}$-space. The network looks like a perceptron (Chapter 5). But here the hidden layers work in a different way. Perceptrons have hidden McCulloch-Pitts neurons that compute *non-local* outputs $\sigma(\boldsymbol{w}_j \cdot \boldsymbol{x} - \theta)$. The output of radial basis functions $u_j(\boldsymbol{x})$, by contrast, is *localised* in input space [Figure 10.15 (left)]. We saw in Section 7.1 how to make localised basis functions out of McCulloch-Pitts neurons with sigmoid activation functions $\sigma(b)$, but one needs two hidden layers to do that [Figure 10.15 (right)].

Radial basis functions produce localised outputs with a single hidden layer, they divide up input space into localised regions, each corresponding to one radial basis

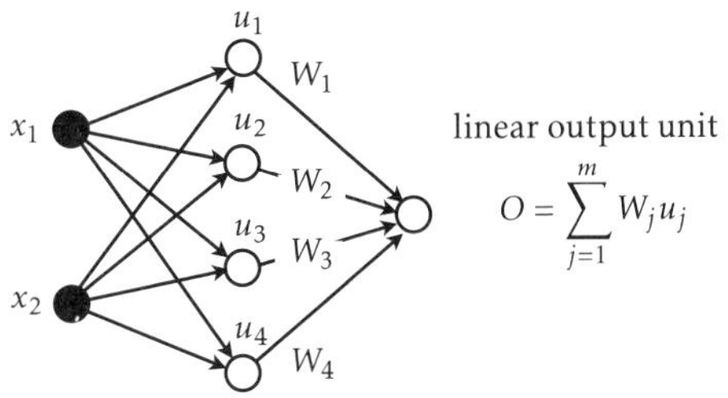

Figure 10.14 Radial basis-function network for $N = 2$ inputs and $m = 4$ radial basis functions (10.38). The output neuron has a linear activation function, weights $\boldsymbol{W}$, and zero threshold

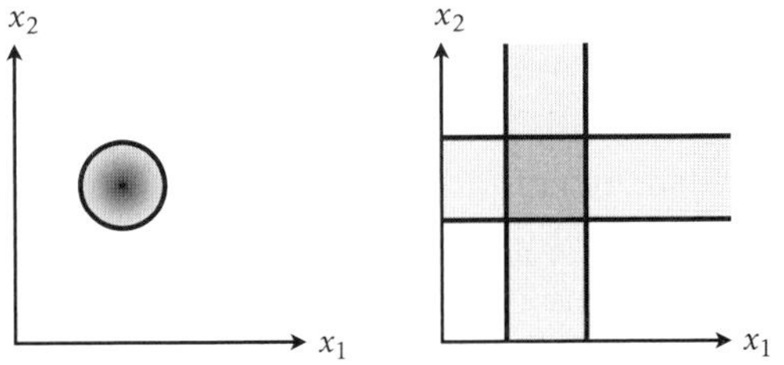

Figure 10.15 Comparison between radial-basis function network and perceptron. Left: the output of a radial basis function is localised in input space. Right: to achieve a localised output with sigmoid units one needs two hidden layers (Figure 7.4)

function. Imagine for a moment that we have as many radial basis functions as input patterns. In this case, we can simply take $\boldsymbol{w}_\nu = \boldsymbol{x}^{(\nu)}$ for $\nu = 1, \ldots, p$. The linear output in Figure 10.14 computes $O^{(\mu)} = \boldsymbol{W}\cdot\boldsymbol{u}(\boldsymbol{x}^{(\mu)})$, and so the classification problem in $\boldsymbol{u}$-space takes the form

$$\sum_{\mu=1}^{p} W_\mu U_{\mu\nu} = t^{(\nu)} \tag{10.40}$$

with $U_{\mu\nu} = u_\nu(\boldsymbol{x}^{(\mu)})$. If all patterns are pairwise different, $\boldsymbol{x}^{(\mu)} \neq \boldsymbol{x}^{(\nu)}$ for $\mu \neq \nu$, then the matrix $\mathbb{U}$ is invertible [2]. In this case, the solution of the classification problem reads

$$W_\mu = \sum_{\nu=1}^{p} t^{(\nu)}[\mathbb{U}^{-1}]_{\nu\mu}\,, \tag{10.41}$$

where $\mathbb{U}$ is the symmetric $p \times p$ matrix with elements $U_{\mu\nu}$.

In practice, one can get away with fewer radial basis functions by choosing their weights to point in the directions of clusters of input data. To this end, one uses unsupervised competitive learning (Algorithm 10), where the index j_0 of the *winning neuron* is defined to be the one with the largest u_j. How are the widths s_j determined? The width s_j of radial basis function $u_j(\boldsymbol{x})$ is taken to be equal to the minimum distance between $\boldsymbol{w}_j$ and the centres of the surrounding radial basis functions. Once weights and widths of the radial basis functions are found, the weights of the output neuron are determined by minimising

$$H = \frac{1}{2}\sum_{\mu}\left(t^{(\mu)} - O^{(\mu)}\right)^2 \tag{10.42}$$

with respect to $\boldsymbol{W}$. This works even if $\mathbb{U}$ is not invertible. An approximate solution can be obtained by stochastic gradient descent on H, keeping the parameters of the radial basis functions fixed. Cover's theorem indicates that the problem is more likely to be separable if the embedding dimension m is higher.

Radial basis-function networks are similar to the perceptrons described in Chapters 5 to 7, in that they are feed-forward networks designed to solve classification problems. A fundamental difference is that the parameters of the radial basis functions are determined by unsupervised learning, whereas perceptrons are trained using supervised learning for *all* units. While McCulloch-Pitts neurons compute weights to minimise their output from given targets, the radial basis functions compute weights by maximising the output u_j as a function of j. The algorithm for finding the weights of the radial basis functions is summarised in Algorithm 10. Further, as opposed to the deep networks from Chapter 7, radial basis-function networks have only one hidden layer and a linear output neuron.

Algorithm 10 Radial basis functions

initialise the weights w_{jk} independently randomly from $[-1, 1]$;
set all widths to $s_j = 0$;
for $t = 1, \ldots, T$ **do**
  feed randomly chosen pattern $\boldsymbol{x}^{(\mu)}$;
  determine winning neuron j_0: $u_{j_0} \geq u_j$ for all values of j;
  update widths: $s_j = \min_{j \neq k} |\boldsymbol{w}_j - \boldsymbol{w}_k|$;
  update only winning neuron: $\delta \boldsymbol{w}_{j_0} = \eta(\boldsymbol{x}^{(\mu)} - \boldsymbol{w}_{j_0})$;
end for

In summary, radial basis-function networks learn using a *hybrid* scheme: unsupervised learning for the parameters of the radial basis functions and supervised learning for the output weights.

10.6 Autoencoders

Multilayer perceptrons, layered feed-forward networks, were developed for supervised learning, as described in Part II. Such layouts can also be used for unsupervised learning. Examples are *autoencoders* and *generative adversarial networks.*

Autoencoders employ layered feed-forward networks for unsupervised learning of an unlabelled data set of input patterns, using the inputs as targets, $\boldsymbol{t}^{(\mu)} = \boldsymbol{x}^{(\mu)}$. The layout is illustrated in Figure 10.16. The network consists of two main parts, an *encoder* (on the left), and a *decoder* (on the right). The encoder consists for instance of several fully connected or convolutional layers and maps the inputs to a bottleneck layer with a small number M of neurons, significantly smaller than the input dimension, $M \ll N$. We denote the states of the bottleneck neurons by z_j. The encoder corresponds to a non-linear mapping $\boldsymbol{z} = \boldsymbol{f}_{\mathrm{e}}(\boldsymbol{x})$. The decoder maps the bottleneck (or *latent*) variables back to the inputs, $\boldsymbol{x} = \boldsymbol{f}_{\mathrm{d}}(\boldsymbol{z})$. One adjusts weights and thresholds by backpropagation until the network learns to approximate the inputs as

$$\boldsymbol{x} = \boldsymbol{f}_{\mathrm{d}}[\boldsymbol{f}_{\mathrm{e}}(\boldsymbol{x})]\,. \tag{10.43}$$

The energy function reads

$$H = \frac{1}{2}\sum_{\mu} \left|\boldsymbol{x}^{(\mu)} - \boldsymbol{f}_{\mathrm{d}}[\boldsymbol{f}_{\mathrm{e}}(\boldsymbol{x}^{(\mu)})]\right|^2, \tag{10.44}$$

where $|\boldsymbol{x}|^2 = \boldsymbol{x}^{\mathsf{T}}\boldsymbol{x}$. In other words, the autoencoder learns the identity function. The point is that the identity is represented in terms of two non-linear functions, the encoder $\boldsymbol{f}_{\mathrm{e}}$ and the decoder $\boldsymbol{f}_{\mathrm{d}}$. While the identity function is trivial, the encoding

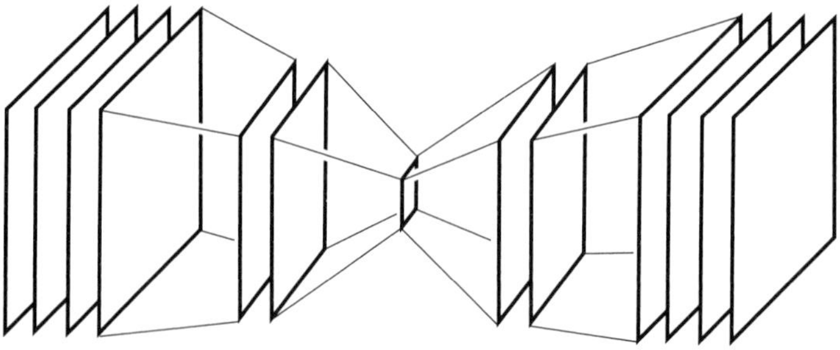

Figure 10.16 Autoencoder (schematic). Both encoder and decoder consist of a number of fully connected or convolutional layers (depicted as squares). In the layout shown, the bottleneck consists of a layer with very few neurons. Sparse autoencoders have bottlenecks with many neurons, but only few are activated

and decoding functions need not be. The bottleneck ensures that the network does not simply learn $\boldsymbol{f}_{\mathrm{e}}(\boldsymbol{x}) = \boldsymbol{f}_{\mathrm{d}}(\boldsymbol{x}) = \boldsymbol{x}$.

The latent variables z may encode interesting properties of the input patterns. If the number of neurons is much smaller than the number of pattern bits, as indicated by the term *bottleneck*, the encoder is a low-dimensional (*compressed*) representation of the input data. In this way, autoencoders can perform non-linear dimensionality reduction, like self-organising maps (Section 10.3). If both encoder and decoder are linear functions with zero thresholds, then $H = \frac{1}{2}\sum_{\mu} |\boldsymbol{x}^{(\mu)} - \mathbb{W}_{\mathrm{d}}\mathbb{W}_{\mathrm{e}}\boldsymbol{x}^{(\mu)})|^2$. In this case, $z_1(\boldsymbol{x}), \ldots, z_M(\boldsymbol{x})$ are simply the first M principal components of zero-mean input data [164] (Exercise 10.14).

Sparse autoencoders [165] have a large number of neurons in the bottleneck, possibly more than the number of pattern bits. But only a small number of bottleneck neurons are allowed to be active at the same time. The idea is that sparse representations of input data are more robust than dense ones and generalise more reliably. At least high-dimensional but sparse representations of binary classification problems are more likely to be linearly separable (Section 5.4). There are different ways of enforcing sparsity, for instance using L_1- or L_2-regularisation (Section 7.6.1). An alternative [165] is to ensure that the average activation of each bottleneck neuron with sigmoid activation function,

$$a_j = \frac{1}{p}\sum_{\mu=1}^{p} \sigma(b_j^{(\mu)}), \tag{10.45}$$

remains small. This is achieved by adding the term

$$\lambda \sum_j a \log \frac{a}{a_j} + (1-a) \log \frac{1-a}{1-a_j} \tag{10.46}$$

to the energy function, with Lagrange multiplier λ (Section 6.3). This term penalises any deviation of a_j from $a_j = a \ll 1$, where $a > 0$ is a sparsity parameter. Each term in the sum is non-negative and vanishes when $a_j = a$ for all j, because each term can be interpreted as the Kullback-Leibler divergence (Section 4.4) between two Bernoulli distributions with parameters a and a_j.

Variational autoencoders [166–168] have layouts similar to the one shown schematically in Figure 10.16, but their purpose is quite different. Variational autoencoders are generative models (Section 4.5): just like restricted Boltzmann machines, they approximate a data distribution of inputs $P_{\rm data}(\boldsymbol{x})$, and allow one to sample from it. As an example consider the MNIST data set of handwritten digits. The patterns define a data distribution that encodes the properties of the digits in terms of covariances and higher-order correlations. The question is how to generate new digits from this distribution, different from those in the data set, yet with the defining properties of handwritten digits. In other words, how can a machine learn to generate images that look like handwritten digits?

The idea of variational autoencoders is to represent the data distribution in terms of a Gaussian distribution $P_{\rm L}(z)$ of latent variables z, using the fact that one can approximate any given data distribution $P_{\rm data}(\boldsymbol{x})$ in terms of $P_{\rm L}(z)$ by a suitable non-linear transformation $\boldsymbol{f}(z))$ [167]. Variational autoencoders are trained not unlike neural networks, but an essential difference to the algorithms described in Part II is that variational autoencoders learn probabilities rather than deterministic input-output mappings.

Given the Gaussian distribution $P_{\rm L}(z)$ of the latent variables, the goal is to maximise the log-likelihood (Section 4.5):

$$\mathscr{L} = \log P(\boldsymbol{x}) = \log \int {\rm d}z\, P(\boldsymbol{x}|z) P_{\rm L}(z)\,. \tag{10.47}$$

Here $P(\boldsymbol{x}|z)$ is the probability to generate $\boldsymbol{x}$ given z. In the simplest case, this distribution is assumed to be Gaussian with mean $\boldsymbol{\mu}_P(z) = \boldsymbol{f}(z)$, and with correlation matrix $\mathbb{C}_P(z)$. The decoder represents these functions in terms of a multilayer perceptron or a convolutional neural net. Weights and thresholds are determined to maximise $\mathscr{L}$ by gradient ascent. To this end, we must find an efficient way of computing $\mathscr{L}$ and its gradients. One possibility is Monte-Carlo sampling (Section 4.2), but this is not very efficient because most values of z drawn from $P_{\rm L}(z)$ result in unlikely patterns $\boldsymbol{x}$, with only negligible contributions to $\mathscr{L}$. To get around this problem, one needs to know which values of z are likely to produce a given pattern $\boldsymbol{x}$. The idea is to learn a second approximate distribution $Q(z|\boldsymbol{x})$ of z given $\boldsymbol{x}$. We can think of $Q(z|\boldsymbol{x})$ as an encoder. So $Q(z|\boldsymbol{x})$ corresponds to the encoder $\boldsymbol{f}_{\rm e}$ discussed above, while $P(\boldsymbol{x}|z)$ corresponds to the decoder $\boldsymbol{f}_{\rm d}$. An important difference is that P and Q are probabilities, not deterministic functions.

To determine a good approximation $Q(z|\boldsymbol{x})$, we minimise the difference between $Q(z|\boldsymbol{x})$ and the unknown distribution $P(z|\boldsymbol{x})$,

$$D_{\mathrm{KL}}[Q(z|\boldsymbol{x}), P(z|\boldsymbol{x})] = \langle \log Q(z|\boldsymbol{x}) - \log P(z|\boldsymbol{x}) \rangle_Q \,. \tag{10.48}$$

A first trick is to rewrite this expression using Bayes' theorem [29],

$$P(z|\boldsymbol{x}) = P(\boldsymbol{x}|z) P_{\mathrm{L}}(z) / P(\boldsymbol{x}) \,. \tag{10.49}$$

This gives

$$\mathscr{L} - D_{\mathrm{KL}}[Q(z|\boldsymbol{x})|P(z|\boldsymbol{x})] = \langle \log P(\boldsymbol{x}|z) - D_{\mathrm{KL}}[Q(z|\boldsymbol{x})|P_{\mathrm{L}}(z)] \rangle_Q \,. \tag{10.50}$$

The second trick is to note that the l.h.s. of Equation (10.50) is a suitable target function to maximise. We want to maximise $\mathscr{L}$ subject to the constraint that the unknown function $Q(z|\boldsymbol{x})$ approximates the probability $P(z|\boldsymbol{x})$ of z encoding the pattern $\boldsymbol{x}$. Usually one takes $Q(z|\boldsymbol{x})$ to be a Gaussian with mean $\boldsymbol{\mu}_Q$ and correlation matrix $\mathbb{C}_Q$. The task is then to determine the functions $\boldsymbol{\mu}_P(z)$, $\mathbb{C}_P(z)$, $\boldsymbol{\mu}_Q(z)$, and $\mathbb{C}_Q(z)$ by adjusting the weights of two neural networks, encoder and decoder.

This task, maximising the r.h.s of Equation (10.50), is not as straightforward as it may seem, because the target function (10.50) involves an average over the distribution Q which in turn depends on the weights. The question is how to move derivative $\partial/\partial_{w_{mn}}$ inside the average $\langle \cdots \rangle_Q$, to obtain an unbiased expression for the weight updates δw_{mn}. In other words, the goal is to ensure that the average weight increments are proportional to the gradients of the target function (10.50). A related problem occurs when training binary stochastic neurons (Section 11.1). Similarities and differences are described in Ref. [169].

One solution is to use *stochastic backpropagation* [168]. In its simplest form, this algorithm makes us of a relation for the gradient of the average of a test function $F(z)$ with a Gaussian probability $Q(z; w_{ij})$ that depends on the weights w_{ij}:

$$\frac{\partial}{\partial w_{mn}} \langle F(z) \rangle_Q = \langle \boldsymbol{b} \cdot \tfrac{\partial}{\partial w_{mn}} \boldsymbol{\mu}_Q + \frac{1}{2} \mathrm{tr} \mathbb{A} \tfrac{\partial}{\partial w_{mn}} \mathbb{C}_Q \rangle_Q \,. \tag{10.51}$$

Here $\boldsymbol{b}$ and $\mathbb{A}$ are the gradient and the Hessian of the function $F(z)$. A challenge is that these derivatives tend to be difficult to compute reliably. Suitable approximations are described in Ref. [168]. The expression inside the average in Equation (10.51) is the unbiased weight increment. Iterating the learning rule allows one to determine the parameters of $P(\boldsymbol{x}|z)$ and $\boldsymbol{Q}(z|\boldsymbol{x})$. This allows to efficiently sample $\boldsymbol{x}$ by sampling the latent variables and then applying the decoder.

Generative adversarial networks [170] are generative models based on learning rules similar to that described above for variational autoencoders, but there are some differences in detail. Generative adversarial networks consist of two multilayer perceptrons, a generator and a discriminator. The generative network produces new outputs from a given data distribution (*fakes*), and the task of the discriminator is to classify these outputs into two classes: real or fake data. Generator and discriminator are trained together. The weights of the generator are adjusted to maximise the classification error of the discriminator, while those of the discriminator are trained to minimise this error [171].

10.7 Summary

The unsupervised-learning algorithms described in Sections 10.1 and 10.2 are based on Hebb's rule. These algorithms can learn different features of unlabelled input data: they can detect the familiarity of inputs, perform principal-component analysis, and identify clusters in the input data.

Self-organising maps also rely on Hebb's rule. An important difference is that the outputs are arranged in an array and that output neurons that are close by in the output array are updated in similar ways. Self-organising maps can therefore represent topographic and semantic maps, where close-by or similar inputs are mapped to nearby outputs. When the dimension of the output array is much lower than the input dimension, self-organising maps perform non-linear dimensional reduction.

Radial basis-function networks are classifiers, just like multilayer perceptrons. Their output neurons are trained in the same way, using labelled input data. However, the decision boundaries of radial basis-function networks are polynomial functions (not just hyperplanes), and their parameters are determined by unsupervised learning.

Autoencoders are multilayer perceptrons. They can learn to encode non-linear features of unlabelled input data by using the input patterns as targets. Finally, generative adversarial networks do not require labelled inputs, so they can be considered unsupervised-learning machines. They are used to generate synthetic data in order to expand training sets for supervised learning, and pose an ethical dilemma because they can be used to generate *deep fakes* [172], manipulated videos where someone's facial expression and speech are replaced by another person's.

In summary, the simple algorithms described in this chapter provide a proof of concept: how machines can learn without labels.

10.8 Further Reading

The primary source for Sections 10.1 and 10.2 is the book by Hertz, Krogh, and Palmer [1]. A good reference for self-organising maps is Kohonen's book [159]. Radial basis-function networks are discussed by Haykin in Chapter 5 of his book [2]. It has been argued that radial-basis function networks do not generalise as well as perceptrons do [173]. To solve this problem, Poggio and Girosi [174] suggested to determine the parameters $\boldsymbol{w}_j$ of the radial basis function by supervised learning, using stochastic gradient descent.

Autoencoders can generate non-linear, low-dimensional representations of an input distribution. The relation to principal-component analysis is discussed in Refs. [164, 175].

The recommended introduction to variational autoencoders is the tutorial by Doersch [167]. He also mentions that the underlying mathematics for variational autoencoders is similar to that of Helmholtz machines (Section 4.7), although the two machines learn in quite different ways.

Variational autoencoders are used for a number of different purposes. Ref. [176] suggests employing a variational autoencoder for active learning (Section 7.8). The idea is to represent the input distribution in terms of lower-dimensional latent variables, and to use K-means clustering (Section 10.4) to identify groups of patterns that should be labelled. Variational autoencoders have also been used for outlier detection [177] and language generation [178].

10.9 Exercises

10.1 Continuous Oja's rule. Using the ansatz $\boldsymbol{w} = \boldsymbol{q}/|\boldsymbol{q}|$, show that Equations (10.4) and (10.5) describe the same angular dynamics. The difference is just that $\boldsymbol{w}$ remains normalised to unity, whereas the norm of $\boldsymbol{q}$ may increase of decrease. See Ref. [156].

10.2 Data-covariance matrix. Determine the data-covariance matrix and the principal direction for the data shown in Figure 10.4.

10.3 Oja's rule. The aim of unsupervised learning is to construct a network that learns the properties of a distribution $P_{\rm data}(\boldsymbol{x})$ of input patterns $\boldsymbol{x} = [x_1, \ldots, x_N]^{\sf T}$. Consider a linear output unit that computes $y = \sum_{j=1}^{N} w_j x_j$. Show that Oja's learning rule $\delta w_j = \eta y (x_j - y w_j)$ has the stable steady state $\boldsymbol{w}^*$ corresponding to the leading eigenvector of the matrix $\mathbb{C}'$ with elements $C'_{ij} = \langle x_i x_j \rangle$. Here $\langle \cdots \rangle$ denotes the average over $P_{\rm data}(\boldsymbol{x})$.

10.4 Linear stability analysis for Oja's rule. Iterate the stochastic dynamics (10.3) near a fixed point $\boldsymbol{w}^*$, linearise, and average the result over a random sequence of patterns $\boldsymbol{x}$. Expand the result to leading order in the learning rate η to show that the linear stability of $\boldsymbol{w}^*$ to this order is determined by Equation (10.9).

10.5 Competitive learning for binary patterns. A competitive learning rule for binary patterns with 0/1 bits reads $\delta w_{ij} = \eta (x_j / \sum_{k=1}^{N} x_k - w_{ij})$ for the winning neuron $i = i_0$, and $\delta w_{ij} = 0$ otherwise. Show that the steady-state weight vectors $\boldsymbol{w}_{i_0}$ have positive components and are normalised as $\sum_{k=1}^{N} w_{i_0,k} = 1$.

10.6 Self-organising map. Write a computer program that implements Kohonen's algorithm with a two-dimensional output array, to learn the properties of a two-dimensional input distribution that is uniform inside an equilateral triangle with sides of unit length, and zero outside. *Hint*: to generate this distribution, sample at least 1,000 points uniformly distributed over the smallest square that contains the triangle, and then accept only points that fall inside the triangle. Increase the number of weights and study how the two-dimensional density of weights near the boundary depends on the distance from the boundary.

10.7 Principal manifolds. Create a data set similar to the one shown in Figure 10.6, using $x_2 = x_1^2 + r$, where r is a Gaussian random number with mean zero and variance $\sigma_r^2 = 0.01$. Determine the principal component (solid line) of the data set (Section 6.3). Use Kohonen's algorithm to find a better approximation to the data, the *principal manifold* (dashed line). For both cases, determine the variance of data that remains unexplained.

10.8 Iris data set. Write a computer program that combines a two-dimensional self-organising map with a simple classifier to classify the Iris data set (Figure 5.1).

10.9 Steady state of two-dimensional Kohonen algorithm. Repeat the analysis of Equation (10.23) for a two-dimensional self-organising map. Derive the equivalent of Equation (10.27) and determine a relation between the weight density ϱ and P_{data} assuming that the data distribution factorises $P_{\text{data}}(\boldsymbol{w}) = f(w_1)g(w_2)$ [161]. Assume that $\boldsymbol{w}(\boldsymbol{r}) = u + iv$ can be written as an analytic function of $\boldsymbol{r} = x + iy$ and derive a relation between ρ and P_{data}.

10.10 Radial basis functions for XOR. Show that the two-dimensional Boolean XOR problem with 0/1 inputs can be solved using the two radial basis functions $u_1(\boldsymbol{x}^{(\mu)}) = \exp(-|\boldsymbol{x}^{(\mu)} - \boldsymbol{w}_1|^2)$ and $u_2(\boldsymbol{x}^{(\mu)}) = \exp(-|\boldsymbol{x}^{(\mu)} - \boldsymbol{w}_2|^2)$ with $\boldsymbol{w}_1 = (1, 1)^{\mathsf{T}}$ and $\boldsymbol{w}_2 = [0, 0]^{\mathsf{T}}$. Draw the positions of the four input patterns in the transformed space with coordinates u_1 and u_2.

10.11 Radial basis functions. Table 10.1 describes a classification problem. Show that this problem can be solved as follows. Transform the inputs $\boldsymbol{x}$ to two-dimensional coordinates u_1, u_2 using radial basis functions:

$$u_1 = \exp\big(-\tfrac{1}{4}|\boldsymbol{x} - \boldsymbol{w}_1|^2\big), \text{ with } \boldsymbol{w}_1 = [\text{-1,1,1}]^{\mathsf{T}}\,, \tag{10.52}$$

$$u_2 = \exp\big(-\tfrac{1}{4}|\boldsymbol{x} - \boldsymbol{w}_2|^2\big), \text{ with } \boldsymbol{w}_2 = [\text{1,1,-1}]^{\mathsf{T}}\,. \tag{10.53}$$

Plot the positions of the eight input patterns in the u_1-u_2-plane. *Hint*: to compute u_j use the following approximations: exp(-1) $\approx$ 0.37, exp(-2) $\approx$ 0.14, exp(-3) $\approx$ 0.05. The transformed data is used as input to a simple perceptron

Table 10.1 *Inputs and targets for Exercise 10.11*

x_1	x_2	x_3	t
−1	−1	−1	1
−1	−1	1	1
−1	1	−1	1
−1	1	1	−1
1	−1	−1	1
1	−1	1	1
1	1	−1	−1
1	1	1	1

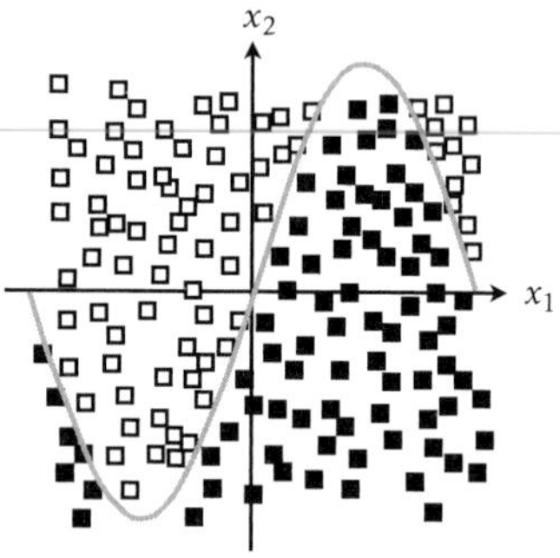

Figure 10.17 Non-linear decision boundary (gray line) $x_2 = \sin(2\pi x_1)$ for a non-linearly separable binary classification problem defined on the square $-1 \leq x_1 \leq 1$ and $-1 \leq x_2 \leq 1$. See Exercise 10.12.

$O^{(\mu)} = \operatorname{sgn}\big(\sum_{i=1}^{2} W_j u_j^{(\mu)} - \Theta\big)$. Draw a decision boundary in the u_1-u_2-plane and determine the corresponding weight vector $\boldsymbol{W}$, as well as the threshold Θ.

10.12 A two-dimensional binary classification problem. Figure 10.17 illustrates a binary classification problem defined on the square $-1 \leq x_1 \leq 1$ and $-1 \leq x_2 \leq 1$ with decision boundary $x_2 = \sin(2\pi x_1)$. Make your own input data set by distributing 1,000 inputs in the two regions shown, half of them with target $t = +1$ (■), the other half with $t = -1$ (□). Find approximate decision boundaries using radial-basis function networks with m radial basis functions, for $m = 5$, 10, 20, and 100. Plot the corresponding decision boundaries in the input plane and determine the classification errors.

10.13 Autoencoder for MNIST. Train the autoencoder network shown in Figure 10.18 on the MNIST data set. Analyse which properties of the data set the latent variables z_1 and z_2 encode.

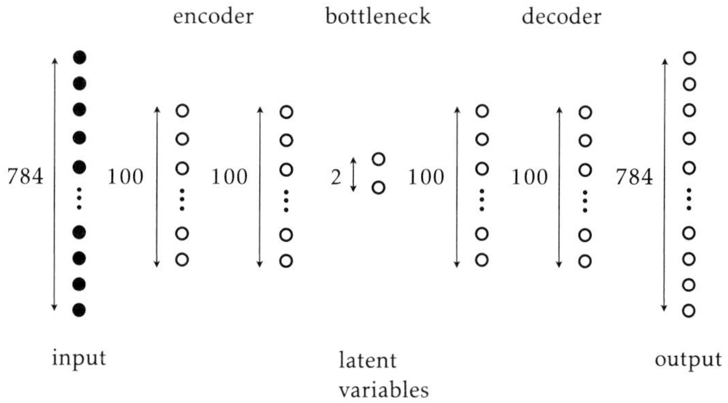

Figure 10.18 Layout for an autoencoder [Equations (10.43) and (10.44)] to analyse the MNIST data set. The layers are fully connected, and the number of neurons are given in the figure. See Exercise 10.13.

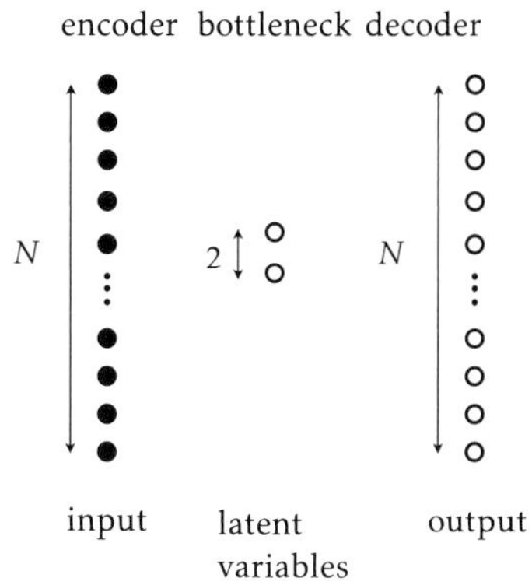

Figure 10.19 Layout of an autoencoder [Equations (10.43) and (10.44)]. The layers are fully connected. The bottleneck consists of two hidden neurons with states z_1 and z_2 (latent variables). Input and output dimensions are equal to N. See Exercise 10.14.

10.14 Autoencoder with linear units. Figure 10.19 shows the layout of an autoencoder [Equations (10.43) and (10.44)]. All neurons are linear units. Assume that the input data has zero mean, so that all thresholds can be set to zero. Prove that the latent variables $z = [z_1, z_2]^{\mathsf{T}}$ are the two principal components of the input data after training the autoencoder with backpropagation. Repeat the proof for inputs with non-zero means.

11

Reinforcement Learning

Supervised learning requires labelled data, where each input comes with a target the network is supposed to learn. Unsupervised learning, by contrast, does not require labelled data. *Reinforcement learning* lies between these extremes. The term reinforcement describes the principle of learning by means of a *reward function*. This function assigns *penalties* or *rewards* to the network output, depending on how the output relates to the learning goal. For a neural network with a vector of outputs, the reward function could be

$$r = \begin{cases} +1 & \text{reward if all outputs correct,} \\ -1 & \text{penalty otherwise.} \end{cases} \tag{11.1}$$

The goal is to learn to produce outputs that receive a reward more frequently than those that trigger a penalty. We say that rewarded outputs are *reinforced*. The feedback may be random, given by a distribution initially unknown to the network.

The reward function reflects the learning goal. The training process as well as the learning outcome depend crucially on this function. Suppose one replaces the reward function (11.1) by the more lenient alternative: $r = 1$ if at least one output is correct, and $r = -1$ if all outputs are wrong. Naturally, this leads to more errors, possibly not a good idea if the goal is to teach a robot to fly.

One distinguishes two different types of reinforcement problems, *associative* and *non-associative* tasks [179]. An example for a non-associative task is the N-armed bandit problem [16]. Imagine N slot machines with different reward distributions, initially unknown to the player. Given a finite amount of money, the question is in which order to play the machines so as to maximise the overall profit. The dilemma is whether to stick with a machine that yields a fairly good reward, or whether to try out other machines that may yield a low reward initially, but could give much higher rewards eventually (*exploit-versus-explore* dilemma). In this type of problem, the player receives only the reinforcement signal, no other inputs.

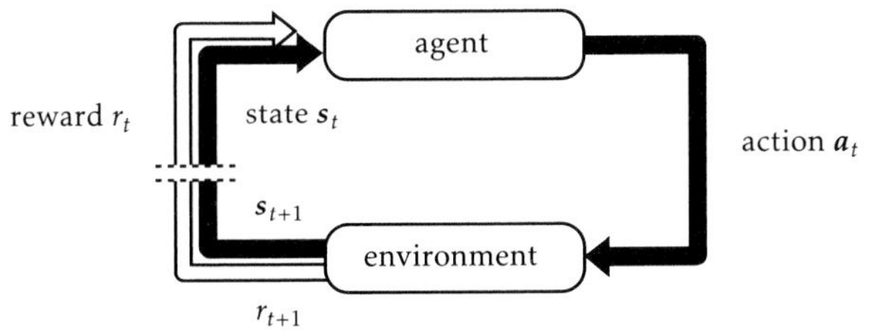

Figure 11.1 Sequential decision process (schematic). After Figure 3.1 in Ref. [16]

In associative tasks, by contrast, the agent receives inputs, or *stimuli*, and it should learn to *associate* with each stimulus the output that yields the highest reward. Such tasks occur for instance in behavioural psychology, where the problem is to discriminate between different stimuli, and to associate the right behaviour with each stimulus.

In general, such associative tasks can be described as *sequential decision processes* (Figure 11.1), where an *agent* explores a sequence of states $\boldsymbol{s}_0, \boldsymbol{s}_1, \boldsymbol{s}_2, \ldots$ through a sequence of actions $\boldsymbol{a}_0, \boldsymbol{a}_1, \boldsymbol{a}_2, \ldots$. Consider for instance a motile microorganism in the turbulent ocean that should swim to the water surface as quickly as possible [180]. It determines its state by observing the local environment. The microorganism might measure the local strain and the vorticity of the flow. The environment provides a reinforcement signal (the distance to the surface, for example), and the organism determines which action to take, given its state and the reinforcement signal. Should it turn, stop to swim, or accelerate? The organism learns to associate actions with certain states that maximise the reward. This sounds quite similar to associating optimal outputs with stimuli. A conceptual difference is that the action of the agent modifies the environment: its actions take it to a different place in the turbulent flow, with different vorticity and different strain. A second point is that the reward to an action may not be immediate. In this case, the challenge is to credit actions that optimise the expected future reward, given the information collected so far. This is the *credit-assignment* problem [181].

There are two different kinds of associative tasks: *continuous* and *episodic* ones. In continuous tasks, the intertwined sequences of states and actions have no natural end, so that one must either terminate the sequence in an *ad-hoc* fashion or introduce a weighting factor to ensure that the expected future reward remains finite. In episodic tasks, by contrast, the learning is divided into episodes that terminate after a finite number of steps. An example is to learn a strategy for winning a board game. In this case, each episode corresponds to a round of the game, and the reward is incurred at the end of each episode. The number of steps per round, the episode length T, may vary from round to round. In order to estimate the expected reward one usually needs many episodes. A second, very simple example is the stimulus problem described above, where the states (stimuli) are independent from the actions. Each episode consists of only one step, so that $T = 1$. In response to a randomly chosen state $\boldsymbol{s}_0$, the agent learns to perform the action $\boldsymbol{a}_0$ that maximises the immediate reward. This can be achieved by the associative reward-penalty algorithm (Section 11.1). It uses stochastic neurons with weights that are trained by gradient ascent to maximise the expected immediate reward.

To estimate the expected future reward when $T > 1$, one must use a different method, usually *temporal difference learning*. It allows one to estimate the expected future reward, after T steps, by breaking up the learning into time steps $t = 1, \ldots, T$. The idea is that it is better to adjust the prediction of a future reward as one iterates, rather than waiting for T iterations before updating the prediction. In temporal difference learning one expresses the reward at time T in terms of differences at time steps $t + 1$ and t (*telescoping sum*) [182].

Temporal difference learning builds up a lookup table that summarises the best actions for each state, the *Q-table* $Q(\boldsymbol{s}, \boldsymbol{a})$. Given $N_{\rm s}$ states $\boldsymbol{s}$ and $N_{\rm a}$ possible actions $\boldsymbol{a}$, $Q(\boldsymbol{s}, \boldsymbol{a})$ is a $N_{\rm s} \times N_{\rm a}$ table. Its elements contain the expected future rewards for each state-action pair. To implement temporal difference learning, one must adopt a *policy* or *strategy*. It specifies for each state which action is taken. In total there are $N_{\rm a}^{N_{\rm s}}$ possibilities of assigning actions to states. When there are many states and actions, it quickly becomes impossible to determine the best strategy by simple sampling, because there are too many to consider.

The advantage of temporal difference learning and related algorithms is that they do not rely on the evaluation of all possible policies. Instead, the policy is updated using iterated estimates of the Q-table. We write Q_t for the estimate at time step t. There are different ways of deriving a policy from a Q-table. The greedy policy is a *deterministic* policy, it corresponds to choosing the action that corresponds to the largest Q-element in a given row of the Q-table: $\boldsymbol{a} = \operatorname{argmax}_{\boldsymbol{a}'} Q_t(\boldsymbol{s}, \boldsymbol{a}')$. This policy maximises the current estimate of the future reward.

Stochastic policies are often better, in particular for non-stationary or stochastic environments, because they allow the agent to explore potentially better alternatives. Also, for a deterministic environment, a deterministic policy may lead to cycles. This can be avoided with a stochastic policy. One example for a stochastic policy is the ε-greedy policy. With probability $1 - \varepsilon$, it chooses the action $\boldsymbol{a} = \operatorname{argmax}_{\boldsymbol{a}'} Q_t(\boldsymbol{s}, \boldsymbol{a}')$, but with a small probability ε, it takes a suboptimal action. Another example is the softmax policy, where argmax is replaced by the softmax function (Section 7.5). The softmax policy can handle actions described in terms of continuous variables.

In general, the policy can change as the algorithm is iterated. A common choice is to reduce the parameter ε in the ε-greedy policy as one iterates, so that the algorithm converges to the optimal deterministic policy.

Q-learning is an approximation to the temporal difference algorithm. In Q-learning, the Q-table is updated assuming that the agent always follows the greedy policy, even though it might actually follow a different policy. Q-learning allows agents to learn to play strategic games [183]. A simple example is the game of tic-tac-toe (Section 11.3). Games such as chess or go require one to keep track of a very large number of states, so large that Q-learning in its simplest form becomes impractical. An alternative is to represent the state-action mapping by a deep neural network [184].

11.1 Associative Reward-Penalty Algorithm

The associative reward-penalty algorithm uses *stochastic neurons* that are trained to maximise the average immediate reward. In Chapters 5 to 9, the output neurons were deterministic functions of their inputs. For reinforcement learning, by contrast, to use stochastic neurons. The idea is the same as in Chapters 3 and 4: stochastic neurons can explore a wider range of possible states which may in the end lead to a better solution. The state y_i of neuron i is given by the *stochastic update rule* (3.1):

$$y_i = \begin{cases} +1 & \text{with probability } p(b_i), \\ -1 & \text{with probability } 1 - p(b_i), \end{cases} \tag{11.2}$$

where $b_i = \boldsymbol{w}_i \cdot \boldsymbol{x}$ is the local field (no thresholds), and $p(b) = (1 + e^{-2\beta b})^{-1}$. Recall that the parameter β^{-1} is the noise level. Since the outputs can assume only two values, $y_i = \pm 1$, Equation (11.2) describes a *binary* stochastic neuron.

To illustrate the associative reward-penalty algorithm for a single binary stochastic neuron, consider an agent experiencing different stimuli $\boldsymbol{x}$ drawn with equal probability from a distribution of inputs. Upon receiving stimulus $\boldsymbol{x}$, the stochastic neuron outputs either $y = 1$ or $y = -1$. Given $\boldsymbol{x}$ and y, the environment provides a stochastic reward $r(\boldsymbol{x}, y) = \pm 1$ drawn from a *reward distribution* $p_{\text{reward}}(\boldsymbol{x}, y)$:

$$r(\boldsymbol{x}, y) = \begin{cases} +1 & \text{with probability } p_{\text{reward}}(\boldsymbol{x}, y), \\ -1 & \text{with probability } 1 - p_{\text{reward}}(\boldsymbol{x}, y). \end{cases} \tag{11.3}$$

The goal is to adjust the weights so that the neuron produces outputs that are rewarded with high probability. Figure 11.2(**a**) shows an example with just two stimuli, $\boldsymbol{x}_1 = [1, 0]^{\mathsf{T}}$ and $\boldsymbol{x}_2 = [1, 1]^{\mathsf{T}}$. The numerical values of $p_{\text{reward}}(\boldsymbol{x}, y)$

p_{reward}	$y=-1$	$y=+1$
$\boldsymbol{x}_1 = [1, 0]^{\mathsf{T}}$	0.6	0.8
$\boldsymbol{x}_2 = [1, 1]^{\mathsf{T}}$	0.3	0.1

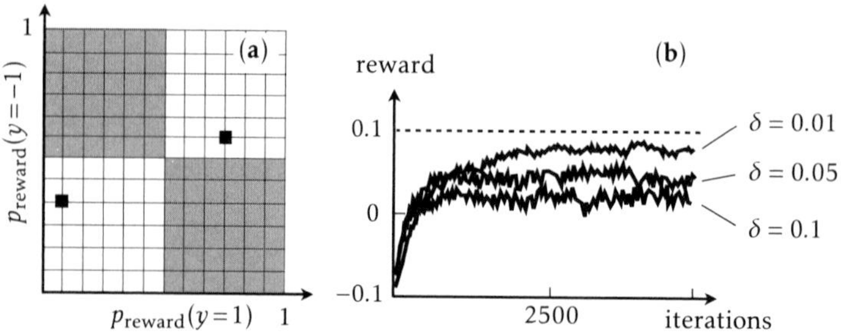

Figure 11.2 Conditioning by reward [185]. A stochastic neuron responds to stimuli $\boldsymbol{x}_1$ and $\boldsymbol{x}_2$ with different outputs, $y=\pm1$ and receives the reward (11.1): $r=+1$ with probability $p_{\text{reward}}(\boldsymbol{x}, y)$, and $r=-1$ with probability $1 - p_{\text{reward}}(\boldsymbol{x}, y)$. The goal is to always respond with the output that maximises the expected reward. Table: reward distribution. Panel (**a**): contingency space [185] of the problem, representing each input $\boldsymbol{x}$ in a plane with coordinates $p_{\text{reward}}(\boldsymbol{x}, +1)$ and $p_{\text{reward}}(\boldsymbol{x}, -1)$. (**b**) Reward versus iteration number of the associative reward-penalty rule (11.10). Schematic, based on simulations by Phillip Graefensteiner averaged over 100 independent realisations

indicate that the expected reward is maximised when the neuron outputs $y=1$ in response to $\boldsymbol{x}_1$, and $y = -1$ in response to $\boldsymbol{x}_2$. Since $\boldsymbol{x}_1$ and $\boldsymbol{x}_2$ occur with equal probability, the maximal expected reward is

$$r_{\max} = \tfrac{1}{2}\big[\langle r(\boldsymbol{x}_1, +1)\rangle_{\text{reward}} + \langle r(\boldsymbol{x}_2, -1)\rangle_{\text{reward}}\big] = 0.1\,. \tag{11.4}$$

Here we used $\langle r(\boldsymbol{x}, y)\rangle_{\text{reward}} = p_{\text{reward}}(\boldsymbol{x}, y) - [1 - p_{\text{reward}}(\boldsymbol{x}, y)]$, as well as the numerical values for $p_{\text{reward}}(\boldsymbol{x}, y)$ given in Figure 11.2(**a**).

Figure 11.2 (**b**) shows the *contingency space* [185] of the problem, representing the inputs $\boldsymbol{x}$ in a plane with coordinates $p_{\text{reward}}(\boldsymbol{x}, +1)$ and $p_{\text{reward}}(\boldsymbol{x}, -1)$. It is easier to learn to associate the correct output with inputs that lie in the shaded regions where $p_{\text{reward}}(\boldsymbol{x}, +1) > \frac{1}{2}$ and $p_{\text{reward}}(\boldsymbol{x}, -1) < \frac{1}{2}$, or vice versa. In this case, one can solve the problem by fixing $y = +1$ and sampling $p_{\text{reward}}(\boldsymbol{x}, +1)$ for all $\boldsymbol{x}$. If $p_{\text{reward}}(\boldsymbol{x}, +1) > \frac{1}{2}$, then $y = +1$ is the optimal output for $\boldsymbol{x}$; otherwise, it is $y = -1$. This strategy cannot be used outside the shaded region. For example, if both reward probabilities are larger than one half, it is necessary to sample both $p_{\text{reward}}(\boldsymbol{x}, -1)$ and $p_{\text{reward}}(\boldsymbol{x}, +1)$ sufficiently often in order to determine which one is larger, one must find *the greater of two goods* according to Barto [185]. This illustrates the fundamental dilemma of reinforcement learning: an output that appears at first to yield a high reward may not be the optimal one in the long run.

To find the optimal output it is necessary to estimate both reward probabilities precisely. This means that one must try all possible outputs frequently, not only the one that appears to be optimal at the moment.

To derive a learning rule we need a cost function. One possibility is to use the average of the immediate reward for a given stimulus $\boldsymbol{x}$,

$$\langle r \rangle = \sum_{y_1=\pm 1,\ldots,y_M=\pm 1} \big\langle r(\boldsymbol{x},\boldsymbol{y}) P(\boldsymbol{y}|\boldsymbol{x}) \big\rangle_{\text{reward}} . \tag{11.5}$$

Here $\langle \cdots \rangle_{\text{reward}}$ is an average over the response of the environment, determined by the reward distribution $p_{\text{reward}}(\boldsymbol{x},\boldsymbol{y})$. It is assumed that the reward distribution is stationary. Furthermore,

$$P(\boldsymbol{y}|\boldsymbol{x}) = \prod_{i=1}^{M} \begin{cases} p(b_i) & \text{for} \quad y_i = 1 , \\ 1 - p(b_i) & \text{for} \quad y_i = -1 \end{cases} \tag{11.6}$$

is the probability that the network produces the output $\boldsymbol{y} = [y_1, \ldots, y_M]^{\mathsf{T}}$ given the local field $b_i = \sum_j w_{ij} x_j$.

To find the maximum of $\langle r \rangle$ one uses gradient ascent on $\langle r \rangle$, analogous to maximising the log-likelihood for Boltzmann machines (Section 4.4), and to gradient descent on the energy function for perceptrons in supervised learning (Chapter 6). The gradient is computed by applying the chain rule, as usual. The calculation is similar to the one for Boltzmann machines (Chapter 4). After some algebra (Exercise 11.2), one finds for given $\boldsymbol{x}$ and $\boldsymbol{y}$ that the derivative of $P(\boldsymbol{y}|\boldsymbol{x})$ with respect to w_{mn} equals $P(\boldsymbol{y}|\boldsymbol{x})\,\beta[y_m - \tanh(\beta b_m)]x_n$. We conclude that

$$\frac{\partial \langle r \rangle}{\partial w_{mn}} = \beta \big\langle r(\boldsymbol{x},\boldsymbol{y}) \big[y_m - \tanh(\beta b_m) \big] \big\rangle x_n \tag{11.7}$$

with $b_m = \sum_j w_{mj} x_j$, as before. The average is taken over the output of the network and over the reward distribution, just as in Equation (11.5).

Now we seek a learning rule that increases the expected immediate reward $\langle r \rangle$. In other words, we require that the weight increment δw_{mn} is an unbiased estimator (Section 10.6) of the gradient of the expected immediate reward [179]:

$$\langle \delta w_{mn} \rangle = \eta \frac{\partial \langle r \rangle}{\partial w_{mn}} . \tag{11.8}$$

Comparison with Equation (11.7) leads to

$$\delta w_{mn} = \alpha r [y_m - \tanh(\beta b_m)] x_n , \tag{11.9}$$

with $\alpha = \eta\beta$. This learning rule belongs to a set of more general rules derived by Williams [179]. It is plausible that the rule (11.9) converges to a steady state, because the weight increments approach zero as the network learns to produce the output $\max_{\boldsymbol{y}}\{p_{\text{reward}}(\boldsymbol{x},\boldsymbol{y})\}$, independent of $\boldsymbol{y}$, so that $\boldsymbol{y}-\langle\boldsymbol{y}\rangle$ averages to zero. But there is no proof of convergence.

An alternative is the *associative reward-penalty* rule [185]:

$$\delta w_{mn} = \alpha \begin{cases} \big[y_m - \tanh(\beta b_m)\big]x_n & \text{for} \quad r=+1\,, \\ -\delta\big[y_m + \tanh(\beta b_m)\big]x_n & \text{for} \quad r=-1\,, \end{cases} \tag{11.10}$$

with $0 < \delta \ll 1$. For $r=1$, the learning rules (11.9) and (11.10) give the same weight increment, but for $r=-1$ the increments are different. With rule (11.10), the agent learns primarily from positive feedback. One advantage of this asymmetric rule is that it can be proven to converge in the limit of $\delta \to 0$ [185]. In general, however, the convergence becomes quite slow when δ is small. Figure11.2(**c**) shows simulation results for the immediate reward, averaged over 100 independent realisations of the learning process, versus the iteration number of the rule (11.10). We see that the average immediate reward approaches a steady state. The steady-state average of the immediate reward is smaller than $r_{\max}=0.1$, but as expected, it approaches $r_{\max}$ as δ decreases.

The averaged learning curves still exhibit substantial fluctuations. They reflect significant variations within and between individual realisations. Furthermore, the convergence proof assumes that the input patterns are linearly independent. This means that the number of patterns cannot exceed the input dimension N. Associative reinforcement problems with linearly dependent inputs can be solved by embedding the input patterns in a higher-dimensional input space (Section 5.4).

The associative reward-penalty rule illustrates how an agent can use a reinforcement signal to maximise the expected immediate reward. The algorithm could for instance be a model for how an animal learns to respond in different ways to different stimuli.

Yet there are many problems where the reward is not immediate. When we play chess, the reward comes at the end of the game, for example $r=+1$ if we won, $r=-1$ if we lost, and $r=0$ if the game ended in a draw. More generally, an agent navigating a complex environment should consider not only immediate rewards but also how a certain action may affect possible future rewards. One way of estimating future rewards for such tasks is temporal difference learning, which is discussed next.

11.2 Temporal Difference Learning

How does temporal difference learning allow an agent to learn to optimise its expected future reward? For an episodic task, given an episode with T steps, the agent visits the finite sequence of states $s_0, \ldots, s_{T-1}$, and collects the rewards $r_1, \ldots, r_T$. The future reward is defined as

$$R_t = \sum_{\tau=t}^{T-1} r_{\tau+1} \,. \tag{11.11}$$

Continuous tasks, by contrast, do not have a defined end point. Since the sum in (11.11) might diverge as $T \to \infty$, it is customary to introduce a weighting factor $0 \le \gamma \le 1$ in the sum over rewards:

$$R_t = \sum_{\tau=t}^{\infty} \gamma^{\tau-t} r_{\tau+1} \,. \tag{11.12}$$

The weighting factor reduces the contribution of the far future to the estimate. Smaller values of γ give more weight to the immediate future, and the limit $\gamma \to 0^+$ corresponds to $R_t = r_{t+1}$. The sum in Equation (11.12) is called *future discounted reward*.

We use a neural network with input $\boldsymbol{s}_t$ to estimate R_t. In general, the network output is a non-linear function of the inputs, parameterised by weights that could be arranged into several layers of hidden neurons (Part II). The simplest choice is to use a single linear unit, just as in Equation (5.19):

$$O(\boldsymbol{s}_t) = \boldsymbol{w} \cdot \boldsymbol{s}_t \,. \tag{11.13}$$

The components w_j of the weight vector $\boldsymbol{w}$ are determined so that the network output $O(\boldsymbol{s}_t)$ approximates R_t. This can be achieved by minimising the energy function

$$H = \tfrac{1}{2} \sum_{t=0}^{T-1} [R_t - O(\boldsymbol{s}_t)]^2 \tag{11.14}$$

using gradient descent. The corresponding learning rule reads

$$\delta w_m = \alpha \sum_{t=0}^{T-1} [R_t - O(\boldsymbol{s}_t)] \frac{\partial O}{\partial w_m} \,. \tag{11.15}$$

The idea of temporal difference learning [182] is to express the error $R_t - O(\boldsymbol{s}_t)$ as a sum of temporal differences:

$$R_t - O(\boldsymbol{s}_t) = \sum_{\tau=t}^{T-1} [r_{\tau+1} + O(\boldsymbol{s}_{\tau+1}) - O(\boldsymbol{s}_\tau)] \,, \tag{11.16}$$

where $O(\boldsymbol{s}_T)$ is defined to be zero, $O(\boldsymbol{s}_T) \equiv 0$. Using the gradient-descent rule (11.15) one obtains

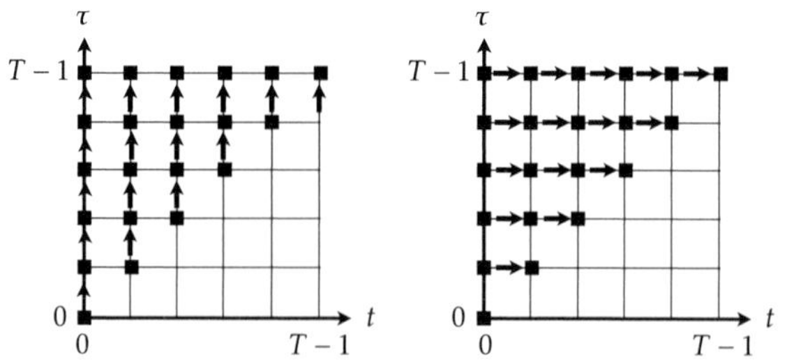

Figure 11.3 The double sum in Equation (11.17) extends over the terms indicated in black. The corresponding terms can be summed in two ways, as illustrated in the two panels, for $T = 6$

$$\delta \boldsymbol{w} = \alpha \sum_{t=0}^{T-1} \sum_{\tau=t}^{T-1} [r_{\tau+1} + O(\boldsymbol{s}_{\tau+1}) - O(\boldsymbol{s}_\tau)]\boldsymbol{s}_t \,. \tag{11.17}$$

The terms in this double sum can be summed in a different way, as illustrated in Figure 11.3:

$$\delta \boldsymbol{w} = \alpha \sum_{\tau=0}^{T-1} \sum_{t=0}^{\tau} [r_{\tau+1} + O(\boldsymbol{s}_{\tau+1}) - O(\boldsymbol{s}_\tau)]\boldsymbol{s}_t \,. \tag{11.18}$$

Exchanging the summation variables and introducing a weighting factor $0 \le \lambda \le 1$ gives

$$\delta \boldsymbol{w} = \alpha \sum_{t=0}^{T-1} [r_{t+1} + O(\boldsymbol{s}_{t+1}) - O(\boldsymbol{s}_t)] \sum_{\tau=0}^{t} \lambda^{t-\tau} \boldsymbol{s}_\tau \,. \tag{11.19}$$

The purpose of the weighting factor is to reduce the weight of past states in the sum [186]. Alternatively, one may update $\boldsymbol{w}$ (and hence O) at each time step, with increment [182]

$$\delta \boldsymbol{w}_t = \alpha [r_{t+1} + O(\boldsymbol{w}_t, \boldsymbol{s}_{t+1}) - O(\boldsymbol{w}_t, \boldsymbol{s}_t)] \sum_{\tau=0}^{t} \lambda^{t-\tau} \boldsymbol{s}_\tau \,. \tag{11.20}$$

This is the temporal difference learning rule, also called TD(λ) [186]. Temporal difference learning allows a machine to learn the board game backgammon [17, 187], using a deep layered network and backpropagation (Section 6.1) to determine the weights.

The rule TD(0) is similar to the learning rule (6.6a) with target $r_{t+1}+O(\boldsymbol{w}_t, \boldsymbol{s}_{t+1})$. It allows one to learn one-step prediction of the time series. Using Equation (11.13), we see that the TD(0)-learning rule corresponds to the following learning rule for the output O:

$$O_{t+1}(\boldsymbol{s}_t) = O_t(\boldsymbol{s}_t) + \alpha [r_{t+1} + O_t(\boldsymbol{s}_{t+1}) - O_t(\boldsymbol{s}_t)] \,. \tag{11.21}$$

The subscript t in O_t emphasises that the output function is updated iteratively. The learning rule (11.21) applies to estimating the future reward (11.11) for episodic tasks. If the environment is stationary, one may average over many consecutive episodes, using the final weights from episode k as initial weight values for episode $k+1$. For continuous tasks, the corresponding rule for estimating the future discounted reward (11.12) reads:

$$O_{t+1}(\boldsymbol{s}_t) = O_t(\boldsymbol{s}_t) + \alpha[r_{t+1} + \gamma O_t(\boldsymbol{s}_{t+1}) - O_t(\boldsymbol{s}_t)]\,. \tag{11.22}$$

Returning to the problem outlined in the beginning of this chapter, consider an agent exploring a complex environment. The task might be to get from location A to location B as quickly as possible, or expending as little energy as possible. At time t the agent is at position $\boldsymbol{x}_t$ with velocity $\boldsymbol{v}_t$. These variables as well as the local state of the environment are summarised in the state vector $\boldsymbol{s}_t$. Given $\boldsymbol{s}_t$, the agent can act in certain ways: it might slow down, speed up, or turn. These possible actions are summarised in a vector $\boldsymbol{a}_t$. At each time step, the agent takes the action $\boldsymbol{a}_t$ that optimises the expected future discounted reward (11.12), given its present state $\boldsymbol{s}_t$. The estimated expected future reward for any state-action pair is summarised in a table: the *Q-table* with elements $Q_t(\boldsymbol{s}_t, \boldsymbol{a}_t)$ is the analogue of $O_t(\boldsymbol{s}_t)$. Different rows of the Q-table correspond to different states, and different columns correspond to different actions. The TD(0) rule for the Q-table reads:

$$Q_{t+1}(\boldsymbol{s}_t, \boldsymbol{a}_t) = Q_t(\boldsymbol{s}_t, \boldsymbol{a}_t) + \alpha_t\big[r_{t+1} + \gamma Q_t(\boldsymbol{s}_{t+1}, \boldsymbol{a}_{t+1}) - Q_t(\boldsymbol{s}_t, \boldsymbol{a}_t)\big]\,. \tag{11.23}$$

This algorithm is called SARSA, because one needs $\boldsymbol{s}_t$, $\boldsymbol{a}_t$, r_{t+1}, $\boldsymbol{s}_{t+1}$, and $\boldsymbol{a}_{t+1}$ to update the Q-table (Figure 11.4). A difficulty with the rule (11.23) is that it depends not only on the present state-action pair $[\boldsymbol{s}_t, \boldsymbol{a}_t]$ but also on the next action $\boldsymbol{a}_{t+1}$, and thus indirectly upon the policy. Sometimes this is indicated by writing Q_π for the Q-table given policy π.

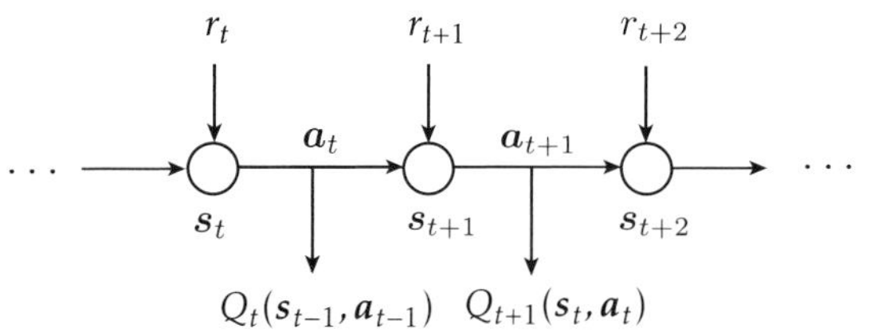

Figure 11.4 Sequence of states $\boldsymbol{s}_t$ in sequential reinforcement learning. The action a_t leads from $\boldsymbol{s}_t$ to $\boldsymbol{s}_{t+1}$ where the agent receives reinforcement r_{t+1}. The Q-table with elements $Q_{t+1}(\boldsymbol{s}_t, \boldsymbol{a}_t)$ estimates the future discounted reward

Algorithm 11 Q-learning for episodic task with the ε-greedy policy

initialise Q;
for $k = 1, \ldots, K$ **do**
  initialise s_0;
  for $t = 0, \ldots, T_k - 1$ **do**
    choose a_t from $Q(a_t, s_t)$ according to ε-greedy policy;
    compute s_{t+1} and record r_{t+1};
    update $Q(s_t, a_t) \leftarrow Q(s_t, a_t) + \alpha[r_{t+1} + \max_a Q(s_{t+1}, a) - Q(s_t, a_t)]$;
  end for
end for

11.3 Q-Learning

The Q-learning rule [188] is an approximation to Equation (11.23) that does not depend on a_{t+1}. Instead one assumes that the next action, a_{t+1}, is the optimal one:

$$Q_{t+1}(s_t, a_t) = Q_t(s_t, a_t) + \alpha_t\big[r_{t+1} + \gamma \max_{a} Q_t(s_{t+1}, a) - Q_t(s_t, a_t)\big], \quad (11.24)$$

regardless of the policy that is currently followed. Although Equation (11.24) does not refer to any policy, the learning outcome nevertheless depends on it, because the policy determines the sequence of states and actions $[s_t, a_t]$. For the greedy policy, Equation (11.24) is equivalent to (11.23), but in general, the two algorithms differ and converge to different solutions. While the Q-table converges to the expected future reward of the greedy policy in Q-learning, for SARSA it converges to the expected future reward corresponding to the policy used in training. This can be an advantage when performance during training is important, for example when training an expensive robot that should not crash too often, or for a small bird that learns flying by doing. Q-learning is simpler and can be used for problems where the final strategy based on $\text{argmax}_{a'} Q(s, a')$ matters, but where the reward during training is less important. Examples are board games where only the quality of the final strategy counts, not how often one loses during training. In summary, Q-learning is simpler, and if one takes $\varepsilon \to 0$ during training, it yields the optimal strategy as well as SARSA, but it may give lower rewards during training.

The Q-learning algorithm is summarised in Algorithm 11. Usually one sets the initial entries in the Q-table to large positive values (*optimistic* initialisation), because this prompts the agent to explore many different actions, at least in the beginning. If the agent is in state s_t, it chooses the action a_t from $Q_t(s_t, a_t)$ according to the given policy. For the ε-greedy policy, for example, the agent picks a random action from the corresponding row of the Q-table with probability ε. With probability $1 - \varepsilon$, it chooses the action

$\boldsymbol{a}_t$ that yields the largest[1] $Q_t(\boldsymbol{s}_t, \boldsymbol{a}_t)$ given $\boldsymbol{s}_t$. The choice of action $\boldsymbol{a}_t$ determines the next state $\boldsymbol{s}_{t+1}$, and this in turn allows one to update the Q-table: given the new state $\boldsymbol{s}_{t+1}$ resulting from the action $\boldsymbol{a}_t$, one updates $Q_t(\boldsymbol{s}_t, \boldsymbol{a}_t)$ using Equation (11.24). For episodic tasks one puts $\gamma = 1$ in (11.24), and one averages over many episodes using the outcome Q_{T_k} from episode k as initial condition for the Q-table for episode $k+1$. Each new episode can start with a new initial state $\boldsymbol{s}_0$. It helps the exploration process if $\boldsymbol{s}_0$ is one of the states that are rarely visited by the learning algorithm.

When the sequence $\boldsymbol{s}_0, \boldsymbol{s}_1, \boldsymbol{s}_2, \ldots$ is a Markov chain (Section 4.2), then the Q-learning algorithm can be shown [186, 189] to converge if one uses a time-dependent learning rate α_t that satisfies

$$\sum_{t=0}^{\infty} \alpha_t = \infty \quad \text{and} \quad \sum_{t=0}^{\infty} \alpha_t^2 < \infty \,. \tag{11.25}$$

Often Q-learning is implemented in combination with the ε-greedy policy. This policy shares an important property with the associative reward-penalty algorithm with stochastic neurons: stochasticity allows for a wider range of responses, some of which may turn out beneficial in the long run. When ε is very small, the agent picks the action that appears optimal. As a consequence, suboptimal Q-elements are sampled less frequently and are therefore subject to larger errors. Therefore it is advantageous to begin with a relatively large value of ε. It is customary to decrease ε as the algorithm is iterated, because this accelerates convergence to the greedy policy.

It is important to bear in mind that the learning outcome depends on the reward function, as mentioned above. In general, it is a good idea to analyse how the optimal strategy changes as one varies the reward function. Sometimes we are faced with the inverse problem: consider how a microorganism swimming in the turbulent ocean responds to different stimuli. How was this behaviour shaped by genetic evolution? Which quantity was optimised? Is it most important to reduce the energy cost for propulsion? Or is it more important to avoid predation?

Another challenge is to determine suitable states and actions. An agent navigating a complex environment may have a continuous range of positions and velocities, and may experience continuous-valued signals from the environment. To represent the corresponding states in a Q-table it is necessary to discretise. To this end, one must determine suitable ranges and resolutions of these variables, and for the actions. If there are too many states and actions, Q-learning becomes inefficient. This is referred to as the *curse of dimensionality* [16, 190].

[1] If several elements in the relevant row have the same maximal value, then any one of them is chosen with equal probability.

Let us see how Q-learning works for a very simple example, for the associative task described in Figure 11.2. One episode corresponds to computing the output of the neuron given its initial state, so $T = 1$. There is no sequence of states, and the task is to estimate the immediate reward. In this case, the learning rule (11.24) simplifies to

$$\delta Q(\boldsymbol{s}, a) = \alpha\big[r(\boldsymbol{s}, a) - Q(\boldsymbol{s}, a)\big]. \tag{11.26}$$

Since each episode consists only of a single time step, we dropped the subscript t. Also, the term $\max_{\boldsymbol{a}} Q(\boldsymbol{s}_{t+1}, \boldsymbol{a})$ from Equation (11.24) does not appear in (11.26) since Q estimates the immediate reward. There are only two states in this problem, $\boldsymbol{s} = \boldsymbol{x}_1$ and $\boldsymbol{s} = \boldsymbol{x}_2$, and the possible actions are $a = \pm 1$. In other words, $N_{\rm s} = N_{\rm a} = 2$ in this case. In each round, one of the states is chosen randomly, with equal probability. The action is determined from the current estimate of the immediate reward as $\operatorname{argmax}_a Q(\boldsymbol{s}, a)$ with probability $1 - \varepsilon$, and uniformly randomly otherwise. These steps are iterated over many iterations (episodes), using the outcome of episode k as an initial condition for episode $k + 1$. The rule (11.26) describes exponential relaxation to the target for small learning rate α. In this limit, Equation (12.26) is approximated by the stochastic differential equation

$$\tfrac{\rm d}{{\rm d}k} Q_k(\boldsymbol{s}, a) = \alpha\, f_\varepsilon(\boldsymbol{s}, a)\, [r(\boldsymbol{s}, a) - Q_k(\boldsymbol{s}, a)]\,, \tag{11.27}$$

where $f_\varepsilon(\boldsymbol{s}, a)$ is the stationary frequency with which the state-action pair $[\boldsymbol{s}, a]$ is visited using the ε-greedy policy:

$$f_\varepsilon(\boldsymbol{s}, a) = \frac{1}{N_{\rm s}} \begin{cases} 1 - \varepsilon + \frac{\varepsilon}{N_{\rm a}} & \text{if } a = \operatorname{argmax}_{a'} Q_k(\boldsymbol{s}, a')\,, \\ \frac{\varepsilon}{N_{\rm a}} & \text{otherwise.} \end{cases} \tag{11.28}$$

The frequency is normalised to unity, $1 = \sum_{\boldsymbol{s},a} f_\varepsilon(\boldsymbol{s}, a)$. Averaging the solution of Equation (11.27) with initial condition $Q_0(\boldsymbol{s}, a) = 1$ over the reward distribution gives

$$\langle Q_k(\boldsymbol{s}, a)\rangle = \exp[-\alpha f_\varepsilon(\boldsymbol{s}, a)k] + \alpha f_\varepsilon(\boldsymbol{s}, a) \int_0^k {\rm d}k'\, \langle r(\boldsymbol{s}, a)\rangle \exp[f_\varepsilon(\boldsymbol{s}, a)\alpha(k' - k)]\,. \tag{11.29}$$

For $\varepsilon > 0$, $Q_k(\boldsymbol{s}, \boldsymbol{a})$ converges on average to

$$\begin{bmatrix} Q^*(\boldsymbol{x}_1, -1) & Q^*(\boldsymbol{x}_1, +1) \\ Q^*(\boldsymbol{x}_2, -1) & Q^*(\boldsymbol{x}_2, +1) \end{bmatrix} = \begin{bmatrix} 0.2 & 0.6 \\ -0.4 & -0.8 \end{bmatrix}, \tag{11.30}$$

as $k \to \infty$. Here we used that $\langle r(\boldsymbol{x}, y)\rangle = 2p_{\rm reward}(\boldsymbol{x}, y) - 1$, as well the reward probabilities in Figure 11.2. Figure 11.5 illustrates how the rate of convergence

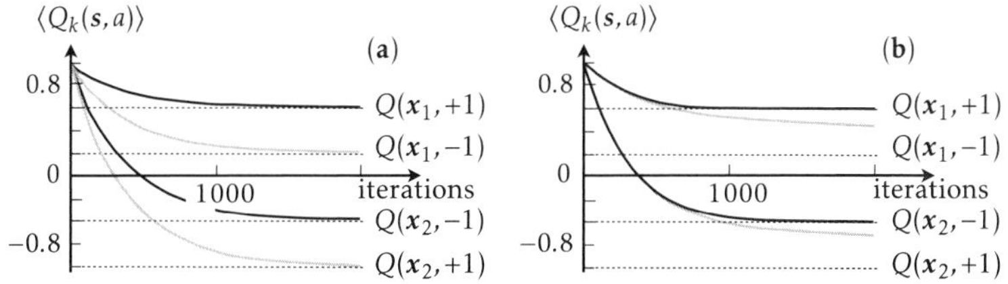

Figure 11.5 Q-learning for the task described in Figure 11.2. (**a**) Entries of the Q-table versus the number of iterations of Equation (11.26) for $\alpha = 0.01$ and $\varepsilon = 1$. (**b**) Same, but for $\varepsilon = 0.05$. Schematic, based on simulations by Navid Mousavi, averaged over 5,000 independent realisations of the learning curve

depends on the value of the parameter ε. For $\varepsilon = 1$, all state-action pairs are visited and evaluated equally often, independently of the present elements of the Q-table. Equations (11.28) and (11.29) show that all elements of the Q-table converge to their steady-state values at the same rate, equal to $\frac{\alpha}{4}$. For small values of ε, by contrast, the algorithm tends to take optimal actions, $\text{argmax}_{a'} Q_k(\boldsymbol{s}, a')$. Therefore it finds the optimal elements of the Q-table more quickly, at the rate $\frac{\alpha}{2}$. However, the other elements converge much more slowly. Initially, the theory (11.29) does not apply because optimal and suboptimal Q-elements in each row of the Q-table are not well separated. As a result, the decay rates of all elements are similar at first. But once optimal and suboptimal elements are significantly different, the suboptimal ones decay at the rate $\varepsilon\alpha/4$, as predicted by the theory.

This example illustrates the strength of Q-learning with the ε-greedy policy. For small values of ε, the algorithm tends to converge to the optimal strategy more quickly than a brute-force algorithm that visits every state-action pair equally often. Equation (11.28) shows that this advantage becomes larger for larger values of N_{a}. The price one pays is that the suboptimal entries of the Q-table converge more slowly.

The gain is even more significant for episodic tasks with $T > 1$ steps. In this case, the learning rule for the Q-table depends on $\text{argmax}_{\boldsymbol{a}'} Q_t(\boldsymbol{s}, \boldsymbol{a}')$. As mentioned above, there are $N_{\text{a}}^{N_{\text{s}}}$ ways in which the largest elements can be distributed over the rows of the Q-table. To find the optimal deterministic strategy by complete enumeration, one must run a large number of episodes for each of the $N_{\text{a}}^{N_{\text{s}}}$ possibilities, to find out which one gives the largest reward. This is impractical when either N_{a} or N_{s} or both become too large. Q-learning with a small value of ε, by contrast, tends to explore actions that appear to yield the largest expected future reward given the current estimates of the Q-values. Using experience in this way allows the algorithm to simultaneously improve its policy and the Q-values towards optimality. As a consequence, Q-learning can find optimal, or at least good strategies when complete enumeration of all possibilities fails.

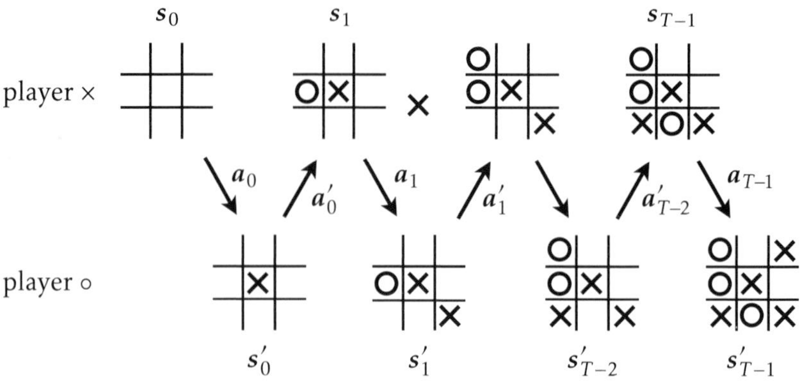

Figure 11.6 Tic-tac-toe. Two players, $\times$ and $\circ$, take turns in placing a piece on an empty field of 3×3 board. The goal is to be the first to complete a row, column, or diagonal consisting of three of one's own pieces. In the example shown, player $\times$ starts and ends up winning the game. The states encountered by player $\times$ are denoted by $\boldsymbol{s}_t$, those encountered by player $\circ$ by $\boldsymbol{s}_t'$. Their actions are denoted by $\boldsymbol{a}_t$ and $\boldsymbol{a}_t'$

A second example for Q-learning is illustrated in Figure 11.6, the board game tic-tac-toe. It is a very simple game where two players take turns in placing their pieces on a 3×3 board. The player who manages to first obtain three pieces in a row, column, or diagonal wins and receives the reward $r = +1$. A draw gives $r = 0$, and the player receives $r = -1$ when the round is lost. The goal is to win as often as possible, to maximise the expected future reward. However, there is a strategy for both players to ensure that they do not lose. If both players try to maximise their expected future reward, then they end up following this strategy. As a consequence, every game must end in a draw [191]. As a result, the game is quite boring.

Nevertheless, it is instructive to ask how the players can learn to find this strategy using Q-learning with the ε-greedy policy. To this end we let two agents play many rounds against each other. The state space is the collection of all board configurations. Player $\times$ starts, and thus always sees a board with an even number of pieces, while the number of pieces is odd for player $\circ$. Since the players encounter different sets of states, each must keep track of their own Q-table. The task is episodic, and the number T of steps may vary from round to round. Feedback is only obtained at the end of each round.

We use Equation (11.24) with a constant learning rate α. We can set $\gamma = 1$ since the number of steps in each round is finite. The Q-table is a $2 \times n$ table where each entry is a 3×3 array. Here n is the number of states the player encountered so far. The first row lists the states, each a 3×3 array with entries -1 ($\circ$), 1 ($\times$), or 0 (empty). The second row contains the Q-values. Given a certain state of the board, a player can play a piece on any empty field. The corresponding estimate of the expected future reward is stored in the corresponding 3×3 array in the second row. Since one cannot place a piece onto an occupied field, the corresponding entries in the Q-table are assigned NaN.

During a round, the Q-tables of the players are updated in turns, always the one of the player who places a piece. After a round, the Q-tables of both players are updated. During the first round, the elements of Q-tables encountered are initialised to zero and remain zero. The first change to Q occurs in the last step of this round. If player $\times$ wins, for example, the element $Q_{T-1}(s_{T-1}, a_{T-1})$ corresponding to the state-action pair that led to the final state s'_{T-1} is updated for the winning player, and $Q_{T-2}(s'_{T-2}, a'_{T-2})$ is set to -1 for player $\circ$.

Both players follow the ε-greedy policy. With probability $1 - \varepsilon$ they take the optimal move (if the maximal Q-element in the relevant row is degenerate, then one of the maximal elements is chosen randomly). With probability ε, a random action is chosen. As the players continue to play rounds against each other, the rewards spread to other elements of Q. Suppose that the state s_{T-1} is encountered once more, the one that allowed player $\times$ to win the first round with a_{T-1}. Then the term $\max_a Q_{T-1}(s_{T-1}, a)$ causes a Q-element for the previous state to change, the one from which s_{T-1} was reached the second time. However, as time goes on, this process slows down, because later updates are multiplied with higher powers of the learning rate α. Also, if the opponent lost in the previous round, it will try different actions that may block winning moves for the other player.

Figure 11.7 illustrates how the players learn, after playing many rounds against each other. Since both players try to maximise their expected future reward in the steady state of the Q-learning algorithm, all games end in a draw in this case.

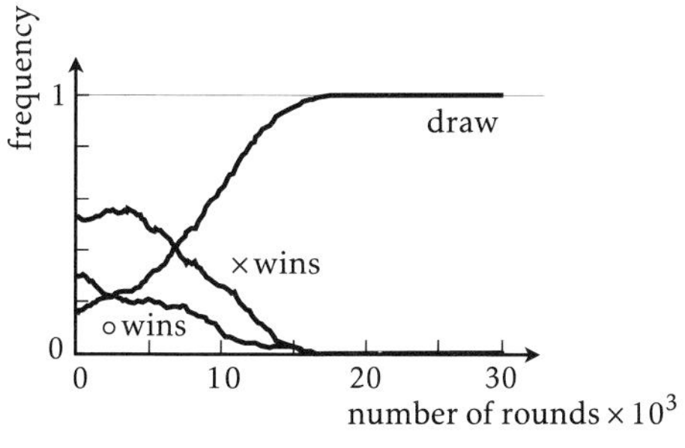

Figure 11.7 Learning curves for two players learning to play tic-tac-toe with Q-learning and the ε-greedy policy. Shown are the frequencies that the game ends in a draw, that player $\times$ wins, and that player $\circ$ wins. Similar curves are obtained using a learning rate $\alpha = 0.1$. The parameter ε was equal to unity for the first 10^4 rounds, and then decreased by a factor of 0.9 after each 100 rounds, and averaging each curve over a running window of 30 rounds. Schematic, based on simulations performed by Navid Mousavi

The corresponding Q-tables contain the strategies each player should adopt to maximise their reward. Suppose player $\circ$ places the first piece as shown here:

board

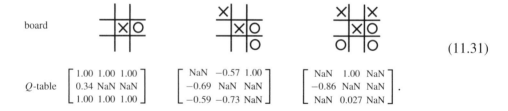

$$
Q\text{-table}\quad \begin{bmatrix} 1.00 & 1.00 & 1.00 \\ 0.34 & \text{NaN} & \text{NaN} \\ 1.00 & 1.00 & 1.00 \end{bmatrix} \quad \begin{bmatrix} \text{NaN} & -0.57 & 1.00 \\ -0.69 & \text{NaN} & \text{NaN} \\ -0.59 & -0.73 & \text{NaN} \end{bmatrix} \quad \begin{bmatrix} \text{NaN} & 1.00 & \text{NaN} \\ -0.86 & \text{NaN} & \text{NaN} \\ \text{NaN} & 0.027 & \text{NaN} \end{bmatrix}. \tag{11.31}
$$

How does this game continue? There are several different ways in which player $\times$ may try to win. The left Q-table in Equation (11.31) shows that one possibility is to place the piece in the top or bottom row, because this creates the opportunity of creating a *bridge* in the next move, a configuration that cannot be blocked by the opponent, allowing $\times$ to win. The right Q-table shows that player $\times$ could still lose or end up with a draw if he makes the wrong move. The corresponding Q-entries have not quite converged to -1 and 0, respectively. Q-entries corresponding to suboptimal states are not estimated as precisely because they are visited less frequently. Here $\varepsilon = 0.3$ was chosen quite large. Smaller values of ε give even less accurate estimates for the suboptimal Q-elements compared with those in Equation (11.31), after training for the same number of rounds.

As pointed out above, the learning outcome depends on the reward function. If one increases the reward for winning, to $r = +2$, for instance, the optimal strategy appears to be to take turns in winning. The same learning outcome is expected if one imposes a penalty for a draw, $r = +1$ (win), $r = -1$ (draw, lose), Exercise 11.8. More examples of reinforcement-learning problems in robotics and in the natural sciences are described in Ref. [192].

The Q-learning algorithm described above is quite efficient when the number of states and actions is not too large. For very large Q-tables, the algorithm becomes quite slow. In this case, it may be more efficient to replace the Q-table by an approximate Q-function that maps states to actions. As explained in Section 11.2, one can use a neural network to represent the Q-function [184] (*deep reinforcement learning*). An application of this method is AlphaGo, a machine-learning algorithm that learnt to play the game of go [183]. The Q-function is represented in terms of a convolutional neural network. This makes it possible to use a variant of Q-learning, despite the fact that the number of states is enormous.

In recent years, many proof-of-principle studies have demonstrated the possibilities of reinforcement learning in a wide range of scientific problems. Recent advances in deep reinforcement learning hold promise for the future, for real-world control problems in the engineering sciences.

11.4 Summary

Reinforcement learning lies between unsupervised learning (Chapter 10) and supervised learning (Chapters 5 to 9). In reinforcement learning, there are no labelled data sets. Instead, the neural network or agent learns through feedback from the environment in the shape of a reward or a penalty. The goal is to find a strategy that maximises the expected reward. Reinforcement learning is applied in a wide range of fields, from psychology to mechanical engineering, using a large variety of algorithms. The associative reward-penalty algorithm and many versions of temporal difference learning were originally formulated using neural networks. Q-learning is an approximation to temporal difference learning for sequential decision processes. In its simplest form, it does not rely on neural networks. However, when the number of states and actions is large, this algorithm becomes slow. In this case, it may be more efficient to approximate the Q-function by a neural network.

11.5 Further Reading

The standard reference for reinforcement learning is *Reinforcement Learning: An Introduction* by Sutton and Barto [16]. The original reference for the convergence of the Q-learning algorithm is Ref. [189]. A more mathematical introduction to reinforcement learning is given in Ref. [186]. Examples for reinforcement learning in statistical and non-linear physics are summarised in Ref. [192].

An open question is when and how *symmetries* can be exploited to simplify a reinforcement problem. For a small microorganism learning to navigate a turbulent flow, some aspects are discussed in Ref. [193], but little is known in general. Another open question concerns the convergence of the Q-learning algorithm. Convergence to the optimal policy is assured if the sequence of states is a Markov chain. However, most real-world problems are not Markovian, so that convergence is not guaranteed. The algorithm appears to perform well nevertheless (a recurring theme in this book), but it is an open question under which circumstances it may fail.

11.6 Exercises

11.1 Binary stochastic neurons. Consider binary stochastic neurons y_i with update rule $y_i = +1$ with probability $p(b_i)$ and $y_i = -1$ otherwise. Here $b_i = \sum_j w_{ij} x_j$, and $p(b) = (1 + \mathrm{e}^{-2\beta b})^{-1}$. The parameter β^{-1} is the noise level, x_j are inputs, and w_{ij} are weights. The energy function reads $H = \frac{1}{2}\sum_{i\mu}\big(t_i^{(\mu)} - y_i^{(\mu)}\big)^2$, with targets $t_i^{(\mu)} = \pm 1$. Stochastic neurons can be trained by gradient descent on the energy function $H' = \frac{1}{2}\sum_{i\mu}\big(t_i^{(\mu)} - \langle y_i^{(\mu)}\rangle\big)^2$, defined in terms of the average

Table 11.1 *Stochastic XOR problem from Ref. [185]. See Exercise 11.4.*

p_{reward}	$y=-1$	$y=+1$
$\boldsymbol{x}_1 = [0, 0]^{\mathsf{T}}$	0.8	0.1
$\boldsymbol{x}_2 = [0, 1]^{\mathsf{T}}$	0.1	0.8
$\boldsymbol{x}_3 = [1, 0]^{\mathsf{T}}$	0.1	0.8
$\boldsymbol{x}_4 = [1, 1]^{\mathsf{T}}$	0.8	0.1

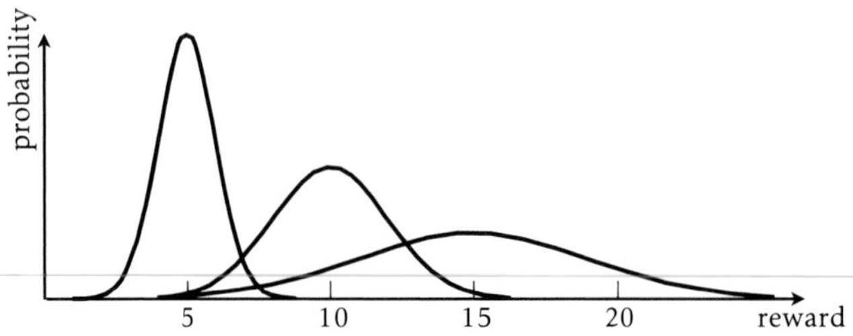

Figure 11.8 Three-armed bandit problem [16]. Three slot machines have Gaussian reward distributions shown, with means and standard deviations $\mu_1 = 7.5, \sigma_1 = 2$, $\mu_2 = 10, \sigma_2 = 1$, and $\mu_3 = 15, \sigma_3 = 5$

outputs $\langle y_i^{(\mu)} \rangle$. The error $\delta_m^{(\mu)}$ is defined by $\delta w_{mn} \equiv -\eta \frac{\partial H'}{\partial w_{mn}} = \eta \sum_\mu \delta_m^{(\mu)} x_n^{(\mu)}$. Show that $\delta_m^{(\mu)} = (t_m^{(\mu)} - \langle y_m^{(\mu)} \rangle)\beta(1 - \langle y_m^{(\mu)} \rangle^2)$. Show that this rule does not necessarily minimise $\langle H \rangle$.

11.2 Gradient of average reward. Derive Equation (11.7). An outline of the derivation is given in Section 7.4 of Hertz, Krogh, and Palmer [1].

11.3 Klopf's self-interested neuron. Klopf's self-interested neuron [194] is a binary stochastic neuron with outputs 0 and 1. Derive a learning rule that is equivalent to Equation (11.9).

11.4 Associate reward-penalty algorithm. Barto [185] explains how to solve association tasks with linearly dependent inputs using hidden neurons. One of his examples is the XOR problem, as shown in Table 11.1. Verify by numerical simulation that the task can be solved by a single binary stochastic neuron if you embed the input data in four-dimensional input space.

11.5 Three-armed bandit problem. Implement the Q-learning algorithm for the three-armed bandit problem with reward distributions shown in Figure 11.8. Analyse the convergence of Q-learning for different values of ε. Discuss the exploitation-exploration dilemma. Illustrate with results of your computer simulations.

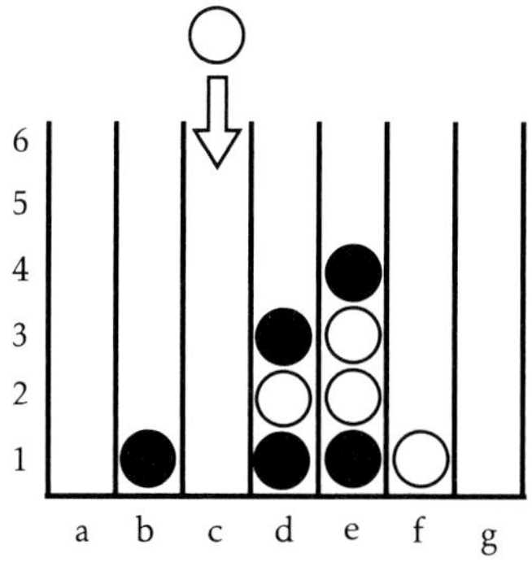

Figure 11.9 Connect Four is a game for two players who take turns dropping their pieces into one of k columns of height l ($k = 7$ and $l = 6$ in the figure). The player first completing a horizontal, vertical, or diagonal row of four pieces wins. The red player started

11.6 Psychology of rock-paper-scissors. Learn to exploit idiosyncrasies in the strategy of your opponent. Suppose that your opponent tends to repeat his action if he won the previous round [195], but changes his action if he lost. Let us say that your opponent randomly chooses between two strategies. With probability p, he repeats his action if he won but chooses one of the other two actions with equal probability if he lost. With probability $1 - p$ he picks one of the three actions, rock with probability $1 - 2q$, and paper or scissors with probability q (your opponent has a preference for rock, so $q < \frac{1}{3}$), and this is also the strategy for the first round. Implement the Q-learning algorithm to determine your optimal strategy as a function of q and p after N rounds.

11.7 Tic-tac-toe. Implement the Q-learning algorithm for two agents learning to play tic-tac-toe (Figure 11.6), with reward function $r = +1$ (win), $r = 0$ (draw), and $r = -1$ (lose). Crowley and Siegler [191] described how a perfect player should play to never lose. Their table 1 summarises how to play given a certain configuration of the board. Determine the Q-table of a perfect player, to verify table 1 of Crowley and Siegler.

11.8 Different reward function for tic-tac-toe. For the tic-tac-toe problem (Figure 11.6) investigate, with Q learning, how the optimal strategy depends on the reward function. Determine the optimal strategy for $r = +2$ (win), $r = 0$ (draw), $r = -1$ (lose), and for $r = +1$ (win), $r = -1$ (draw, lose).

11.9 Connect Four. Implement the Q-learning algorithm for two agents learning to play Connect Four (Figure 11.9) on a 6×6 board. Show that the second player can always achieve a draw against a perfect player [196].

11.10 Eat or save the chocolate. Suppose you get a piece of chocolate every morning. Either you save your chocolate for the next day or you eat all of it during the day, including all pieces you may have saved from previous days. Each day, before you go to bed, you receive a reward: for each piece of chocolate you ate during the day, you get $+2$. If you save the chocolate instead, you get $+1$ for each piece of chocolate in stock. But your brother likes chocolate too, and he searches for your stock while you are asleep. Suppose he finds it with probability p and eats all the chocolate. What is your strategy to optimise your future reward over N days?

Bibliography

[1] Hertz, J., Krogh, A., & Palmer, R. 1991. *Introduction to the Theory of Neural Computation*. Boston: Addison-Wesley.

[2] Haykin, S. 1999. *Neural Networks: A Comprehensive Foundation*, 2nd ed. New Jersey: Prentice Hall.

[3] Horner, H. Neuronale Netze. www.tphys.uni-heidelberg.de/~horner (last accessed 8 November 2018).

[4] Goodfellow, I. J., Bengio, Y., & Courville, A. Deep learning. www.deeplearningbook.org (last accessed 5 September 2018).

[5] Nielsen, M. Neural networks and deep learning. http://neuralnetworksanddeeplearning.com (last accessed 13 August 2018).

[6] McCulloch, W., & Pitts, W. 1943. A logical calculus of the ideas immanent in nervous activity. *Bull. Math. Biophys.* **5**, 115–133.

[7] Carnap, R. 1937. *The Logical Syntax of Language*. London: K. Paul, Trench, Trubner & Co.

[8] McCulloch, W., & Pitts, W. 1947. How we know universals: The perception of auditory and visual forms. *Bull. Math. Biophys.* **9**, 127–147.

[9] Hebb, D. O. 1949. *The Organization of Behavior: A Neuropsychological Theory*. New York: Wiley.

[10] Rosenblatt, F. 1958. The perceptron: A probabilistic model for information storage and organization in the brain. *Psychological Rev.* **65**, 386–408.

[11] Minsky, M., & Papert, S. 1969. *Perceptrons: An Introduction to Computational Geometry*. Cambridge, MA: MIT Press.

[12] Rumelhart, D. E., Hinton, G. E., & Williams, R. J. 1986. Learning internal representations by error propagation. In *Parallel Distributed Processing: Explorations in the Microstructure of Cognition* (ed. D. E. Rumelhart & J. L. McClelland). Cambridge, MA: MIT Press.

[13] Hopfield, J. J. 1982. Neural networks and physical systems with emergent collective computational abilities. *Proceedings of the National Academy of Sciences* **79**, 2554–2558.

[14] Hinton, G. E. 2010. Boltzmann machines. In *Encyclopedia of Machine Learning* (ed. C. Sammut & G. I. Webb), 132–136. Boston: Springer US.

[15] Hinton, G. E., & Sejnowski, T. J. 1986. *Learning and Relearning in Boltzmann Machines*, 282–317. Cambridge, MA: MIT Press.

[16] Sutton, R. S., & Barto, A. G. 2018. *Reinforcement Learning: An Introduction*, 2nd ed. Boston: MIT Press.

[17] Tesauro, G. 1995. Temporal difference learning and TD-Gammon. *Communications of the ACM* **38**, 58–68.
[18] Kohonen, T. 1990. The self-organizing map. *Proceedings of the IEEE* **78**, 1464–1480.
[19] Senior, A. W., Evans, R., Jumper, J., et al. 2020. Improved protein structure prediction using potentials from deep learning. *Nature* **577**, 706–710.
[20] The Nobel Prize in Physiology or Medicine, 1906. www.nobelprize.org (last accessed 1 October 2020).
[21] Newman, E. A., Araque, A., & Dubinsky, J. M., ed. 2017. *The Beautiful Brain: The Drawings of Santiago Ramón y Cajal*. New York: Abrams.
[22] Gabbiani, F., & Metzner, W. 1999. Encoding and processing of sensory information in neuronal spike trains. *Journal of Experimental Biology* **202**, 1267–1279.
[23] Kanal, L. 2001. Perceptrons. In *International Encyclopedia of the Social & Behavioral Sciences* (ed. N. J. Smelser & P. B. Baltes), 11218–11221. Oxford: Pergamon.
[24] Little, W. 1974. The existence of persistent states in the brain. *Mathematical Biosciences* **19**, 101–120.
[25] Fischer, A., & Igel, C. 2014. Training restricted Boltzmann machines: An introduction. *Pattern Recognition* **47**, 25–39.
[26] Sherrington, D. Spin glasses: A perspective. arxiv.org/abs/cond-mat/0512425 (last accessed 5 December 2020).
[27] Sherrington, D., & Kirkpatrick, S. 1975. Solvable model of a spin-glass. *Phys. Rev. Lett.* **35**, 1792–1796.
[28] Lippmann, R. 1987. An introduction to computing with neural nets. *IEEE ASSP Magazine* **4**, 4–22.
[29] Mathews, J., & Walker, R. L. 1964. *Mathematical Methods of Physics*. New York: W.A. Benjamin.
[30] Feller, W. 1968. *An Introduction to Probability Theory and Its applications*, 3rd ed. New York: John Wiley & Sons.
[31] Weisstein, E. W. WolframMathWorld – a Wolfram web resource. http://mathworld.wolfram.com/Erf.html (last accessed 17 September 2019).
[32] Kadanoff, L. P. More is the Same: Phase transitions and mean field theories. http://arxiv.org/abs/0906.0653 (last accessed 3 September 2020).
[33] Amit, D. J., Gutfreund, H., & Sompolinsky, H. 1985. Spin-glass models of neural networks. *Phys. Rev. A* **32**, 1007.
[34] Amit, D. J., & Gutfreund, H. 1987. Statistical mechanics of neural networks near saturation. *Ann. Phys.* **173**, 30–67.
[35] Hopfield, J. J. 1984. Neurons with graded response have collective computational properties like those of two-state neurons. *Proceedings of the National Academy of Sciences* **81**, 3088–3092.
[36] Hinton, G. E., & Sejnowski, T. J. 1983. Optimal perceptual inference. In *Proceedings of the IEEE Conference on Computer Vision and Pattern Recognition*, 444–453. New York: IEEE.
[37] Müller, B., Reinhardt, J., & Strickland, M. T. 1999. *Neural Networks: An Introduction*. Heidelberg: Springer.
[38] Geszti, T. 1990. *Physical Models of Neural Networks*. Singapore: World Scientific.
[39] Steffan, H., & Kühn, R. 1994. Replica symmetry breaking in attractor neural network models. *Zeitschrift für Physik B Condensed Matter* **95**, 249–260.
[40] Volk, D. 1998. On the phase transition of Hopfield networks – another Monte Carlo study. *Int. J. Mod. Phys. C* **9**, 693–700.

[41] Löwe, M. 1998. On the storage capacity of Hopfield models with correlated patterns. *Ann. Prob.* **8**, 1216–1250.
[42] Engel, A. & Van den Broeck, C. 2001. *Statistical Mechanics of Learning*. Cambridge: Cambridge University Press.
[43] Watkin, T. L. H., Rau, A., & Biehl, M. 1993. The statistical mechanics of learning a rule. *Rev. Mod. Phys.* **65**, 499–556.
[44] Kirkpatrick, S., Gelatt, C. D., & Vecchi, M. P. 1983. Optimization by simulated annealing. *Science* **220**, 671–680.
[45] Hinton, G. E. Boltzmann machine. http://scholarpedia.org/article/Boltzmann_machine (last accessed 21 September 2019).
[46] Hinton, G. E. A practical guide to training restricted Boltzmann machines. http://cs.toronto.edu/~hinton/absps/guideTR.pdf (last accessed 18 September 2019).
[47] MacKay, D. J. C. 2003. *Information Theory, Inference and Learning Algorithms*. New Jersey: Cambridge University Press.
[48] Van Kampen, N. G. 2007. *Stochastic Processes in Physics and Chemistry*. Amsterdam: Elsevier.
[49] Sokal, A. 1997. Monte Carlo methods in statistical mechanics: Foundations and new algorithms. In *Functional Integration: Basics and Applications* (ed. C. DeWitt-Morette, P. Cartier, & A. Folacci), 131–192. Boston: Springer US.
[50] Mehlig, B., Heermann, D. W., & Forrest, B. M. 1992. Hybrid Monte Carlo method for condensed-matter systems. *Phys. Rev. B* **45**, 679–685.
[51] Metropolis, N., Rosenbluth, A. W., Rosenbluth, M. N., Teller, M., & Teller, E. 1953. Equation of state calculations by very fast computing machine. *Journal of Chemical Physics* **21**, 1087–1092.
[52] Binder, K., ed. 1986. *Monte-Carlo Methods in Statistical Physics*, 2nd ed. Berlin: Springer.
[53] Press, W. H., Teukolsky, S. A., Vetterling, W. T., & Flannery, W. P. 1992 *Numerical Recipes in C: The Art of Scientific Computing*, 2nd ed. New York: Cambridge University Press.
[54] Hopfield, J. J., & Tank, D. W. 1985. Neural computation of decisions in optimisation problems. *Biol. Cybern.* **52**, 141.
[55] Waterman, M. S. 1995. *Introduction to Bioinformatics*. Boston: Prentice Hall.
[56] Lander, E., Linton, L., Birren, B, et al. 2001. Initial sequencing and analysis of the Human genome. *Nature* **409**, 860–921.
[57] Smolensky, P. 1987. *Information Processing in Dynamical Systems: Foundations of Harmony Theory*. Cambridge, MA: MIT Press.
[58] Le Roux, N., & Bengio, Y. 2008. Representational power of restricted Boltzmann machines and deep belief networks. *Neural Computation* **20**, 1631–1649.
[59] Le Roux, N., & Bengio, Y. 2010. Deep belief networks are compact universal approximators. *Neural Computation* **22**, 2192–2207.
[60] Montúfar, G. F., & Ay, N. 2011. Refinements of universal approximation results for deep belief networks and restricted Boltzmann machines. *Neural Computation* **23**, 1306–1319.
[61] Montúfar, G. F., Rauh, J., & Ay, N. 2011. Expressive power and approximation errors of restricted Boltzmann machines. In *Advances in Neural Information Processing Systems* (ed. J. Shawe-Taylor, R. Zemel, P. Bartlett, F. Pereira, & K. Q. Weinberger), vol. 24, 415–423. Red Hook: Curran.
[62] Montúfar, G. F., Rauh, J., & Ay, N. 2013. Maximal information divergence from statistical models defined by neural networks. In *Geometric Science of Information*

(ed. F. Nielsen & F. Barbaresco), 759–766. Berlin, Heidelberg: Springer Berlin Heidelberg.

[63] Carleo, G., & Troyer, M. 2017. Solving the quantum many-body problem with artificial neural networks. *Science* **355**, 602–606.

[64] Gubernatis, J. E. 2005. Marshal Rosenbluth and the Metropolis algorithm. *Physics of Plasmas* **12**, 057303.

[65] Murphy, K. P. 2012. *Machine Learning: A Probabilistic Perspective*. Cambridge, MA: MIT Press.

[66] Fischer, A., & Igel, C. 2012. An introduction to restricted Boltzmann machines. In *Progress in Pattern Recognition, Image Analysis, Computer Vision, and Applications* (ed. L. Alvarez, M. Mejail, L. Gomez, & J. Jacobo), 14–36. Berlin, Heidelberg: Springer Berlin Heidelberg.

[67] Bengio, Y. 2009. Learning deep architectures for AI. *Foundations and Trends in Machine Learning* **2**, 1–127.

[68] Dayan, P., Hinton, G. E., Neal, R. M., & Zemel, R. S. 1995. The Helmholtz machine. *Neural Computation* **7**, 889–904.

[69] Dayan, P., & Hinton, G. E. 1996. Varieties of Helmholtz machine. *Neural Networks* **9**, 1385–1403.

[70] Dua, D., & Graff, C. UCI machine learning repository. http://archive.ics.uci.edu/ml (last accessed 18 August 2018).

[71] Fisher, R. A. 1936. The use of multiple measurements in taxonomic problems. *Ann. Eugenics* **7**, 179–188.

[72] Cover, T. M. 1965. Geometrical and statistical properties of systems of linear inequalities with applications in pattern recognition. *IEEE Trans. on Electronic Computers* **14**, 326–334.

[73] Sompolinsky, H. Introduction: The perceptron. http://web.mit.edu (last accessed 9 October 2018).

[74] Sloane, N. J. A. Online encyclopedia of integer sequences. http://oeis.org/A000609 (last accessed 9 November 2020).

[75] Greub, W. 1981. *Linear Algebra*. New York: Springer.

[76] LeCun, Y., Bottou, L., Orr, G. B., & Müller, K.-R. 1998. Efficient back prop. In *Neural Networks: Tricks of the Trade* (ed. G. B. Orr & K.-R. Müller). Berlin: Springer.

[77] Nesterov, Y. 1983. A method of solving a convex programming problem with convergence rate $o(1/k^2)$. *Soviet Mathematics Doklady* **27**, 372–376.

[78] Sutskever, I. 2013. Training recurrent neural networks. PhD thesis, University of Toronto.

[79] Hornik, K., Stinchcombe, M., & White, H. 1989. Multilayer feedforward networks are universal approximators. *Neural Networks* **2**, 359–366.

[80] Lapedes, A., & Farber, R. 1988. How neural nets work. In *Neural Information Processing Systems* (ed. D Anderson), 442–456. New York: American Institute of Physics.

[81] Franco, L., & Cannas, S.A. 2001. Generalization properties of modular networks: implementing the parity function. *IEEE Transactions on neural networks* **12**, 1306–1313.

[82] Crisanti, A., Vulpiani, A., & Paladin, G. 1993. *Products of Random Matrices in Statistical Physics*. Berlin: Springer.

[83] Cvitanovic, P., Artuso, G., Mainieri, R., Tanner, G., & Vattay, G. Lyapunov exponents. http://chaosbook.org/chapters/Lyapunov.pdf (last accessed 30 September 2018).

[84] Eckmann, J. P., & Ruelle, D. 1985. Ergodic theory of chaos and strange attractors. *Rev. Mod. Phys.* **57**, 617–656.

[85] Strogatz, S. H. 2000. *Nonlinear Dynamics and Chaos: With Applications to Physics, Biology, Chemistry and Engineering*. Boulder: Westview Press.

[86] Storm, L. 2020. Unstable gradients in deep neural nets. MSc thesis, Chalmers University of Technology.

[87] Pennington, J., Schoenholz, S. S., & Ganguli, S. 2017. Resurrecting the sigmoid in deep learning through dynamical isometry: Theory and practice. In *Advances in Neural Information Processing Systems* (ed. I. Guyon, U. V. Luxburg, S. Bengio, H. Wallach, R. Fergus, S. Vishwanathan, & R. Garnett), vol. 30, pp. 4785–4795. Red Hook: Curran Associates Inc.

[88] Sutskever, I., Martens, J., Dahl, G., & Hinton, G. E. 2013. On the importance of initialization and momentum in deep learning. In *Proceedings of the 30th International Conference on Machine Learning – Volume 28*, pp. III–1139–III–1147, https://jmlr.org/proceedings/spec.html.

[89] Glorot, X., & Bengio, Y. 2010. Understanding the difficulty of training deep feedforward neural networks. In *Proceedings of the Thirteenth International Conference on Artificial Intelligence and Statistics* (ed. Y. W. Teh & M. Titterington), *Proceedings of Machine Learning Research*, vol. 9, pp. 249–256. Chia Laguna Resort, Sardinia, Italy: JMLR Workshop and Conference Proceedings.

[90] Schoenholz, S. S., Gilmer, J., Ganguli, S., & Sohl-Dickstein, J. Deep information propagation. http://arxiv.org/abs/1611.01232. (last accessed 5 December 2020).

[91] Glorot, X., Bordes, A., & Bengio, Y. 2011. Deep sparse rectifier neural networks. In *Proceedings of the Fourteenth International Conference on Artificial Intelligence and Statistics* (ed. G. Gordon, D. Dunson, & M. Dudík), *Proceedings of Machine Learning Research*, vol. 15, pp. 315–323. Fort Lauderdale, FL: JMLR Workshop and Conference Proceedings.

[92] He, K., Zhang, X., Ren, S., & Sun, J. 2016. Deep residual learning for image recognition. In *2016 IEEE Conference on Computer Vision and Pattern Recognition (CVPR)*, 770–778. New York: IEEE.

[93] Residual neural network, wikipedia.org/wiki/Residual_neural_network (last accessed 25 May 2021).

[94] Kleinbaum, D., Kupper, L., & Nizam, A. 2008. *Applied Regression Analysis and Other Multivariable Methods*, 3rd ed. Belmont: Thomson Higher Education.

[95] Srivastava, N., Hinton, G. E., Krizhevsky, A., Sutskever, I., & Salakhutdinov, R. 2014. Dropout: A simple way to prevent neural networks from overfitting. *Journal of Machine Learning Research* **15**, 1929–1958.

[96] Hanson, S., & Pratt, L. 1989. Comparing biases for minimal network construction with back-propagation. In *Advances in Neural Information Processing Systems* (ed. D Touretzky), vol. 1, pp. 177–185. San Francisco: Morgan-Kaufmann.

[97] Hassibi, B., & Stork, D. 1993. Second order derivatives for network pruning: Optimal brain surgeon. In *Advances in Neural Information Processing Systems* (ed. S. Hanson, J. Cowan, & C. Giles), vol. 5, pp. 164–171. San Francisco: Morgan-Kaufmann.

[98] LeCun, Y., Denker, J., & Solla, S. 1990. Optimal brain damage. In *Advances in Neural Information Processing Systems* (ed. D Touretzky), vol. 2, pp. 598–605. San Francisco: Morgan-Kaufmann.

[99] Frankle, J., & Carbin, M. The lottery ticket hypothesis: Finding small, trainable neural networks. http://arxiv.org/abs/1803.03635 (last accessed 5 December 2020).

[100] Deng, J., Dong, W., Socher, R., Li, L. J., Li, K., & Li, F. F. 2009. ImageNet: A large-scale hierarchical image database. In *2009 IEEE Conference on Computer Vision and Pattern Recognition*, 248–255. New York: IEEE.

[101] Ioffe, S., & Szegedy, C. Batch normalization: Accelerating deep network training by reducing internal covariate shift. http://arxiv.org/abs/1502.03167 (last accessed 5 December 2020).

[102] Santurkar, S., Tsipras, D., Ilyas, A., & Madry, A., How does batch normalization help optimization? (No, it is not about internal covariate shift). http://arxiv.org/abs/1805.11604 (last accessed 5 December 2020).

[103] Kirkpatrick, J., Pascanu, R., Rabinowitz, N., et al. 2017. Overcoming catastrophic forgetting in neural networks. *Proceedings of the National Academy of Sciences* **114**, 3521–3526.

[104] Settles, B. Active learning literature survey. http://burrsettles.com/pub/settles.active learning.pdf (last accessed 5 December 2020).

[105] Choromanska, A., Henaff, M., Mathieu, M., Arous, G. B., & LeCun, Y. 2015. The loss surfaces of multilayer networks. In *Proceedings of the Eighteenth International Conference on Artificial Intelligence and Statistics* (ed. G. Lebanon & S. V. N. Vishwanathan), *Proceedings of Machine Learning Research*, vol. 38, pp. 192–204. San Diego: PMLR.

[106] Fyodorov, Y. V. 2004. Complexity of random energy landscapes, glass transition, and absolute value of the spectral determinant of random matrices. *Phys. Rev. Lett.* **92**, 240601.

[107] Becker, S., Zhang, Y., & Lee, A. A. 2020. Geometry of energy landscapes and the optimizability of deep neural networks. *Phys. Rev. Lett.* **124**, 108301.

[108] Y. Wang, Q. Yao, J. Kwok, and L. M. Ni, 2020. Generalizing from a Few Examples: A Survey on Few-Shot Learning. ACM Computing Surveys **53**, 63.

[109] Krizhevsky, A., Sutskever, I., & Hinton, G. E. 2012. ImageNet classification with deep convolutional neural networks. In *Advances in Neural Information Processing Systems* (ed. F. Pereira, C. J. C. Burges, L. Bottou, & K. Q. Weinberger), vol. 25, pp. 1097–1105. Red Hook: Curran Associates Inc.

[110] Abadi, M., Agarwal, A., Barham, P., et al., TensorFlow: Large-scale machine learning on heterogeneous systems. www.tensorflow.org (last accessed 3 September 2018).

[111] LeCun, Y., Cortes, C., & Burges, C. J. The MNIST database of handwritten digits. http://yann.lecun.com/exdb/mnist (last accessed 3 September 2018).

[112] Smith, L. N. 2017. Cyclical learning rates for training neural networks. In *2017 IEEE Winter Conference on Applications of Computer Vision (WACV)*, 464–472. New York: IEEE.

[113] Deep learning in MATLAB. http://se.mathworks.com (last accessed 14 January 2020).

[114] Ciregan, D., Meier, U., & Schmidhuber, J. 2012. Multi-column deep neural networks for image classification. In *2012 IEEE Conference on Computer Vision and Pattern Recognition*, 3642–3649. New York: IEEE.

[115] Picasso, J. P. Pre-processing before digit recognition for NN and CNN trained with MNIST dataset. http://stackoverflow.com (last accessed 26 September 2018).

[116] Kozielski, M., Forster, J., & Ney, H. 2012. Moment-based image normalization for handwritten text recognition. In *2012 International Conference on Frontiers in Handwriting Recognition*, 256–261. New York: IEEE.

[117] Russakovsky, O., Deng, J., Su, H., Krause, J., Satheesh, S., Ma, S., Huang, Z., Karpathy, A., Khosla, A., Bernstein, M., Berg, A. C., & Li, F. F. 2015. ImageNet large scale visual recognition challenge. *International Journal of Computer Vision* **115**, 211–252.

[118] Li, F. F., Johnson, J., & Yeung, S. CNN architectures. http://cs231n.stanford.edu (last accessed 4 December 2020).

[119] Hu, J., Shen, L., & Sun, G. 2018. Squeeze-and-excitation networks. In *2018 IEEE/CVF Conference on Computer Vision and Pattern Recognition*, 7132–7141. New York: IEEE.

[120] Seif, G. Deep learning for image recognition: Why it's challenging, where we've been, and what's next. http://towardsdatascience.com (last accessed 26 September 2018).

[121] Szegedy, C., Wei Liu, Yangqing Jia, Sermanet, P., Reed, S., Anguelov, D., Erhan, D., Vanhoucke, V., & Rabinovich, A. 2015. Going deeper with convolutions. In *2015 IEEE Conference on Computer Vision and Pattern Recognition (CVPR)*, 1–9. New York: IEEE.

[122] Zeng, X., Ouyang, W., Yan, J., et al. 2018. Crafting gbd-net for object detection. *IEEE Transactions on Pattern Analysis and Machine Intelligence* **40** (9), 2109–2123.

[123] Hern, A. Computers now better than humans at recognising and sorting images. www.theguardian.com (last accessed 26 September 2018).

[124] Karpathy, A. What I learned from competing against a ConvNet on ImageNet. http://karpathy.github.io (last accessed 26 September 2018).

[125] Khurshudov, A. Suddenly, a leopard print sofa appears. http://rocknrollnerd.github.io (last accessed 23 August 2018).

[126] Geirhos, R., Medina Temme, C. R., Rauber, J., Schütt, H. H., Bethge, M., & Wichmann, F. A. 2018. Generalisation in humans and deep neural networks. In *Advances in Neural Information Processing Systems* (ed. S. Bengio, H. Wallach, H. Larochelle, K. Grauman, N. Cesa-Bianchi, & R. Garnett), vol. 31, pp. 7538–7550. Red Hook: Curran Associates Inc.

[127] Szegedy, C., Zaremba, W., Sutskever, I., Bruna, J., Erban, D., Goodfellow, I. J., & Fergus, R., Intriguing properties of neural networks. http://arxiv.org/abs/1312.6199 (last accessed 5 December 2020).

[128] Nguyen, A., Yosinski, J., & Clune, J. 2015. Deep neural networks are easily fooled: High confidence predictions for unrecognizable images. In *2015 IEEE Conference on Computer Vision and Pattern Recognition (CVPR)*, 427–436. New York: IEEE.

[129] Yosinski, J., Clune, J., Nguyen, A., Fuchs, T., & Lipson, H., Understanding neural networks through deep visualization. http://arxiv.org/abs/1506.06579 (last accessed 5 December 2020).

[130] Graetz, F. M. How to visualize convolutional features in 40 lines of code. https://towardsdatascience.com (last accessed 30 December 2020).

[131] Dosovitskiy, A., Beyer, L., Kolesnikov, A., et al. An image is worth 16x16 words: Transformers for image recognition at scale. http://arxiv.org/abs/2010.11929 (last accessed 5 December 2020).

[132] Krizhevsky, A., Learning multiple layers of features from tinyimages. www.cs.toronto.edu/kriz (last accessed 1 November 2020).

[133] Ott, E. 2002. *Chaos in Dynamical Systems*, 2nd ed. Cambridge: Cambridge University Press.

[134] Sutskever, I., Vinyals, O., & Le, Q. V. 2014. Sequence to sequence learning with neural networks. In *Advances in Neural Information Processing Systems* (ed. Z. Ghahramani, M. Welling, C. Cortes, N. Lawrence, & K. Q. Weinberger), vol. 27, pp. 3104–3112. Red Hook: Curran Associates Inc.

[135] Lipton, Z. C., Berkowitz, J., & Elkan, C. A critical review of recurrent neural networks for sequence learning. http://arxiv.org/abs/1506.00019 (last accessed 5 December 2020).

[136] Pascanu, R., Mikolov, T., & Bengio, Y. 2013. On the difficulty of training recurrent neural networks. In *Proceedings of the 30th International Conference on International Conference on Machine Learning – Volume 28*, pp. III-1310–III-1318. https://jmlr.org/proceedings/spec.html.

[137] Hochreiter, S., & Schmidhuber, J. 1997. Long short-term memory. *Neural Computation* **9**, 1735–1780.

[138] Olah, C. Understanding LSTM networks. http://colah.github.io (last accessed 30 September 2020).

[139] Cho, K., van Merriënboer, B., Gulcehre, C., Bahdanau, D., Bougares, F., Schwenk, H., & Bengio, Y. 2014. Learning phrase representations using RNN encoder–decoder for statistical machine translation. In *Proceedings of the 2014 Conference on Empirical Methods in Natural Language Processing (EMNLP)*, 1724–1734. Doha, Qatar: Association for Computational Linguistics.

[140] Heck, J. C., & Salem, F. M. 2017. Simplified minimal gated unit variations for recurrent neural networks. In *2017 IEEE 60th International Midwest Symposium on Circuits and Systems (MWSCAS)*, 1593–1596. New York: IEEE.

[141] Wu, Y., Schuster, M., Chen, Z., et al. Google's neural machine translation system: Bridging the gap between human and machine translation. http://arxiv.org/abs/1609.08144 (last accessed 5 December 2020).

[142] Papineni, K., Roukos, S., Ward, T., & Zhu, W.-J. 2002. BLEU: A method for automatic evaluation of machine translation. In *Proceedings of the 40th Annual Meeting of the Association for Computational Linguistics*, 311–318.

[143] Lukosevicius, M., & Jaeger, H. 2009. Reservoir computing approaches to recurrent neural network training. *Computer Science Review* **3**, 127.

[144] Pathak, J., Hunt, B., Girvan, M., Lu, Z., & Ott, E. 2018. Model-free prediction of large spatiotemporally chaotic systems from data: A reservoir computing approach. *Phys. Rev. Lett.* **120**, 024102.

[145] Jaeger, H., & Haas, H. 2004. Harnessing nonlinearity: Predicting chaotic systems and saving energy in wireless communication. *Science* **304**, 78–80.

[146] Lim, S. H., Giorgini, L. T. T., Moon, W., & Wettlaufer, J. S., Predicting critical transitions in multiscale dynamical systems using reservoir computing. http://arxiv.org/abs/:1908.03771 (last accessed 5 December 2020).

[147] Lukosevicius, M. 2012. A practical guide to applying echo state networks. In *Neural Networks: Tricks of the Trade* (ed. G. Montavon, G. Orr, & K. Müller). Berlin, Heidelberg: Springer.

[148] Tanaka, G., Yamane, T., Héroux, J. B., et al. 2019. Recent advances in physical reservoir computing: A review. *Neural Networks* **115**, 100–123.

[149] Doya, K. 1993. Bifurcations of recurrent neural networks in gradient descent learning. *IEEE Transactions on Neural Networks* **1**, 75–80.

[150] Williams, R. J., & Zipser, D. 1995. Gradient-based learning algorithms for recurrent networks and their computational complexity. In *Back-propagation: Theory, Architectures and Applications* (ed. Y. Chauvin & D. E. Rumelhart), 433–486. Hillsdale, NJ: Erlbaum.

[151] Doya, K. 1995. Recurrent networks: Supervised learning. In *The Handbook of Brain Theory and Neural Networks* (ed. M. A. Arbib), 796–799. Cambridge: MIT Press.
[152] Karpathy, A. The unreasonable effectiveness of recurrent neural networks. http://karpathy.github.io (last accessed 4 October 2018).
[153] Ikeda, K., Daido, H., & Akimoto, O.1980. Optical Turbulence: Chaotic Behaviour of Transmitted Light from a Ring Cavity. *Physical Review Letters.* **45**, 709–712.
[154] Kantz, H., & Schreiber, T., 2004. *Nonlinear Time Series Analysis.* Cambridge: Cambridge University Press.
[155] Oja, E. 1982. A simplified neuron model as a principal component analyzer. *J. Math. Biol.* **15**, 267–273.
[156] Wilkinson, M., Bezuglyy, V., & Mehlig, B. 2009. Fingerprints of random flows? *Phys. Fluids* **21**, 043304.
[157] Weliky, M., Bosking, W. H. & Fitzpatrick, D. 1996. A systematic map of direction preference in primary visual cortex. *Nature* **379**, 1476–4687.
[158] Kohonen, T. 2013. Essentials of the self-organizing map. *Neural Networks* **37**, 52–65.
[159] Kohonen, T. 1995. *Self-Organizing Maps*. Berlin: Springer.
[160] Martin, R., & Obermayer, K. 2009. Self-organizing maps. In *Encyclopedia of Neuroscience* (ed. L. R. Squire), 551. Oxford: Academic Press.
[161] Ritter, H., & Schulten, K. 1986. On the stationary state of kohonen's self-organizing sensory mapping. *Biological Cybernetics* **54**, 99–106.
[162] Jackson, J. D. 1999. *Classical Electrodynamics*, 3rd ed. New York: Wiley.
[163] Snyder, W., Nissman, D., Van den Bout, D., & Bilbro, G. 1991. Kohonen networks and clustering: Comparative performance in color clustering. In *Advances in Neural Information Processing Systems* (ed. R. P. Lippmann, J. Moody, & D. Touretzky), vol. 3, pp. 984–990. San Francisco: Morgan-Kaufmann.
[164] Bourlard, H., & Kamp, Y. 1988. Auto-association by multilayer perceptrons and singular value decomposition. *Biological Cybernetics* **59**, 201.
[165] Ng, A. Sparse autoencoder. http://web.stanford.edu/class/cs294a (last accessed 13 October 2020).
[166] Kingma, D. P., & Welling, M. Auto-encoding variational Bayes. http://arxiv.org/abs/1312.6114 (last accessed 5 December 2020).
[167] Doersch, C. Tutorial on variational autoencoders. http://arxiv.org/abs/1606.05908 (last accessed 5 December 2020).
[168] Jimenez Rezende, D., Mohamed, S., & Wierstra, D. 2014. Stochastic backpropagation and approximate inference in deep generative models. In *Proceedings of the 31st International Conference on Machine Learning* (ed. E. P. Xing & T. Jebara), *Proceedings of Machine Learning Research*, vol. 32, pp. 1278–1286. Bejing: PMLR.
[169] Jankowiak, M. & Obermeyer, F. 2018. Pathwise Derivatives beyond the Reparameterization Trick. http://arxiv.org/abs/1806.01851 (last accessed 20 April 2021).
[170] Goodfellow, I. J., Pouget-Abadie, J., Mirza, M., Xu, B., Warde-Farley, D., Ozair, S., Courville, A., & Bengio, Y. 2014. Generative adversarial nets. In *Advances in Neural Information Processing Systems* (ed. Z. Ghahramani, M. Welling, C. Cortes, N. Lawrence, & K. Q. Weinberger), vol. 27, pp. 2672–2680. Red Hook: Curran Associates Inc.
[171] Rocca, J. Understanding generative adversarial networks. http://towardsdata science.com (last accessed 15 October 2020).
[172] Sample, I. What are deepfakes – and how can you spot them? http://theguard ian.com (last accessed 30 September 2020).

[173] Wettschereck, D., & Dietterich, T. 1992. Improving the performance of radial basis function networks by learning center locations. In *Advances in Neural Information Processing Systems* (ed. J. Moody, S. Hanson, & R. P. Lippmann), vol. 4, pp. 1133–1140. San Francisco: Morgan-Kaufmann.

[174] Poggio, T., & Girosi, F. 1990. Networks for approximation and learning. *Proceedings of the IEEE* **78** (9), 1481–1497.

[175] Bourlard, H., Auto-association by multilayer perceptrons and singular value decomposition. http://publications.idiap.ch/downloads/reports/2000/rr00-16.pdf (last accessed 16 October 2020).

[176] Pourkamali-Anaraki, F., & Wakin, M. B., The effectiveness of variational autoencoders for active learning. https://arxiv.org/abs/1911.07716 (last accessed 5 December 2020).

[177] Eduardo, S., Nazabal, A., Williams, C. K. I., & Sutton, C. Robust variational autoencoders for outlier detection and repair of mixed-type data. http://arxiv.org/abs/1907.06671 (last accessed 5 December 2020).

[178] Li, C., Gao, X., Li, Y., Peng, B., Li, X., Zhang, Y., & Gao, J., Optimus: Organizing sentences via pre-trained modeling of a latent space. http://arxiv.org/abs/2004.04092 (last accessed 5 December 2020).

[179] Williams, R. J. 1992. Simple statistical gradient-following algorithms for connectionist reinforcement learning. *Machine Learning* **8**, 229–256.

[180] Colabrese, S., Gustavsson, K., Celani, A., & Biferale, L. 2017. Flow navigation by smart microswimmers via reinforcement learning. *Phys. Rev. Lett.* **118**, 158004.

[181] Minsky, M. 1961. Steps toward artificial intelligence. *Proceedings of the IRE*, 8–30. https://ieeexplore.ieee.org/document/4066245.

[182] Sutton, R. S. 1988. Learning to predict by the methods of temporal differences. *Machine Learning* **3**, 9–44.

[183] Silver, D., Huang, A., Maddison, C. J., et al. 2016. Mastering the game of go with deep neural networks and tree search. *Nature* **529**, 484–489.

[184] Mnih, V., Kavukcuoglu, K., Silver, D., et al. 2015. Human-level control through deep reinforcement learning. *Nature* **518**, 1476–4687.

[185] Barto, A. G. 1985. Learning by statistical cooperation of self-interested neuron-like computing elements. *Hum. Neurobiol.* **4**, 229–256.

[186] Szepesvari, C. 2010. Algorithms for reinforcement learning. In *Synthesis Lectures on Artificial Intelligence and Machine Learning* (ed. R. J. Brachmann & T. Dietterich). San Rafael: Morgan and Claypool Publishers.

[187] McClelland, J. L. 2015. *Explorations in Parallel Distributed Processing: A Handbook of Models, Programs, and Exercises*. New Jersey: Prentice Hall.

[188] Watkins, C. J. C. H. 1989. Learning from delayed rewards. PhD thesis, University of Cambridge.

[189] Watkins, C. J. C. H., & Dayan, P. 1992. Q-learning. *Machine Learning* **8**, 279–292.

[190] Bellman, R. E. 1957. *Dynamic Programming*. New York: Dover Publications.

[191] Crowley, K., & Siegler, R. S. 1993. Flexible strategy use in young children's tic-tac-toe. *Cognitive Science* **17**, 531–561.

[192] Cichos, F., Gustavsson, K., Mehlig, B., & Volpe, G. 2020. Machine learning for active matter. *Nature Machine Intelligence* **2**, 94–103.

[193] Qiu, J., Mousavi, N., Gustavsson, K., Xu, C., Mehlig, B., & Zhao, L. 2020. Navigation of a micro-swimmer in steady flow: The importance of symmetries. https://arxiv.org/abs/2104.11303 (last accessed 28 April 2021).

[194] Klopf, A. H. 1982. *The Hedonistic Neuron: Theory of Memory, Learning and Intelligence*. New York: Taylor and Francis.

[195] Morgan, J. How to win at rock-paper-scissors. http://bbc.com/news/science-environment-27228416 (last accessed 7 September 2020).
[196] Allis, V. 1988. A knowledge-based approach of Connect-Four. Report IR-163, Faculty of Mathematics and Computer Science at the Vrije Universiteit Amsterdam.

Author Index

Subject Index

Congratulations, you reached the end of this book. This is the end of the first episode. You are rewarded by +1. Please multiply everything you have learned by $\alpha = 0.01$ *and add to your knowledge. Then start reading from the first page to start a new episode.*